A Tide-Swept Coast of Sand and Marsh

Coastal Geology and Ecology of GEORGIA

Miles O. Hayes and Jacqueline Michel

Joseph M. Holmes

Illustrator

PANDION BOOKS
a Division of Research Planning, Inc.
Columbia, South Carolina

PANDION BOOKS
a division of Research Planning, Inc.
P.O. Box 328
Columbia, South Carolina USA 29202

Email: permissions@researchplanning.com

A CIP catalog record for this book has been applied for from the Library of Congress.

ISBN 978-0-9816618-3-4

All photographs by Miles O. Hayes and Jacqueline Michel unless otherwise indicated.
Illustration and design by Joseph M. Holmes.

The paper used in this book meets the minimum requirements of the American National Standard for Information Sciences – Permanence of Paper for Printed Library Materials, ANSI/NISO Z39.48-1992

Printed in China by Everbest through Four Colour Imports, Ltd., Louisville, KY

Front cover: Southern end of Ossabaw Island during a low spring tide on 28 April 2010.

Back cover: Eroding north end of Jekyll Island during a low tide in March 2010.

DEDICATION

TO THE LEGENDS OF THE GAME

Geological studies of the Georgia coastal systems began in the early 1960s and continued on pretty hot and heavy into the early 1980s and beyond. These first pioneering studies on the Georgia Coast area were carried out by geologists working at the University of Georgia Marine Institute on Sapelo Island. This book is dedicated to those pathfinders, who, at the time, were generating a new field of endeavor now known as coastal geology (or coastal geomorphology). As a matter of fact, co-author Hayes was of that same vintage, as he produced the first major study of the geological effects of hurricanes on coasts of the USA (Hurricane Carla, 1961, on the Texas Coast) that was published in 1965. Of course, we all had a few predecessors, such as W. Armstrong Price in Texas, V. P. Zenkovitch on the Russian shore, and F. P. Shepard, W. Bascom, and D. L. Inman on the California Coast.

As we were preparing this book, Hayes, a lover and student of the game of baseball, began to think of these early pioneer geologists on the Georgia shore as the "Legends of the Game," comparing some of them to the baseball greats. Examples:

1) Orrin H. Pilkey, who did beach profiles and continental shelf work in Georgia at the start of his career in the early 1960s, is the Babe Ruth of the group. The "Babe," of course, set both the season and lifetime home run production records for the major leagues that stood for many years. Pilkey, with his prolific production of not only scientific papers, but a continuing flood of books on coastal topics, is the all time producer of the group.

2) Jim Howard was the Joe DiMaggio of the original geological team. "Joltin' Joe" was not only great at patrolling center field, but he set a record that will never be broken, getting at least one base hit in 56 straight games. Jim Howard was also extremely prolific in his early days, with a string of revolutionary papers on the relationships of invertebrates and other fauna to habitats of sedimentation along the Georgia Coast that are still being used by geologists to interpret ancient rock sequences, another record that will never be broken. Jim's colleague from the Geology Department of the University of Georgia, Robert Frey, participated in many of these projects.

3) John Hoyt, who early on described such features as intertidal bar systems, bubble sand, steeply dipping beds on the beach, and numerous other first observations is the Jackie Robinson of the group, early putting coastal geology into the limelight of scientific studies. Not only did Jackie Robinson break the color barrier and, hence, allowing a flood of skillful black players into the major leagues, he is probably the best second baseman to ever play the game. He had the purest, most level swing of all, rattling long line drives off the left field walls. But neither John nor Jackie were the single pioneers for very long. Jackie had Roy Campanella, the great catcher, who helped him to not only break the color barrier, but to also produce the brilliant Brooklyn Dodger teams of the late 1940s and 1950s. John similarly had a great partner in Jim Henry as they built their world-renowned coastal geology team, but John's untimely death cut his career short. However, Jim Henry continued to be productive until his recent death, especially with his work on the continental shelf.

[NOTE: In April, 1951, Hayes joined several friends and played hooky from high school for the afternoon as they went to town (Asheville, NC) to see the Brooklyn Dodgers play the Asheville Tourists, a class B Dodger farm team. The baseball park, called McCormick Field, that was carved into the side of a mountain, had a seating capacity of approximately 3,500. Almost 7,000 people turned out for that afternoon game on a weekday in April (numbers provided by American Encyclopedia Regional Cultures – The South). The spectators for the game were dominated by large numbers of black men dressed in their Sunday's finest, many of whom spilled outside of the park and up the side of the mountain. Roy Campanella won the game with an RBI and also hit a home run, if memory serves me. Hayes and his colleagues from the Oakley High School baseball team stood on top of the Dodger dugout before the game, getting the autographs of Robinson, Campanella, Gil Hodges, Duke Snider, Pee Wee Reese and the rest of the gang. This day is one Hayes will never forget! These late spring-training tours through the south as the Dodgers headed north to start the new season, presumably partly to expose the population of the south to these new black players, are memorialized in Roger Kahn's book, *The Boys of Summer* (1972).]

Without further application of the baseball analogy, three additional "Legends" of the early work on the Georgia Coast are George Oertel, who did pioneering work on the large tidal inlets along the coast, Chester DePratter, who spearheaded the application of archaeological dating, and John Teal, the original guru of the salt marshes.

TABLE OF CONTENTS

LIST OF FIGURES

PREFACE

In the early 1960s, while I was a graduate student in the Geology Department of the University of Texas (UT), I had a part-time job in the Defense Research Laboratory at UT that allowed me to conduct a study of the coasts (and inner continental shelves) of the world as part of their classified anti-submarine warfare program. As a result of that work, I began to distinguish among coasts like those in Texas (low-relief coastal plains and deltas, which we eventually called *depositional coasts*) on the basis of **tidal range** (vertical distance between high and low tide). I recognized that depositional coasts with a large tidal range were distinctly different from those with a small tidal range. An outside advisor of my dissertation committee, Dr. W. Armstrong Price, had pioneered these ideas, but I then had a statistical basis for extending them.

As I was completing my Ph.D. study of the South Texas Coast, which has a small tidal range, my ambition was to eventually carry out similar studies of coasts around the world with intermediate and large tidal ranges. Therefore, the very first research grant I applied for, with the U.S. Navy research program, was to study the **depositional coast of the Georgia Bight with its intermediate-range tides**. That research proposal was turned down flat. Also, there were no job openings requiring my skills in any of the geology departments of the universities in the southeastern United States.

Fortunately, shortly thereafter I was able to secure a job at the University of Massachusetts. The relatively unstudied coast of New England, which has a tidal range similar to South Carolina and Georgia, was at the disposal of me and my versatile crew of graduate students. It was a great experience and we learned a lot, but the coast of the Georgia Bight still beckoned. Therefore, in the fall of 1972, a group of my students and I migrated to Columbia, South Carolina to join the Geology Department of the University of South Carolina (USC). During the ensuing years at USC, my students produced over 25 M.S. and Ph.D. theses on studies of the coast of the Georgia Bight. In addition to that research, we led geological field seminars on the South Carolina Coast attended by several thousand oil-and-gas geologists from all over the world for 30 years, beginning in the fall of 1976 and finally concluding them in the fall of 2005.

Fortunately, a couple of years after I arrived in Columbia, Jacqui Michel joined our research trips from time to time while completing her Ph.D. in geochemistry at USC under Dr. Billy Moore. Jacqui and I were married in 1977, eventually forming a science technology company called Research Planning, Inc. (RPI), and we have continued to work together on the Georgia Bight coast ever since. Also, during the decades of the 1980s and 1990s, we were able to fulfill my original ambition to study coasts with different tidal ranges, culminating with studies of shorelines with tidal ranges >30 feet in Chile, Alaska, and France. We have personally carried out several projects on the Georgia Coast with its large tides, such as environmental sensitivity mapping and responding to oil spills. As noted in the dedication, however, we were lucky to have an excellent historic scientific basis to work with as we completed this book.

Whereas we won't be taking you on any boat rides, beach walks, or overflights in a small airplane, which was always a part of our field seminars, we hope the descriptions and illustrations in this book will convey a little of the excitement that we always feel when we are out in the field on this beautiful coastline here at home in the southeastern United States.

Miles O. Hayes, Columbia, S.C., 31 October 2012.

ACKNOWLEDGEMENTS

A special thanks to Clark Alexander, Richard A. Davis, Jr., George Oertel, and Orrin Pilkey for their reviews of an earlier draft of this book. All of their suggestions are much appreciated. However, we should add that we take full responsibility for any of the scientific discussions or personal accounts that are given here.

Much of the information presented is based upon research carried out throughout the Georgia Bight funded by a number of sources, including the U.S. Army Corps of Engineers, U.S. Army Research Office, Office of Naval Research, National Science Foundation, South Carolina Sea Grant Consortium, South Carolina Coastal Council, National Oceanic and Atmospheric Administration (NOAA), Geology Department of the University of South Carolina, RPI International, Inc., and Research Planning, Inc. The support of the American Association of Petroleum Geologists and Schlumberger Well Services during the 30 years we taught geological training seminars on the coast of South Carolina is also gratefully acknowledged. The logistics coordinators for those courses over the years included Tim Kana, Aileen Duc, Patrice Cunningham, Leslie Sauter, Wendy Early, and Kristi Suggs.

Associates in the Coastal Research Division of the Geology Department of the University of South Carolina and at RPI, who carried out research on the Georgia Bight coast, much of which is discussed in this book, included Tim Kana, Walter J. Sexton, Pulak Ray, Mike Stephen, Dag Nummedal, Larry Ward, Gary Zarillo, Bob Finley, Duncan FitzGerald, Dennis Hubbard, John Barwis, Tom Moslow, Aileen Duc, Chris Ruby, Stan Humphries, Ray Levey, Cary Fico, Jeff Knoth, Craig Shipp, Pete Reinhart, Bob Tye, Martha Griffin, John Hodge, Dave Nelligan, Mike Svetlichney, Chris Reel, Tom Freeman, Cathy McCreech, C. Y. McCants, Dan Domeracki, Helen Mary Johnson, Doug Imperato, Steve Wilson, Leita Jean Hulmes, Tim Eckard, Colin Plank, Norm Dodson, and Dallon Weathers.

The Pandion Books Division of RPI provided funding, as well as staff support (notably Joe Holmes and Wendy Early) for the final publication of this book.

In the Dedication, we discussed the "Legends of The Game," placing emphasis on the historic coastal geology programs in the state of Georgia. We should point out, however, that coastal research in Georgia is still ongoing at three institutions and universities. For example, at the Skidaway Institute of Oceanography, several programs are underway. The University of Georgia Marine Institute at Sapelo Island is presently conducting research at the *ecosystem level and basic scientific studies of salt marsh processes, estuarine complexes, and near shore coastal ecosystems* (UGAMI, 2010). One of the reviewers of this book, Dr. Clark Alexander, is the Director of the Georgia Southern University Applied Coastal Research Laboratory on Skidaway Island, where he has been carrying out management-related studies of coastal and estuarine processes, as well as human impacts on the natural ecosystems of the Georgia Coast. He also has a program determining the Holocene history of geologic development of the Georgia Coast using OSL (optically stimulated luminescence) and ^{14}C dating. Dr. Chester Jackson, also affiliated with Georgia Southern, is examining inlet and coastal change dynamics from LIDAR (terrestrial and airborne) and other imagery, using new geospatial tools that he has developed.

1 INTRODUCTION

Our goal for this book is to introduce you to the natural setting of the coast of Georgia, one of the most beautiful in all the country and one you can enjoy visiting during all seasons of the year (see satellite image of the entire state in Figure 1). Many aspects of this coast make it a special place to visit and to enjoy, including:

1) The wide, slowly sinking coastal plain that is flooded daily with the largest tides in the southeastern United States that inundate broad expanses of salt and brackish marshes;
2) Extensive areas of essentially untouched wilderness, such as the Cumberland Island National Seashore;
3) Wide sand beaches and tidal flats populated with a variety of birds, sea turtles, and other wildlife; and
4) The pristine and historic Altamaha River Delta area.

> There are two ways you may want to consider using this book:
>
> 1) If you wish to gain an introductory understanding of the coastal processes and land forms, and general ecological trends, you should spend some time right away reading Sections I and II.
> 2) On the other hand, if you are visiting the coast and you want to learn about a specific area, you can look up that area in Section III. Then, if you have questions about some aspect of that particular area, such as why the beach is eroding or what principal bird groups are present, you can read up on those topics in Sections I and II.

It is a place that captures the imagination when one sees how the role the physical development of the coast since the beginning of the Ice Ages (Pleistocene epoch) has dictated the positioning of historic forts, harbors, rice farms, and numerous other relicts of the activities of the European immigrants since the visit by Hernando de Soto in 1540. But most of all, if you are natural-born environmentalists like we are, it is a chance to experience a beautiful natural setting of the type that is rarely preserved on the modern-day coasts of the lower 48 states.

The authors of this book have been studying coasts for many years, with Hayes' analysis of the coasts of the world for the U.S. Navy in 1960-65 being a starting point (Hayes, 1964; 1967a). Since that time, the authors have worked on the coasts of over 50 countries, including almost all of the coasts of North America, the Middle East, parts of Central and South America, and West Africa. This experience includes conducting numerous major projects along the Georgia Bight (North Carolina to Florida) from 1972 to the present. Hayes also supervised Masters and Doctoral research of 72 graduate students, most of which had to do with coasts (including those along the Georgia Bight), while a Professor at the Universities of Massachusetts and South Carolina. Michel's specialty of oil-spill response has enabled her to carry out research in a wide variety of coastal settings, which has included many long days starting on 28 April 2010 through early 2013, while providing scientific support on cleanup of the *Deepwater Horizon* oil spill in the Gulf of Mexico (for NOAA and the U.S. Coast Guard).

The overall approach and underlying theme of this book are based on a description of the **physical characteristics** of the coast, because that physical framework is the basic skeleton upon which other relevant features, such as coastal wetlands and tidal flats, rest. For example, certain fundamental questions, such as why do some barrier islands erode substantially while others do not, cannot be answered without a firm understanding of the physical processes that formed the islands. However, none of the major coastal habitats, including their biological components, are neglected as we explore the coast from the Savannah River to the

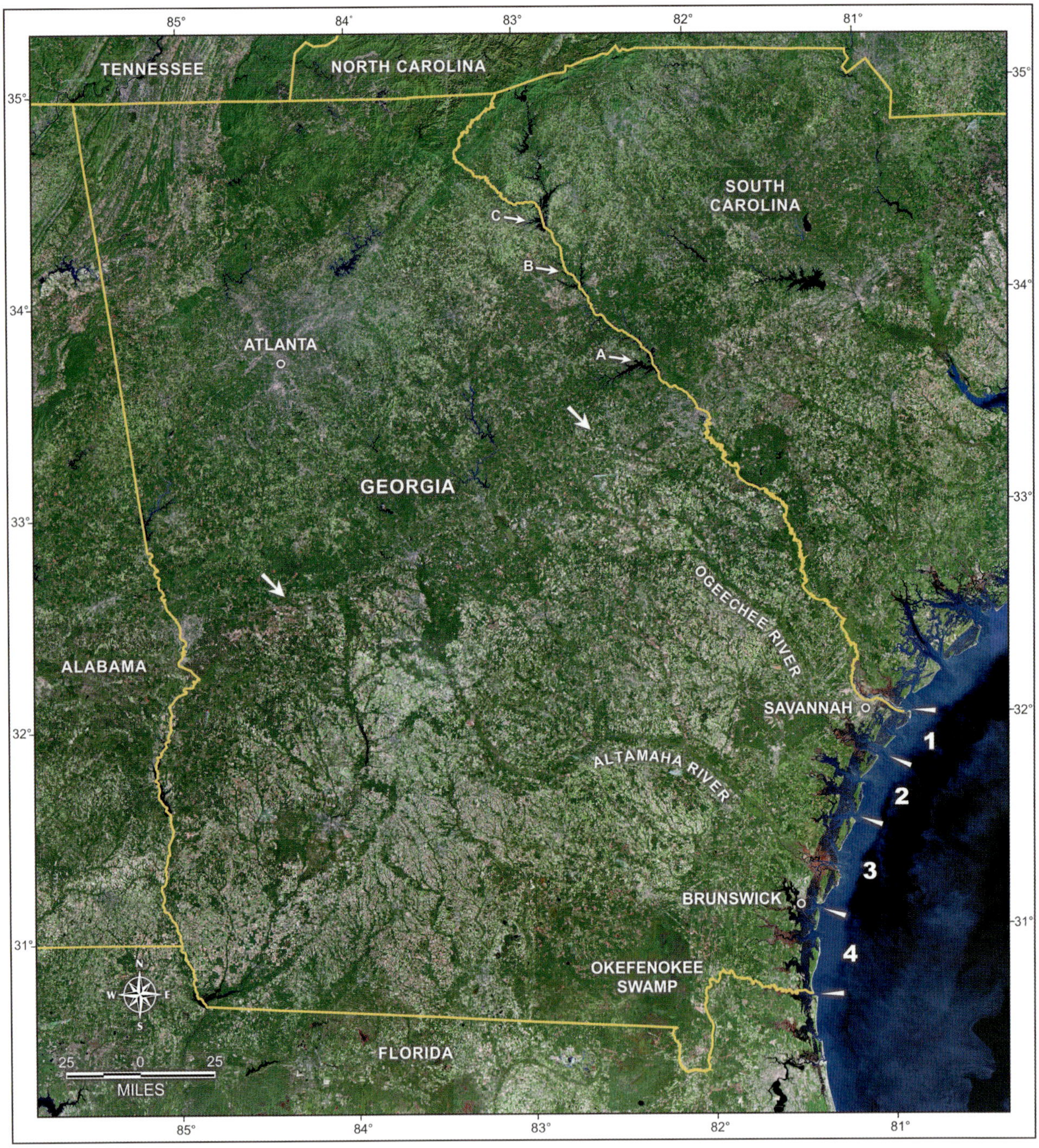

FIGURE 1. Satellite image of Georgia. Note the clear demarcation of the **fall line** (arrows), which marks the boundary between the Piedmont and Coastal Plain Physiographic Provinces. The boundaries of the four major morphological compartments of the coast [Savannah River to Ossabaw Island (1); Ossabaw Island to Blackbeard Island (2); Blackbeard Island to Jekyll Island (3); and Jekyll Island to St. Marys Entrance (4)], which are discussed in detail in the text, are also shown. Landsat image mosaic acquired in 2000, courtesy of the NASA Applied Science & Technology Project Office. Arrow A points to Clarks Hill Lake, Arrow B points to Richard B. Russell Lake, and Arrow C points to Hartwell Lake.

Florida border.

Therefore, this book is designed to provide two somewhat different types of understanding:

1) An abundance of information and diagrams on the geology and ecology of the Georgia Coast; and
2) A guidebook that describes places, environments, plants, and animals all along the coast, with suggestions of places to visit.

By way of introduction, the coast of Georgia borders a wide plain on the trailing edge of the North American tectonic plate (which is slowly moving west; discussed in detail later). In this area, the plate is generally tectonically stable to slightly downwarped. The coast is low-lying and depositional, with Pleistocene and Holocene sediments deposited in barrier islands, salt marshes, a number of large estuaries, and two rather small river deltas blanketing an irregular topography eroded during low stands of sea level during the Pleistocene epoch (the Ice Ages; Hayes, 1985b; 1994). The state's coastline borders the western flank of the Georgia Bight, also called the South Atlantic Bight, a large indentation along the southeast coast of the United States. The outer coast of the Bight is about 100 miles long, but the intricate tidal shorelines behind the barrier islands are several times that length.

Wave heights are typically less than two feet, with the coast being generally dominated by large tides that exceed 10 feet at times. This is in contrast with the more wave-dominated coasts in Florida to the south and North Carolina to the north. The strongest water currents found along the Georgia Coast are generated by rising and falling tides in the entrances to eight large tidal inlets, or sounds, and at the mouths of the Savannah and Altamaha Rivers.

The climate is humid subtropical, with abundant rainfall throughout the year. Hurricanes occasionally cross the area in late summer and early fall, and extratropical cyclonic depressions are common during the non-summer months. Except during the passage of these cyclonic storms, the wind blows prevailingly offshore, from the northwest and southwest, making this a lee coast, in contrast to the California Coast, where the prevailing and dominant winds blow onshore, an important distinction (Hayes and Michel, 2010).

Before getting further into this discussion, we need to introduce the term **geomorphology** (*geo* = earth; *morph* = shape, form; *ology* = to study), a scientific discipline devoted to understanding the origin and three-dimensional shape of the landforms of the Earth. The discipline of **coastal geomorphology** is focused on the understanding of how the different coasts of the world originated, as well as their resulting forms and organization. Both of the authors of this book are coastal geomorphologists, but our interests and professional activities have provided for us a broad experience in other aspects of **coastal ecology**. We are ardent birders and the work of our company, Research Planning, Inc. (RPI) in Columbia, South Carolina, in areas such as beach erosion, oil-spill response, and natural resources mapping, has broadened our understanding of coastal ecosystems. At the time of this writing (31 July 2012), our company is heavily involved in the response to BP's *Deepwater Horizon* oil spill in the Gulf of Mexico, with Michel having been on scene in Louisiana for many months.

The book begins in Section I with a general discussion of the origin of the Georgia Coast, starting as far back as over 300 million years ago when the rocks that presently underlie the coastal areas at depth were part of the supercontinent **Pangaea**. This history of the evolution of the coast ends with a discussion of the factors that have shaped the modern shore since the last rise in sea level that resulted from the melting of the glaciers and ice fields of the last **Ice Age**, which peaked about 20,000 years ago. Next, this coast is compared with others around the world, identifying the key factors that give the Georgia Coast its distinguishing characteristics.

In the rest of Section I, the dynamic physical processes that shape the present shore are considered, namely waves and tides. The origin and influence of hurricanes is also introduced in some detail. Next, the major landforms of the Georgia Coast are described, specifically – barrier islands, tidal inlets, estuaries, major freshwater river systems, and deltas. The outer coast is mostly sand beaches, thus there is a detailed discussion of the cycle of erosion and deposition on them. Specific habitats of the major landforms, such as coastal dunes, marshes, tidal flats, and tidal channels, are also treated at length.

In Section II, we briefly describe the dominant animals that you will see during your visits. In Section III, we discuss and identify opportunities for you to learn about and enjoy the diverse ecosystems that exist within the **four major geomorphological compartments** of the coastal zone that are delineated in Figure 1. Each of the major compartments has several subdivisions, for which discussions of the geomorphology, beach erosion (where appropriate), and so on, as well as some information on their human history, are presented. A list and discussion of the **key places to visit** conclude the coverage of each of the subdivisions of the four major compartments that are introduced below:

- **Compartment 1. Savannah River to Ossabaw Island** – Beginning in the north with the Savannah River Delta, this compartment contains: a) One of the four developed islands on the coast, Tybee Island; b) Two major sounds, Wassaw and Ossabaw; c) The only large, purely Holocene prograding barrier island in Georgia, Wassaw Island; and d) The largest coastal plain (black-water) river in Georgia, the Ogeechee.
- **Compartment 2. Ossabaw Island to Blackbeard Island** – Beginning at the north end of Ossabaw Island and terminating at the north end of Blackbeard Island, this compartment contains: a) Two large "combination" barrier islands which contain cores of the Pleistocene Silver Bluff barrier island system and a wide rim of Holocene deposits, Ossabaw and St. Catherines Islands; b) Two major sounds, St. Catherines and Sapelo; and c) The Harris Neck National Wildlife Refuge.
- **Compartment 3. Blackbeard Island to Jekyll Island** – The most complex of all the compartments, this one, which begins at the north end of Blackbeard Island and extends to the north end of Jekyll Island, contains: a) Another "combination" barrier island, Sapelo, made famous in the geological literature, and Wolf Island (a low-profile overwash island); b) Two major sounds, Doboy and St. Simons; c) The truly unique Altamaha River Delta; d) Two of the state's four developed barrier islands, Sea and St. Simons Islands; and e) Two National Wildlife Refuges, Blackbeard and Wolf Island.
- **Compartment 4. Jekyll Island to St. Marys Entrance** – Beginning at the north end of Jekyll Island and ending at the jettied mouth of the St. Marys River, this compartment contains: a) The last of the four major developed barrier islands, Jekyll (doubling as a state park), which is also one of the "combination" islands; b) Two major sounds, St. Andrew and Cumberland; c) The spectacular Cumberland Island National Seashore; and d) The partially infilled valley of the St. Marys River with its wide salt marsh complex.

Section III concludes with a brief discussion of the Okefenokee Swamp, as well as some thoughts on the possible future of the barrier islands along the coast.

As geologists, we use a lot of geological terms, though we have tried to minimize the jargon. If you come across words or phrases that you don't understand, please look them up in the Glossary.

SECTION I
Coastal Processes and Landforms

The intertidal flats on the south side of the ebb-tidal delta at the entrance to Ossabaw Sound during a low spring tide on 28 April 2010.

2 ORIGIN OF THE COAST

HOW OLD IS IT?

Before giving details of the origin of the Georgia Coast, we will first review the most elemental question in the interpretation of the origin of rock and sedimentary sequences, namely, how old are they? This information is necessary before the evolution of any significant portion of the Earth's crust can be properly interpreted. Three of the most commonly used methods to answer this question with regard to major rock sequences are: 1) Applying the "law of superposition;" 2) Paleontology correlations; and 3) Radiometric dating.

In the 11th century, a Persian Muslim philosopher named Avicenna came up with an idea for determining the age of rock sequences that was later formulated more clearly in the 17th century by the Danish scientist Nicolaus Steno. This concept later became known as Steno's Law or The Law of Superposition, which states: *Sedimentary layers are deposited in a time sequence, with the oldest at the bottom and the youngest at the top.*

In other words, layers of sedimentary rocks are laid down in a sequence from bottom to top like you would do if baking a cake composed of three distinct layers. The first layer you put in the pan is "oldest" and the third one, which is placed on top of the second layer, is "youngest." The difference is that, while it takes only a few seconds to create the three layers in the cake, geological processes may take thousands of years to create stacked layers of rocks. However, the end product, a sequence of vertically stacked layers ranging from the oldest at the bottom to the youngest at the top, is the same. The problem becomes much more complicated if the rock layers have been tilted, or even completely overturned, as is true for many rock sequences, especially those located in areas that have been subject to severe tectonic processes, such as in the Appalachian Mountains or on the California Coast.

The rock layers exposed in the sides of the Grand Canyon in Arizona are a classic example of relatively simple "layer-cake geology." Rocks at the bottom of the canyon consist of granites and schists (metamorphosed fine-grained rocks) overtopped by some tilted rock layers in places, all of which are Precambrian in age (over 542 million years ago; see Table 1). These rocks were scoured off into a relatively flat surface during a major erosional event. The resultant eroded surface is called an unconformity and, because many of the rocks under the eroded surface in the Grand Canyon are tilted, that unconformity is more specifically referred to as an angular unconformity.

The major angular unconformity near the bottom of the Grand Canyon marks a significant gap in the record of geological time provided by the rocks of the canyon. This unconformity was later covered by over 5,000 feet of horizontal layers of sedimentary rocks, beginning with Cambrian sandstones more than 500 million years old at the base. Coming up the sides of the canyon, younger and younger relatively horizontal layers of sedimentary rocks are encountered until one reaches the top of the canyon, where rocks of Late Triassic age (around 225 million years old) are found (Bues and Morales, 1990). These Triassic rocks, named the Chinle Formation, consist of abundant siltstones and some sandstones that were originally deposited mostly in river floodplain and lake environments. Notable examples of petrified forests and dinosaur remains are also contained within the Chinle Formation.

There are some time intervals missing from the sedimentary layers in the Grand Canyon, but this is a remarkably complete record of 300 million years of relatively steady deposition of horizontal rock layers, oldest at the bottom and youngest at the top. This technique of examining layered rocks has been used to interpret the history of the Earth's rocks since the earliest geological maps were constructed, even as far back as the 11th century!

[NOTE: Many important economic products, such as oil, gas, and coal, have accumulated in sedimentary rock

TABLE 1. Geologic time. Time intervals are listed in millions of years before the present [from a chart published by the International Commission on Stratigraphy (2008)].

Era/Period/Epoch			Time (Million yr ago)
Cenozoic era "Recent Life"	Quarternary period	Holocene epoch	0.012-0
		Pleistocene epoch	2.6-0.012
	Neogene period	Pliocene epoch	5.3-2.6
		Miocene epoch	23.0-5.3
	Paleogene period	Oligocene epoch	33.9-23.0
		Eocene epoch	55.8-33.9
		Paleocene epoch	65.5-55.8
Mesozoic era	Cretaceous period		145.5-65.5
	Jurassic period		199.6-145.5
	Triassic period		251.0-199.6
Paleozoic era	Permian period		299.0-251.0
	Carboniferous (Mississippian/Pennsylvanian) period		359.2-299.0
	Devonian period		416.0-359.2
	Silurian period		443.7-416.0
	Ordovician period		488.3-443.7
	Cambrian period		542.0-488.3
Precambrian	Proterozoic era		2,500-542
	Archaeozoic (Archean) era		4,600-2,500

layers similar in some respects to those exposed in the sides of the Grand Canyon. Consequently, the branch of geology known as **stratigraphy**, the study of sedimentary rock layers or strata, particularly their distribution, mode of deposition, and age, has provided professional careers for more geologists than any other branch of the geological sciences.]

Once the sequence of sedimentary rock layers had been worked out sufficiently enough to build a rudimentary time scale, it became apparent that over time many thousands of organisms had come and gone through evolutionary trends and extinctions. The somewhat limited life spans of some of the most critical species were related to rock sequences around the world early in the pursuit of the science of geology. Therefore, it became possible to determine the relative ages of the sedimentary rock sequences based upon their fossil content, or their **paleontology**. The discoveries of sequential vertical changes in these fossil sequences throughout the geological rock column (as well as observations on living organisms) eventually led scientists, most notably Charles Darwin as revealed in his transforming book *Origin of Species* (1859), to derive some basic information relevant to the law of biological evolution and natural selection.

Neither the law of superposition nor paleontological correlation works for determining the ages of intrusive masses of rocks like granite (because they are not layered nor do they contain significant fossil remains), highly metamorphosed rocks, or markedly tilted and overturned strata, such as many of the rock sequences in the Blue Ridge Physiographic Province in northeastern Georgia. Therefore, radiometric, or radioisotope, dating is commonly used to date igneous and metamorphic rocks, as well as organic remains associated with geologically young materials (e.g., by archaeologists). Some of the basic elements and concepts of this process, outlined by Gore (1999), include:

- Naturally occurring radioactive elements break down into other elements at known rates (a process known as **radioactive decay**).
- Radioactive parent elements eventually decay to stable daughter elements. Each radioactive isotope has its own **half-life**, the time it takes for half of the

radioactive parent element to decay to the daughter product.

- Each radioactive element decays at its own nearly constant rate. Examples include Potassium-40 (radioactive parent) to Argon 40 (stable daughter) with a half-life of 1.25 billion years, and Uranium-235 (radioactive parent) to Lead-207 (stable daughter) with a half-life of 700 million years.
- Therefore, measuring (in a rock sample) the amount of radioactive parent element and the amount of stable daughter elements allows the geologist to estimate the length of time over which the decay has been occurring.
- A method called Carbon-14 dating is used for preserved formerly living plant or animal remains less than 50,000 years old. All living things contain a constant ratio of Carbon-14 to Carbon-12. After death, any Carbon-14 in the tissues decays to Nitrogen-14 with a half-life of 5,730 years. The specimen is dated on the basis of the change in the Carbon-14 to Carbon-12 ratio.
- Thermoluminescence (TL) is another dating method by means of measuring the accumulated radiation dose, of the time elapsed since material containing crystalline minerals was either heated (lava, ceramics) or exposed to sunlight (sediments). As the material is heated, thermoluminescence, a weak light signal, is emitted proportional to the radiation dose absorbed by the material. Thermoluminescence dating is used for material where radiocarbon dating is not available, such as sediments. This method has applications in solving issues related to beach nourishment with sand (Ronline et al., 2008; Liritz, 2011).
- A method presently being used extensively by researchers on the Georgia Coast, including reviewer Clark Alexander, is called optically stimulated luminescence (OSL), which was defined as follows by Luminescence Laboratory (2012): *OSL has emerged in the past decade as valuable Quaternary dating method for a variety of eolian, fluvial, marine, and colluvial sediments. The OSL clock, like in TL dating, is reset by exposure of sediment to sunlight prior to deposition. A key advantage of OSL dating is that the luminescence of quartz and feldspar grains is reduced to a low definable level after a few minutes of sunlight exposures versus hours for the corresponding thermoluminescence response. This level, called the residual level, is the point from which the geological luminescence accumulated after burial. OSL dating uses light of a particular wavelength or range of wavelengths, usually either blue, green or infrared light, releasing rapidly the most light sensitive trapped electrons from the crystal lattice. Astonishingly, the decrease in the OSL after a 20 second exposure to sunlight is equivalent to the reduction in TL after a 20-hour light exposure.*
- Burdette et al. (2012) described another method called resonance optical dating (ESROD) that *utilizes light-sensitive radiation-sensitive defects at silicon sites that have been replaced by aluminum and titanium atoms to give rise to a time-dependent signal. These defects saturate at higher levels of radiation dose, compared to optically stimulated luminescence (OSL), and therefore extend the optical dating range back into millions of years.* In their work on Pleistocene sediments in Florida, Burdette et al. found that *every sample's OSL signal was found to be saturated, indicating that the samples were beyond the dating capacity of OSL. Based on dose rates and behavior of Florida quartz, we estimate the maximum OSL age range at approximately 200,000 years.*
- Within the past 30 years or so, understanding the history of the Quaternary period (Pleistocene and Holocene epochs, Table 1), especially with regard to changes in sea level, has been greatly enhanced by other developments in Quaternary dating methods. Two primary methods are: a) Measurements of uranium-series disequilibrium, particularly of preserved coral deposits; and b) The fact that certain configurations of amino acids within the proteins of living organisms (such as marine mollusks) convert to different configurations upon the death of the organism (Muhs et al., 2004).
- A method of dating that has been particularly well developed on the Georgia Coast is a merging of archaeological with geological studies. For example, during much of the past 4,500 years, Native Americans inhabiting the Georgia coastal area subsisted in large part on oysters and clams, as evidenced by large shell middens which mark former habitation sites that have been studied in detail (DePratter, 1977; DePratter and Thompson, 2013). Furthermore, DePratter and Howard (1980) observed that *Indian populations in the coastal area produced copious amounts of ceramics, beginning about 4500 years ago. Through time, characteristics of the pottery changed rapidly as new methods of tempering and decoration*

were developed. In this paper, they illustrated a vertical ceramic sequence that included eight periods of human development, starting with the St. Simons period about 3,300 years ago and ending with the Altamaha period starting about 400 years ago. This vertical sequence resulted from a somewhat basic application of the *Law of Superposition*, along with which the dates for each period were determined by Carbon-14 dating methods.

EVOLUTION OF THE COAST

Once upon a time at the spot where the authors of this book have a mountain house in western North Carolina, in place of the now serene, forested Blue Ridge Mountains, stood a stark, snow-capped mountain chain. Eventually, a significant part of that high mountain range was eroded and, in the process, provided the bulk of the sediment that now comprises the Coastal Plain of Georgia, including the light gray sand on the beaches. Unraveling the long history of how the present coast originated has fascinated geologists for many decades. The story is rather complicated, and there are still many unanswered questions regarding details, such as the exact age and timing of the different geological episodes. However, the following brief listing of this chain of events, based on a variety of sources (sciencedaily.com; earthquakes.usgs.gov; news.softpedia.com; sciencedaily.com; geology.er.usgs.gov; Glen, 1975; Hatcher, 2005), provides a general introduction to the subject. See also the time line in Figure 2.

There is still some question about the exact timing of all this, but conventional wisdom has it that about 400 million years ago, there were two main masses of land on the Earth separated by the Rheic Ocean. In the south sat **Gondwana**, a supercontinent consisting of what is now South America, Africa, India, Australia, and Antarctica, and in the north sat **Laurasia**, made up of North America, Greenland, Europe, and part of Asia (Figure 3A). There is still some dispute about the exact makeup of these two landmasses.

[NOTE: In the winter of 1964/65, Hayes led a geological expedition in Antarctica to examine a sequence of sedimentary rocks called the Beacon Sandstone of Permian age (around 250 million years old). This expedition is discussed in some detail in a book by Hayes (2011), titled *Coastal Heroes.* These rocks were very similar to other Permian units in South Africa and India, some containing huge petrified trees and coal beds, among other artifacts. The correlation of these very similar sandstones by the geologists at that time was supporting evidence for the original makeup of the supercontinent Gondwana. We also found evidence of a major glaciation episode during the Permian period (references: Matz and Hayes, 1966; Pinet, Matz, and Hayes, 1967; Pinet, Matz, and Hayes, 1971)].

Around 280 million years ago, during the Carboniferous period of the Paleozoic era (see geological time scale in Table 1), the two landmasses of Gondwana and Laurasia collided to form a single large continent called **Pangaea** (see Figure 3A). As a result of this collision, the Appalachian Mountains were formed (Dietz and Holden, 1970). These mountains are said by some to have been as high as the Himalayas and by others at least as high as the modern Rocky Mountains. During this time interval, the ancient igneous and metamorphic rocks that presently compose much of the Piedmont area of Georgia (zone between the fall line and the mountains – Figure 1) were intensely folded and faulted.

[NOTE: The three major classes of rocks–igneous, sedimentary, and metamorphic–were first defined by the pioneering Scottish geologist James Hutton in the mid 1700s. He recognized a class of rocks called **igneous rocks** that crystallized from an extremely hot molten mass of material (magma) that was formed as if by fire (*ignis* = fire). One class of igneous rocks, which includes granite, develops by slow cooling of the molten mass at great depths within the Earth's crust. As a result of the slow cooling, large crystals of individual minerals form that can be as much as an inch or so in diameter. These minerals, typically including feldspar and quartz, eventually coalesce to form a rock mass, which may later be pushed up to the Earth's surface by mountain-building processes. A second class of igneous rocks, called volcanic rocks (e.g., basalt and rhyolite), crystallizes rapidly as a result of molten lava being extruded suddenly upward toward and eventually onto the Earth's surface. The crystals of these minerals that form so rapidly are, for the most part, too small to be seen by the naked eye. On the other hand, **sedimentary rocks** are most commonly formed by weathering and erosion of preexisting rocks on the Earth's surface. This process creates sediment that can be transported by water, or wind in some cases, to be deposited in large masses on river deltas, beaches, and so on. Once deposited, these most commonly fine-grained sediments (sand and mud) may become buried

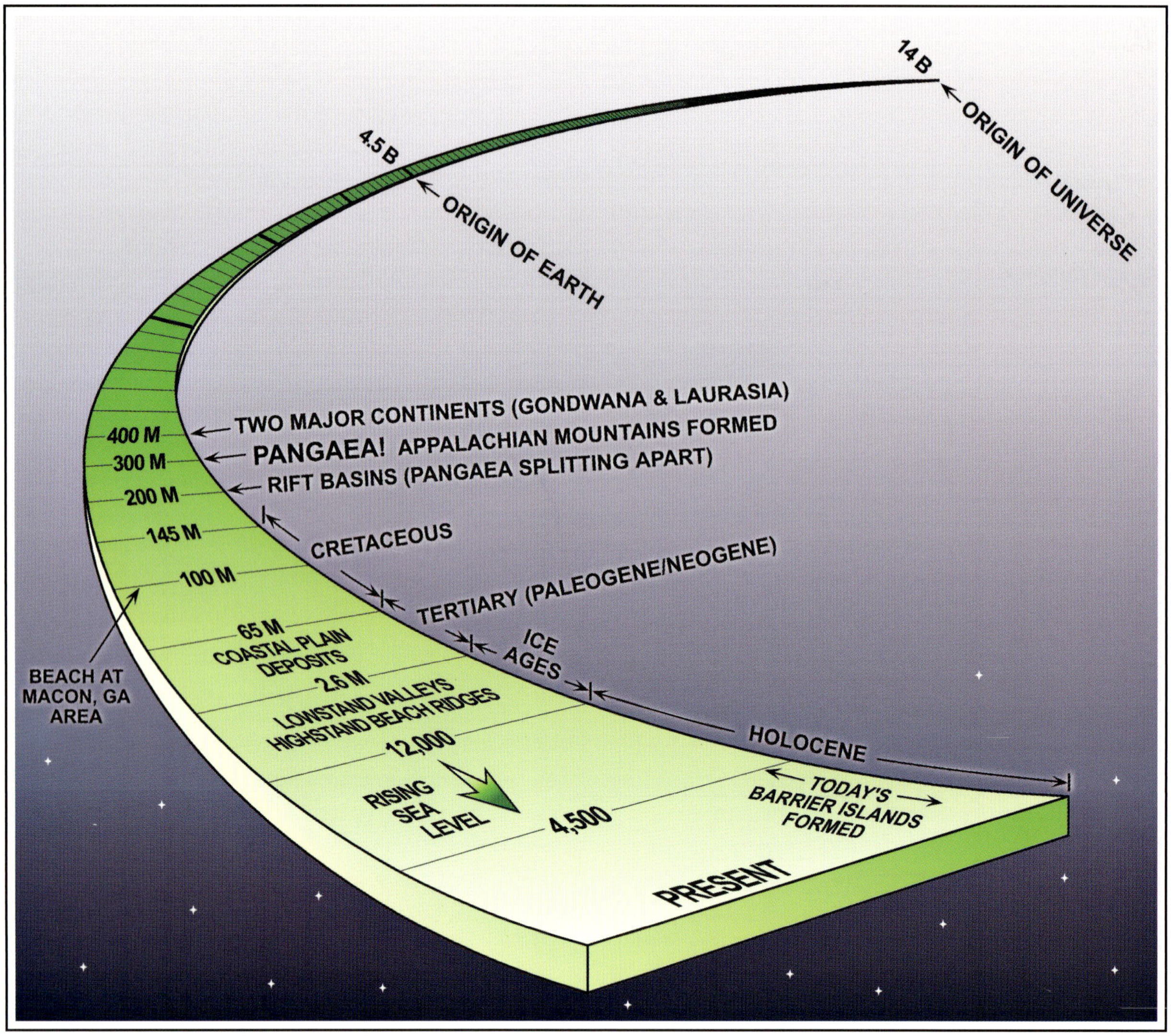

FIGURE 2. Time line for origin of the coast of the Georgia Bight. See Table 1 for nomenclature and timing of the geological events.

where they are consolidated by chemical cementation and other processes. The abundant sedimentary rocks formed by this process are called sandstone, mudstone, and shale. Limestone, which is also a sedimentary rock, usually develops in marine water by a combination of chemical precipitation and accumulation of the hard parts of some marine organisms, such as seashells. The third class, **metamorphic rocks,** results from dramatic changes in igneous and sedimentary rocks affected by heat, pressure, and water that usually result in a more compact and more crystalline condition. These changes usually take place at significant depths below the Earth's surface. Examples of metamorphic rocks include: 1) Slate, derived from a sedimentary rock called shale (originally composed of silt and clay); 2) Marble, derived from limestone; and 3) Gneiss, a banded, micaceous rock typically derived from granite.]

Between 225-190 million years ago, during the Triassic Period, the major landmass of Pangaea started to split apart along the old collision zones, forming several new continents. The present thinking is that *Antarctica has been near or at the South Pole since the formation of Pangaea* (since about 280 million years ago; New World Encyclopedia, 2010). This means that, apparently, while the other continents moved away, Antarctica stayed put. This raises some interesting questions about the ancient climate on Antarctica, but that is beyond the scope of this discussion. However, we now know that Antarctica has

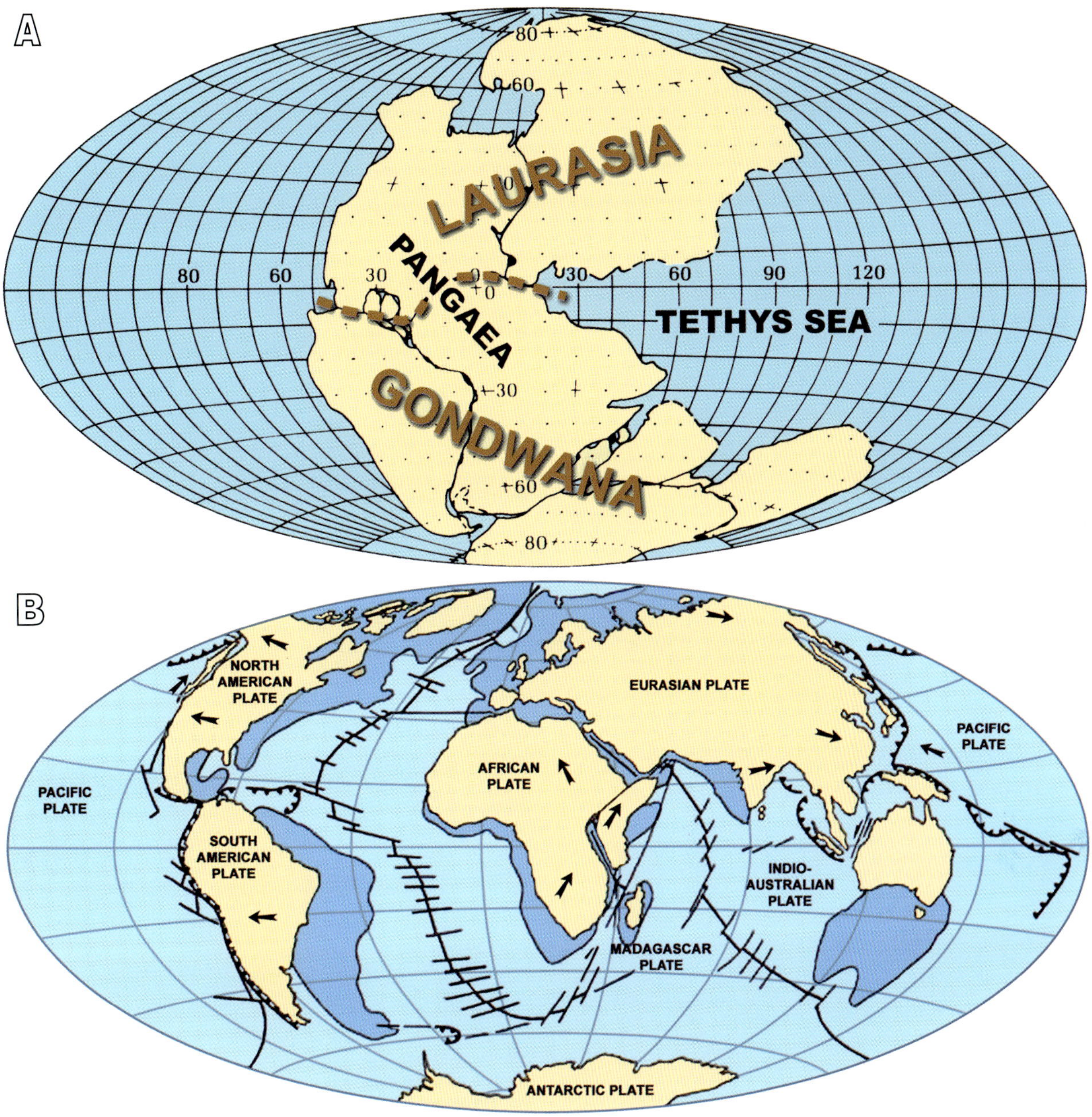

FIGURE 3. Two of the basic elements in the geological history of Georgia as we now know it. (A) Outline of all the continents that assembled to form **Pangaea**, a universal landmass as it existed at the end of the Permian period (Table 1) around 250 million years ago [as defined by Dietz and Holden (1970); modified after Glen (1975: Figure 8.8)]. (B) The major continents in their present position. Arrows indicate the directions the different landmasses have moved since they first started "drifting apart" around 200 million years ago. The blue "shadows" represent previous locations of the continents as they moved along.

been entirely free of ice from time to time back through geological time.

This pulling apart of the different segments of Pangaea created some **rift basins** in parts of the area now known as the eastern United States similar to the ones presently occurring in south-central Africa. These rift valleys filled with sediments eroded from the uplifted blocks along the rift margins. Some of the most noteworthy of these Triassic Rift Basins occur in Massachusetts and North Carolina. However, the map in Figure 4 shows the presence of one of these rifts that underlies much of the southern half of Georgia (South Georgia Rift).

Around 100 million years or so ago as the continents continued to drift apart (see Figure 3B, which shows the final product of this drifting), a huge arm of the world ocean, called the **Cretaceous Seaway,** covered much of what is now North America, extending north to south through the middle of the continent. All of the present United States south of the southern edge of the Appalachian Mountains was flooded by this large sea, which extended to the east into the ocean being created by the continents drifting apart. As shown in Figure 5, during the Late Cretaceous, the shoreline was near Macon, Georgia. The Cretaceous sedimentary rocks in Georgia were deposited in this ancient sea, containing fossils of oysters and other marine animals, as well as dinosaur teeth and bones washed from the shore. These sedimentary rocks can be seen today on the surface at the upper edge of the Coastal Plain. Also, they are important groundwater aquifers in the Coastal Plain region.

Between 65 and 2.6 million years ago, the **Coastal Plain** of the region we know as Georgia was transgressed (flooded) and exposed several times during what was formerly known as the **Tertiary period**. The five main topographic/geologic subdivisions of the state of Georgia, known as physiographic provinces, are shown in Figure 6. A succinct description of the Tertiary sedimentary rocks in the state of South Carolina that applies equally well for those in the state of Georgia, given in the paleoportal.org web site, follows: *Erosion of the Appalachian Mountains supplied much of the material that makes up the Tertiary rocks of South Carolina. Fossils from these rocks indicate that the climate of the Early Tertiary (Paleocene and Early Eocene) was warm and tropical. Worldwide cooling events generated a more moderate climate for South Carolina* (and Georgia) *during the Late Eocene. But by the Middle Tertiary (Oligocene), the climate began to warm again, as evidenced by fossils of whales and large crocodiles. Various marine and terrestrial fossils show that the climate was tropical to subtropical, much like modern Florida, by the Late Tertiary (Pliocene).* Tertiary rocks, which are predominately sandstone and shale, but do contain some significant limestone deposits, are exposed at the surface over much of the middle to upper reaches of the Coastal Plain. They also underlie deposits of Quaternary sediments on the continental shelf (Henry et al., 1981; Henry and Idris, 1992; Popenoe, 1992). Depositional environments of the Tertiary sediments ranged from deltas and other coastal deposits to offshore carbonate mud.

[NOTE: As shown in Table 1, modern stratigraphers have subdivided the formerly designated Tertiary period into two periods: Neogene (2.6 to 23.0 million years ago) and Paleogene (23.0 to 65.5 million years ago).]

A major change in climate that initiated the beginning of the Ice Ages (the **Pleistocene epoch**) occurred around 2.6 million years ago. The Pleistocene epoch ended around 12,000 years ago, after the last of the major glaciations ceased and sea level was already rising at a relatively rapid rate. The peak of this last major glaciation is said to have been around 21,000 years ago. Within the Pleistocene epoch, several major glacial events occurred, during at least four of which ice covered large areas of the North American continent (as far south as the Ohio River in the eastern United States). The approximate median peak times for these four major glacial events, expressed in years before the present (B.P.), have been estimated to be: 650,000 for the Nebraskan glaciation; 450,000 for the Kansan; 150,000 for the Illinoian; and 21,000 for the Wisconsin. During each of these major glaciations, sea level dropped to levels significantly lower than it is today; during the last glaciation period, the Wisconsin, it was 410 feet (+/- 16 feet) lower than it is today, according to Fleming et al. (1998)! Clark and Mix (2002) suggested that the drop of worldwide sea level during the last glacial maximum (Wisconsin) ranged from 410 to 440 feet below the present level. Also, Peltier (2002) proposed a low of 120 meters (394 feet); therefore, all three of these estimates are fairly close.

The determining of these sea levels during the Pleistocene and Holocene epochs is not an easy task, and research on this issue continues at a torrid pace. According to Muhs et al. (2004), *a new framework for sea-level studies began with a seminal paper on U-series*

FIGURE 4. Coast of the southeastern United States, illustrating the major bend in the shoreline known as the Georgia Bight – Cape Hatteras to Cape Canaveral (also called the South Atlantic Bight). Some of the principal tectonic elements, the Cape Fear Arch to the north and the Ocala Uplift (also called the Peninsular Arch) to the south, the South Georgia Rift (of Triassic age; Table 1), and the SE Georgia Embayment, are shown. Tectonic elements were delineated by LeGrand (1961) and Henry and Kellam (1988; modified from Paull and Dillon, 1980). The blue line marks the 500-foot offshore water depth, which is near the outer edge of the continental shelf.

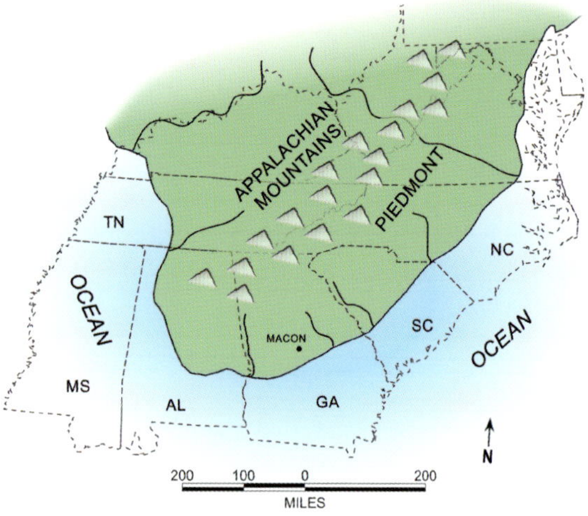

FIGURE 5. Approximate location of the Cretaceous Seaway in the eastern United States around one hundred million years ago. The shoreline of the Seaway was located not far from the present town of Macon, Georgia. Modified after D. T. King, Jr.

dating (discussed later) *that linked the deep-sea oxygen isotope record with uplifted interglacial coral terraces of Barbados (Broecker et al., 1968).* This information was linked to the Milankovitch (1941), or astronomical (orbital forcing), theory of climate change. *This theory postulates that as the Earth travels through space around the Sun cyclical variations in three elements of Earth-Sun geometry combine to produce variations in the amount of solar energy that reaches Earth* (Kaufman, 2002*)*. These three elements that have cyclic variations are eccentricity (the shape of the Earth's orbit around the Sun), obliquity (the tilt of the Earth's axis away from the orbital plane), and precession (change in orientation of the Earth's rotational axis – like the wobble of a spinning top) (Howard, 2002). *Using these three orbital variations, Milankovitch was able to formulate a comprehensive mathematical model that calculated latitudinal differences in insolation* (intensity of incoming solar radiation) *and the corresponding surface temperature for the 600,000 years prior to the year 1800* (Kaufman, 2002). This theory, that its author thought explained the repetitive cycles of glaciation that occurred during the Pleistocene epoch, has been questioned and modified over the years, but some form of it is probably the answer to the question of the cause of the Pleistocene glaciation cycles (Berger et al., 1999). To complicate matters a little more by considering conditions further back in geological time (remember the end of the Pleistocene was only 2.6 million years ago), some other major changes in the Earth's so-called boundary conditions need to be taken into account. These include: 1) Continental geography and topography; 2) Oceanic gateway locations and bathymetry; and 3) The concentration of atmospheric greenhouse gases. *These boundary conditions are controlled largely by plate tectonics, and thus tend to change gradually* (Zachos et al., 2001). Therefore, other major glacial episodes have happened back through geological time, a noteworthy one having occurred during the Permian period.

When sea level was lower during the major glaciations, during what is referred to as **lowstands**, the rivers along the southeast coast of the U.S. carved deep valleys across the coastal plain and out onto the present continental shelf. The locations of the largest of these valleys on the coastal plain of the Georgia Bight are shown in Figure 7. The glaciations were separated by warming periods, called interglacials, during which time sea level rose. In these warming periods, the highest sea levels were higher than it is today, with each succeeding interglacial having lower and lower **highstands**. Sea level at the peak of the last major interglacial, the so-called Sangamon (peak around 120,000 years B.P.), stood over 15 feet higher than it is today (discussed in more detail later). During these highstands, waves eroded scarps and deposited linear beach ridges inland of the present

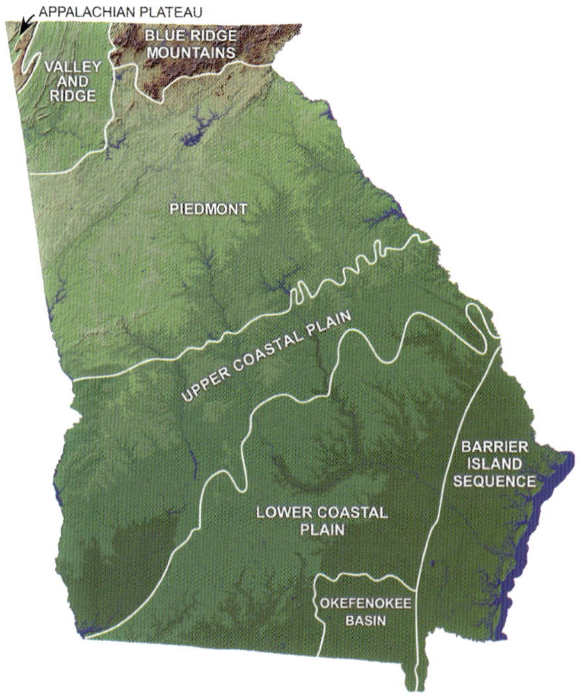

FIGURE 6. Physiographic Provinces of Georgia.

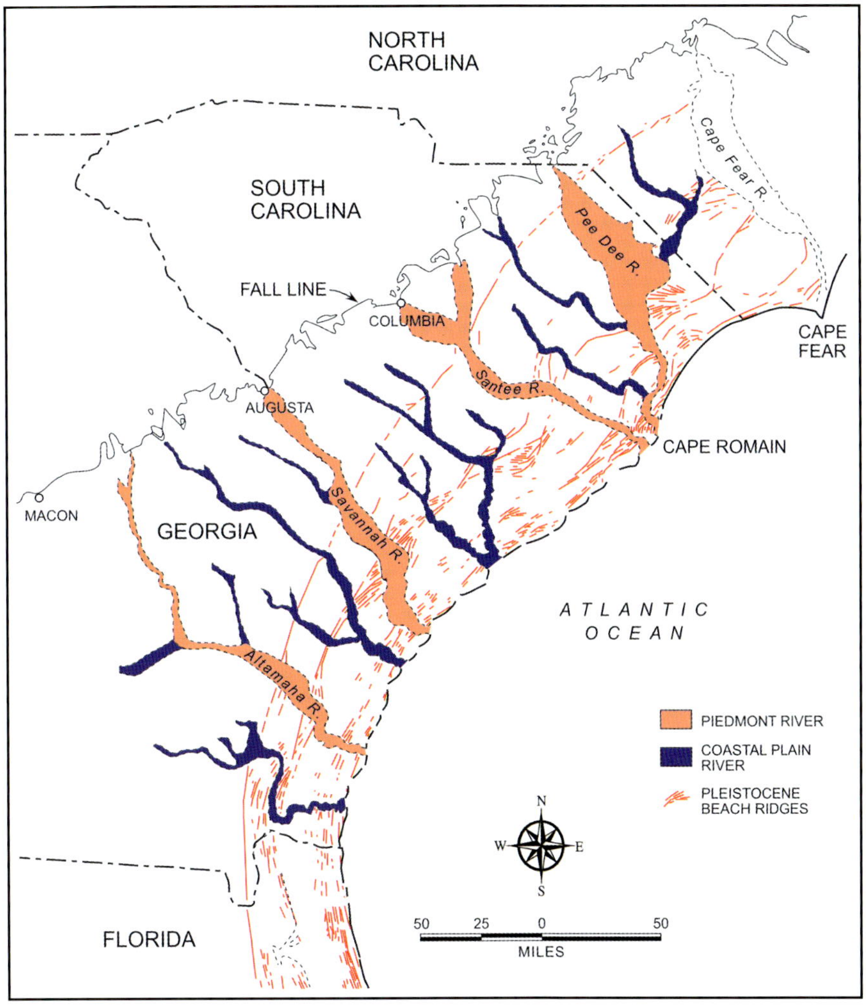

FIGURE 7. Lowstand valleys of the coastal plains of South Carolina and Georgia. The valleys were carved when sea level was lowered during the major glaciations of the Pleistocene epoch. The bulk of the flood plains that now occupy these valleys were formed during the last major rise in sea level, which started about 12-16,000 years ago. The valleys of the four piedmont, or red-water, rivers (Pee Dee, Santee, Savannah, and Altamaha), are shown in orange, and those of the major coastal plain, or black-water, rivers are shown in blue. Valley boundaries after Winker and Howard (1977).

shoreline. Pleistocene sediments blanket the lower portions of the Lower Coastal Plain (Barrier Island Sequence; Figure 6), which is relatively flat. During some of the lowstand episodes, the actual shoreline was near or at the edge of the continental shelf, a distance of 60-70 miles seaward from the present shore. Also, during episodes when sea level was relatively constant for a while during its covering and uncovering of the continental shelf, coastal features, such as river deltas and barrier-island complexes, were deposited; remnants of them are still present on the continental shelf along some parts of the Georgia Bight, creating very complex topography of the shelf in places. DePratter and Howard (1980) made the following comment regarding the character of the shelf off the Georgia Coast: *For the area now comprising the continental shelf, we envision, at any given time as sea level fell, a broad, forested, low coastal plain similar to the present lower Coastal Plain, with associated barrier islands, estuaries or lagoons, and perhaps tidal marshes. Numerous cores and other samples taken on the shelf (Pilkey et al., 1969; Blackwelder et al., 1979; Howard and Frey, 1980) contain faunal and facies evidence of backbarrier environments. Specific evidence of human habitation is lacking, however, because of the reworking that occurred as sea level rose (Bruun, 1962; Schwartz, 1967; Edwards and Merrill, 1977). All that remains of this transgression is a thin cover of reworked material and some fortuitous preservation of estuarine and tidal-inlet facies.*

(NOTE: The term **facies** is one geologists commonly use when referring to suite of rocks or sediments that have similar characteristics that distinguish it from neighboring units. All of the material within an individual facies suite would normally have been formed under somewhat similar conditions.)

The major ice-melting episode that started about 16-18,000 years ago marked the beginning of the end of the Pleistocene epoch (at which time sea level was possibly as much as 400 feet below its present position). The start date for the **Holocene epoch** is currently listed as 12,000 years ago (Table 1). As the melting proceeded, sea level rose rapidly, as much as 2 feet/century, reaching near its present level in Georgia around **4,500-5,000 B.P.** Because a near **stillstand** occurred at that time, the bulk of the major Holocene landforms on the Georgia Coast (e.g., deltas and barrier islands) have formed. However, as shown in Figure 8, there have been several minor sea-level fluctuations, up to four feet or so, in the past 5,000 years. The remainder of this book that relates to features formed during the Holocene epoch, deals with features, such as parts of the barrier islands, extensive salt marshes, and so on, that formed during that last, relatively short time interval.

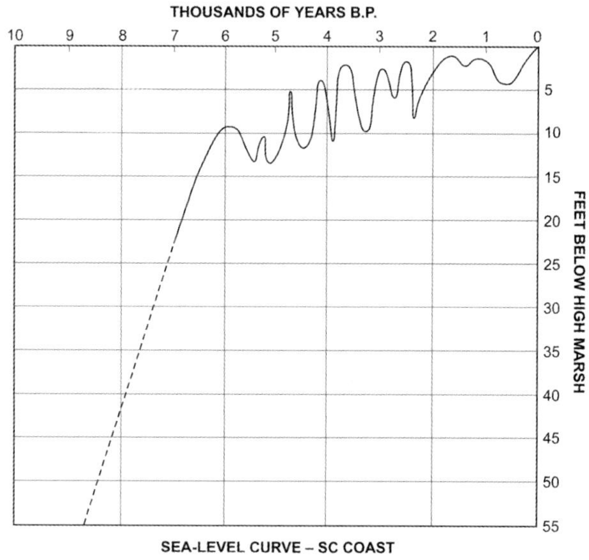

FIGURE 8. Changes in sea level along the South Carolina Coast over the past 6,000 years, based on detailed, combined archaeological and stratigraphical research (after Colquhoun and Brooks, 1986). Some of the lows in the curve between 2,000 and 4,000 years before present had a significant effect on the evolution of the barrier islands in the Georgia Bight.

This long evolution has left a final imprint on the state of Georgia, which consists of the five major Physiographic Provinces shown in Figure 6. Just a sliver of the **Blue Ridge Province** occurs in the extreme northeast portion of the state. This mountainous region has elevations ranging from 1,000 feet in the foothills to 4,784 feet at Brasstown Bald, the highest point in Georgia. The Blue Ridge Mountains of Georgia are but a small portion of a mountain chain that extends to the northeast into northern Virginia and southern Pennsylvania. There are also relatively small fragments of the **Valley and Ridge** and the **Appalachian Plateau Provinces** along the northwestern boundary of the state (Figure 6). The Blue Ridge and Valley and Ridge Provinces are bordered on the southeast by the **Piedmont Province.** Piedmont is derived from a French word meaning "foot of the mountains." This province consists of a northeast-southwest trending band that covers about one third of the state's area. The rock types in the Piedmont are primarily metamorphic, mainly schist, gneiss, and slate, and some granite igneous rocks that intruded into cracks and joints in the existing rocks about 200 million years ago. The Piedmont Province is separated from the **Coastal Plain Province** by the **fall line** (located on Figures 1 and 7), a somewhat imaginary line where, in this case, the stream courses descend abruptly as rapids cutting into the older bedrock of the Piedmont Province (at the fall line) and then flow on into the lowlands of the Coastal Plain Province, to never again erode into material older than Cretaceous sedimentary rocks. The gradients of the streams are also considerably less in the more southerly regions seaward of the fall line. The Coastal Plain Province, which has a maximum width of about 200 miles and covers over 60 percent of the state's area, does have some relief in its upper portions, with hills scattered here and there with a maximum elevation of around 750 feet. However, the lowest portions of the Coastal Plain, with their surface sediments of Pleistocene and Holocene age, are considerably flatter. According to the U.S. Geological Survey's digital atlas of the Physiographic Provinces of Georgia (Sandercock and Alhadeff, 1996), the Upper Coastal Plain is subdivided into the Fall Line Hills and the Fort Valley Plateau. The Lower Coastal Plain, on the other hand, contains six distinctly different subdivisions. However, because of their proximity to the coast and their relatively recent origin (during the Quaternary period), only the Okefenokee Basin (located on Figure 6), an area of low relief and swamps, and the Barrier Island Sequence, composed of marsh, sea islands, and older Pleistocene formations (Hodler and Schretter, 1986), will be treated any further in this discussion.

LOWER COASTAL PLAIN (GENERAL GEOLOGY)

In this discussion, we will concentrate on the so-called Barrier Island Sequence, which consists of: 1) The presently active barrier islands and marsh/estuarine complexes that make up much of the outer coast of the state; and 2) Six elevated Pleistocene barrier island coastlines and associated backbarrier deposits. Hails and Hoyt (1972) described the first (Holocene) unit as follows: *The Georgia Coast is rimmed by a chain of barrier islands separated from the mainland coast by two to six miles of salt marsh which is drained and flooded by a system of branching estuarine tidal channels. These in turn connect with the major tidal channels or sounds between the barrier islands. While most of the islands have a nucleus of Pleistocene material* (Silver Bluff Formation; defined later*), there is a wedge of Holocene deposits on the seaward margin, either in the form of a barrier beach separated from the islands by an extremely narrow area of Holocene salt marsh, or as a beach which overlies the Pleistocene sediments.*

The portion of the Lower Coastal Plain landward of the aforementioned more recently deposited features is another story. As pointed out by Winker and Howard (1977), more than a century ago the famous English scientist Sir Charles Lyell, the foremost geologist of his day, recognized a series of eroded scarps on the coastal plain of Georgia during his travels in North America (Lyell, 1845). These scarps were thought to have been formed during earlier highstands of the sea. Winker and Howard noted further that *the idea of Pleistocene transgressions of the Atlantic coast gained widespread acceptance as marine "terraces" were described in* Maryland, Virginia, North Carolina, and Georgia. Cooke (1931; 1932) *attempted the first regional descriptions by correlating the "terrace-formations" between states. Some recent studies have retained Cooke's terrace nomenclature (Hoyt and Hails, 1974; Colquhoun, 1969; 1974).*

The Barrier Island Sequence of the Lower Coastal Plain is illustrated in Figures 9 and 10. Hails and Hoyt (1972) described this area as follows: *Stratigraphical and morphological evidence shows that below an elevation of approximately 95 to 100 feet, the Georgia Coastal Plain consists of six segmented barrier island coastlines of Pleistocene age which trend parallel to the modern coast (Hoyt and Hails, 1967; Hails and Hoyt, 1968; 1969). These are known as the Wicomico (95 to 100 feet;* its present elevation*), Penholoway (70 to 75 feet), Talbot (40 to 45 feet), Pamlico (24 feet), Princess Anne (13 feet), and Silver Bluff (4.5 feet).* The Wicomico also goes by the name Trail Ridge near the Florida border. It also correlates with the Orangeburg Escarpment in South Carolina.

[NOTE: As more data are collected, the elevations of these Pleistocene barrier island systems given here may be modified, but most numbers suggested are pretty close to these (e.g., GeorgiaInfo, 2010b)].

These presently elevated "barrier island coastlines" were mapped by Hails and Hoyt (1974) as geological "formations," each of which consists of what they called lagoonal-marsh and barrier island facies (Figure 9). The lagoonal-marsh sediments, which are typically fine-grained mud, were said to contain sediments deposited in "estuaries and tidal channels." There are no lagoons *per se* on the present Georgia Coast. On the present-day coasts of Texas and North Carolina, large lagoons occur landward of long barrier islands. Both of these coasts have small tides and few riverine sediment sources, thus sediment has been slow to fill in the area landward of the barrier islands in those areas within the past 4,500 years. In Georgia, however, the large tides plus two piedmont river sources (Savannah and Altamaha Rivers) have carried enough sediment into the backbarrier areas to allow extensive tidal flats to develop that were eventually covered with marshes. In fact, there may have been lagoons landward of the barrier islands when they first formed beginning around 4,500 years ago, which, presumably, is why this term was used by Hails and Hoyt in describing these sediments. On the other hand, the barrier island components of these systems are typically composed of sand deposited on beaches, dunes, the offshore sides of the barrier islands (called the shoreface), and in major tidal inlets.

[NOTE: In his review of this book, Clark Alexander stated that *there is no evidence that barrier islands started to form in GA by 4,500 B.P.* He noted further that *the only evidence we have for barrier existence, DePratter's or*

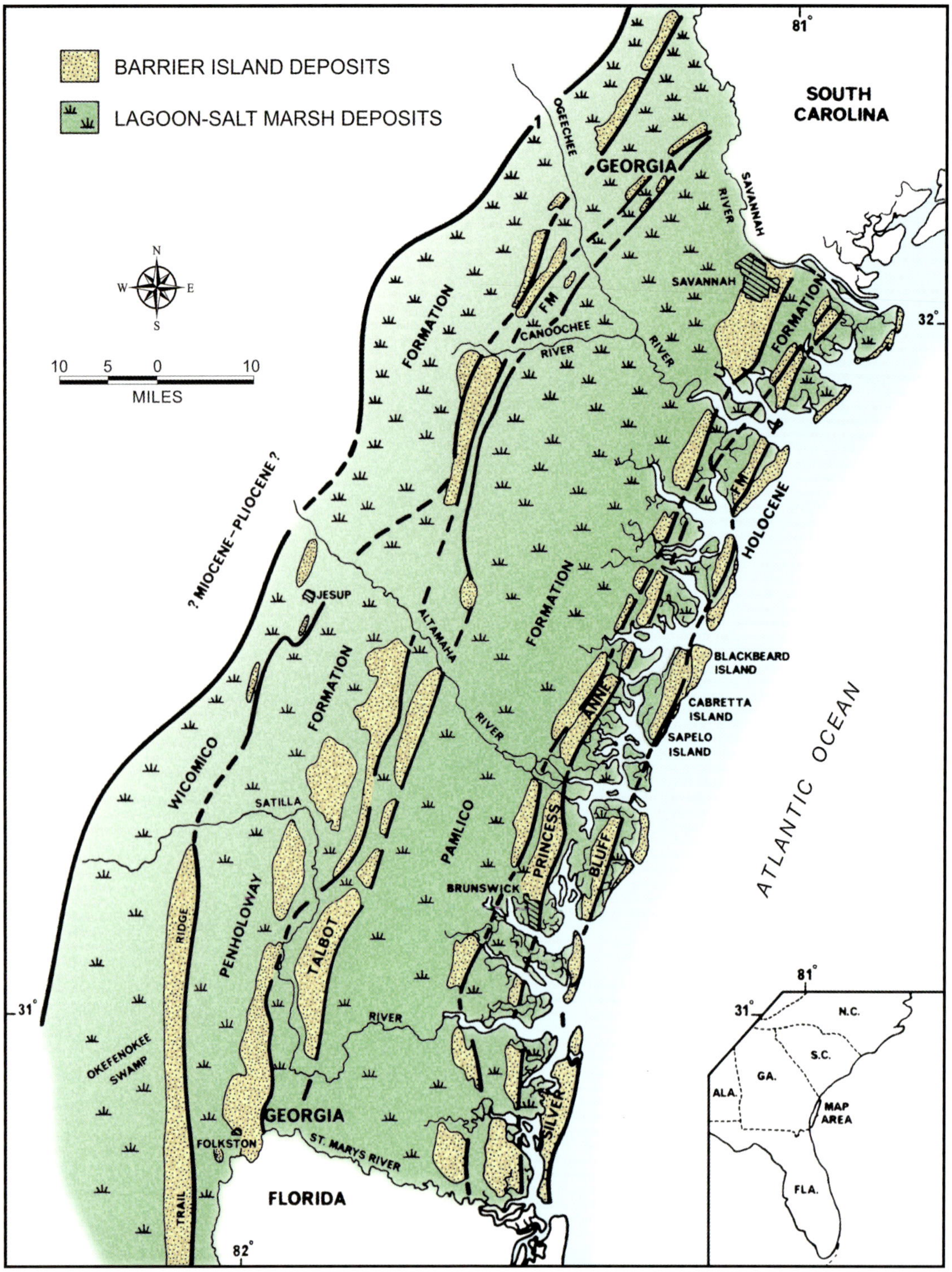

FIGURE 9. Distribution of the Pleistocene barrier island and lagoonal-marsh systems of the Barrier Island Sequence (Figure 6) on the Lower Coastal Plain of Georgia. The heavy lines represent the position of the beachfaces of the most seaward parts of the barrier islands during each of six highstands. At the time of formation, sea level was at 95-100 feet higher than the present sea level during the formation of the Wicomico (Trail Ridge) system and at 4-5 feet higher during deposition of the Silver Bluff system. Modified slightly after Hails and Hoyt (1972). In a map produced later, Huddlestun (1986) gave a different distribution of the Pamlico and Princess Anne units in the St. Marys area (discussed in text).

mine, shows no Holocene barriers prior to 2,000 B.P. Yet, the following two statements are presented in an in-press paper he sent us to read (Turck and Alexander, 2013): 1) *Basal deposits typically consist of relatively coarse, Pleistocene sands, especially where marshes closely flank barrier islands (Edwards and Frey, 1977; Frey and Basan, 1981). Marsh islands in interbarrier areas represent relict beach ridges and dunes, and must have formed within the last 4,500 years given the sea level history in the area*; and 2) *Holocene marsh occurred in the area between Sapelo and Blackbeard Islands sometime after 3,560 B.P. up until about 2,900 B.P. ..... This suggests that a Holocene barrier formed by 3,560 B.P., protecting the area from wave action.* Okay, so 3,560 is not 4,500, but 2,000 isn't either. It is true that data provided by DePratter and Thompson (2013) show that most of the Holocene ridges on the Georgia barrier islands are less than 2,000 years old, but even their data show a precursor to Tybee Island being around 2,900 years old and that beach ridges on the north end of St. Catherines Island are older than 2,400 years. There are two things to keep in mind. There is good evidence that sea level along the coast of the Georgia Bight rose to near its present level by 4,500 B.P. (before present) (Colquhoun and Brooks, 1986) or, possibly, only 4,200 years ago (Gayes et al., 1992). There is also excellent evidence that barrier islands started forming on the South Carolina Coast by 4,500 B.P. (Moslow, 1980; Ruby, 1981). In his comments, Alexander implied that there may not have been enough time for the sediments to migrate to the south away from their primary sources on the Georgia Coast such that the presently existing Holocene barrier islands themselves could not have started to form until around 2,000 B.P. The two areas in South Carolina studied by Moslow and Ruby are located a relatively short distance to the south of the combined Santee/Pee Dee Rivers, the largest potential sediment source in the Bight. Other factors, such as differential rates of uplift or sinking and varying erosional/depositional rates may also be involved in causing the proposed scarcity of older sands in the Georgia barrier islands. Clearly, there is still something to learn about why these dates differ in South Carolina and Georgia. Maybe a couple of MS theses?]

The map in Figure 9 shows the distribution of these six barrier island coastlines, and Figure 10 shows a topographic/geological cross section that illustrates the stair-step nature of them. Of these systems, the Wicomico (or Trail Ridge) is the oldest. Figure 11, a DEM (digital elevation model) re-creation of the coastal area including and landward of St. Simons Island, gives an oblique view of the three most seaward barrier island ridges.

As noted by Johnson et al. (1974), at the time of the Wicomico highstand, *a great barrier island, now called Trail Ridge, formed in front of the area now know as the Okefenokee Swamp which was then a very large salt marsh.* The Trail Ridge system extends further south into Florida; however, the extension of it known as the Wicomico system is now discontinuous further north in the Lower Coastal Plain of Georgia because it has been largely eroded away (Hoyt, 1968). The line indicated as

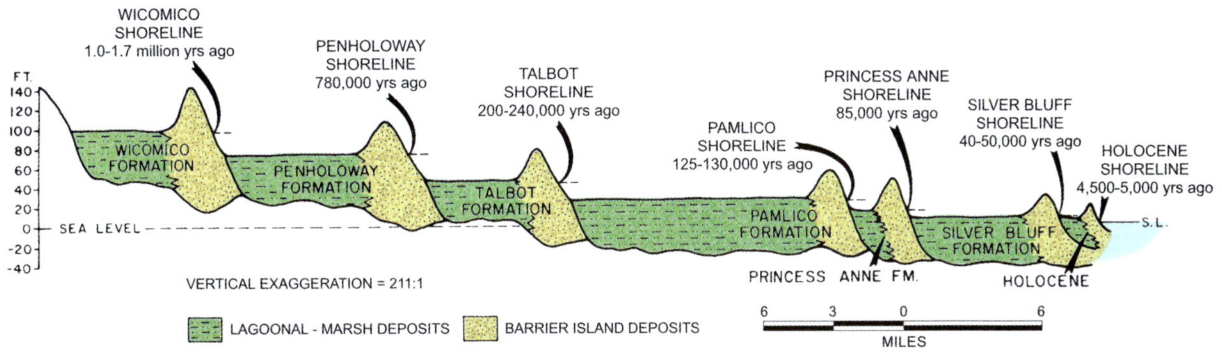

FIGURE 10. General schematized NW-SE geologic/topographic cross-section across the Barrier Island Sequence (Figure 6) on the Lower Coastal Plain of Georgia. Arrows point to approximate mean sea level during the six maximum highstands of the Pleistocene epoch. The designated stratigraphic "formation" for each of the six units includes fine-grained, muddy sediments deposited in backbarrier environments associated with the barrier island (marsh, tidal flat, etc.) and the sandy sediments associated with the barrier island itself. During Wicomico time, sea level stood 95-100 feet above the present sea level, and during Silver Bluff time, it was only 4-5 feet higher. Modified slightly after Hoyt and Hails (1974).

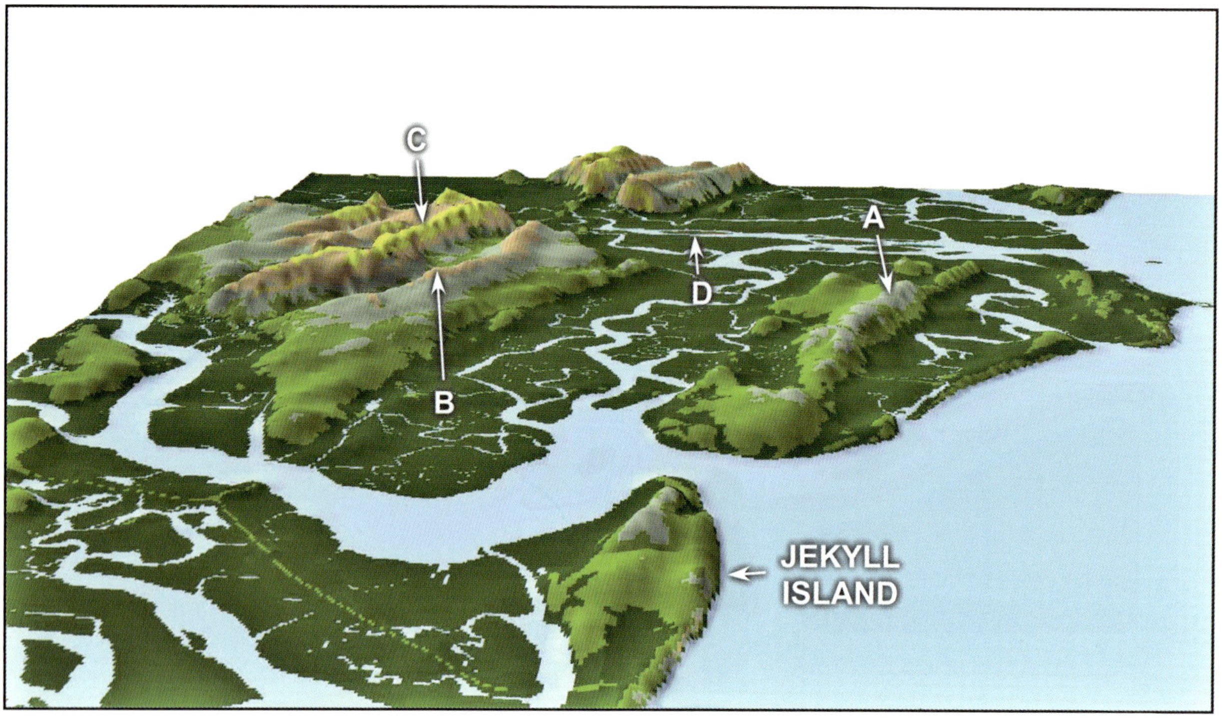

FIGURE 11. DEM (Digital Elevation Model) re-creation of the coastal area including and landward of St. Simons Island. The elevated uplands of remnants of the Pleistocene barrier island systems illustrated in Figures 9 and 10 clearly stand out with high vertical exaggeration. Arrow A points to the Silver Bluff barrier island on St. Simons Island, arrow B points to the Princess Anne barrier island on which Brunswick and the Hofwyl-Broadfield Plantation State Historic Site are located, arrow C points to the Pamlico barrier located immediately to the north of I-95, and arrow D points to the valley of the Altamaha River.

the landward limit of the Wicomico Formation on Figure 9 is presumably the most landward extent of the shoreline during this, the earliest, Pleistocene highstand. Further to the northeast into South Carolina, this line apparently merges with the eroded Orangeburg escarpment.

If one assumes that no major uplift of the Georgia Coast has taken place within the past million years or so, the diagram in Figure 10 would indicate that the height of the maximum sea level during the periods between the major glaciations gradually decreased through time, from an elevation of 95 to 100 feet when the oldest system, the Wicomico (Trail Ridge), developed, to only 4.5 feet when the youngest system, the Silver Bluff, was formed. The geological evidence indicates that, if anything, the land on the central Georgia Coast has gradually sunk, not elevated, during this time; therefore, the elevations just stated for the Pleistocene sequences may have been somewhat higher when they were deposited than they are now (relative to the modern sea level). Details on the dates when these units are thought to have formed will be given in the next section. The elevations of these individual units are fairly laterally consistent in the central portion of the Lower Coastal Plain of Georgia. However, this is not the case with regard to the Pleistocene shorelines of the whole Georgia Bight, as was clearly shown in a revealing study by Winker and Howard (1977).

Winker and Howard combined the six previously described Pleistocene "barrier island" assemblages (illustrated in Figures 9 and 10) into three major Pleistocene coastal sequences, the Chatham, Effington, and Trail Ridge Sequences, which are illustrated in Figure 12. Apparently, the Chatham Sequence includes Hails and Hoyt's (1972) Silver Bluff, Princess Anne, and Pamlico Formations, the Effington Sequence includes the Talbot and Penholoway Formations, and the Trail Ridge Sequence is made up of the Trail Ridge/Wicomico Formations. The components of the Chatham Sequence are similar to the present geomorphology of the coastline of the Georgia Bight, with a "sea island" section in Georgia that *contains broad tidal inlets and, in the Carolinas, a cuspate coast*. As shown in Figure 13,

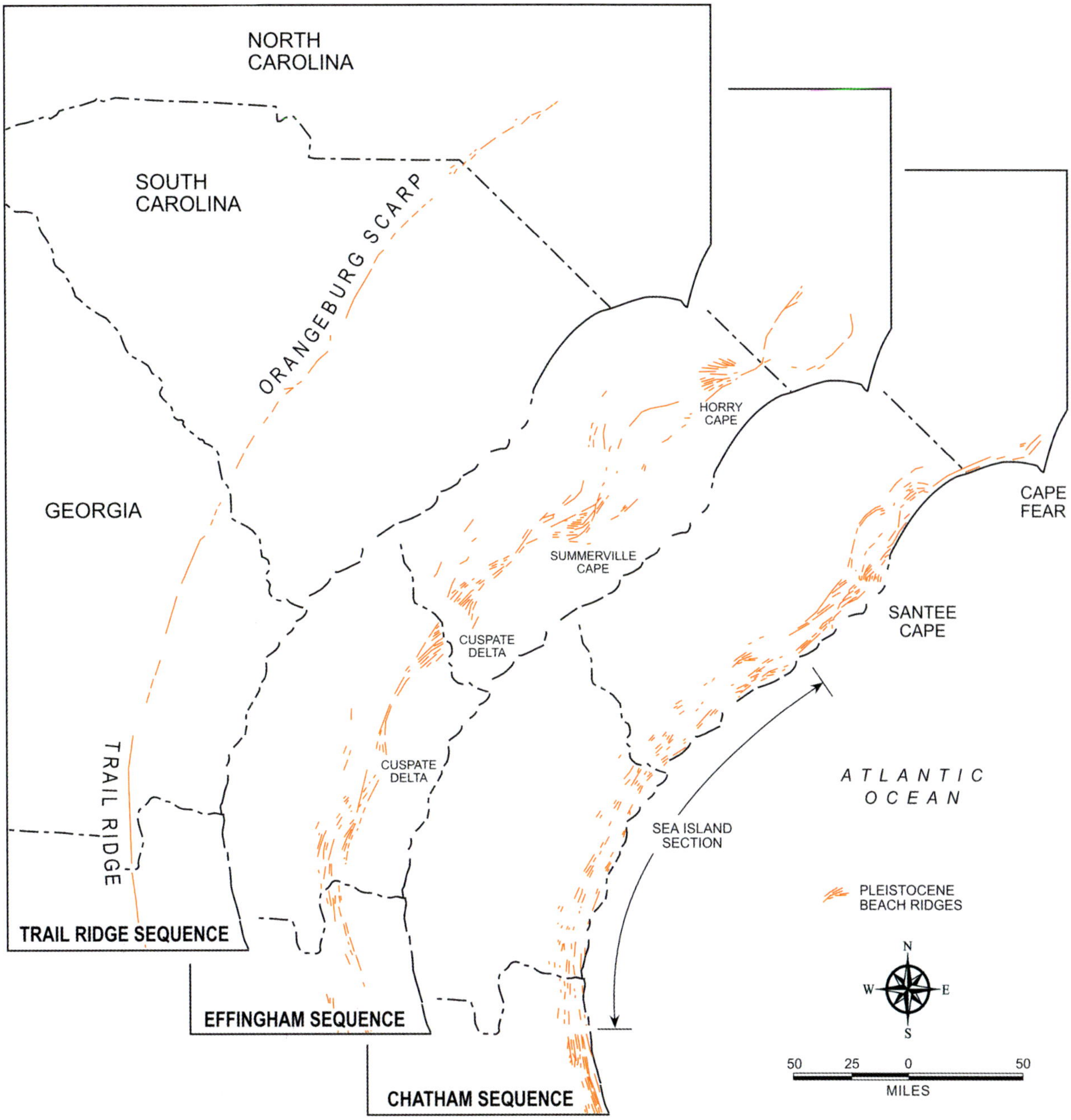

FIGURE 12. Panels representing three different sequences of coastal landforms - the Trail Ridge Sequence, representing the Trail Ridge barrier island system shown in Figures 9 and 10 in the south as well as the Orangeburg escarpment further north (presumably eroded around 1-1.5 million years ago), the Effingham Sequence, which includes the Penholoway and Talbot shorelines shown in Figures 9 and 10 (dates in the literature span from 780,000 to 200,000 years ago), and the Chatham Sequence, which includes the Pamlico, Princess Anne, and Silver Bluff Sequences shown in Figures 9 and 10 (from 125,000 to 40,000 years ago). Modified from Winker and Howard (1977).

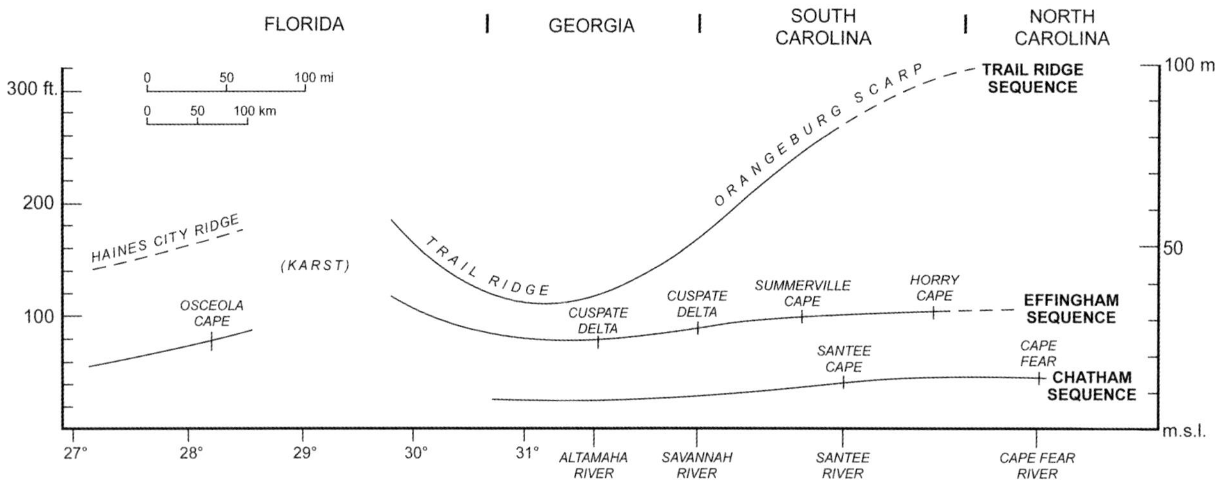

FIGURE 13. Tentative curves of shoreline warping of the Pleistocene sequences of the Lower Coastal Plain of the Georgia Bight. These lines represent the maximum height of transgression during highstands within the Pleistocene epoch for each of the shoreline sequences illustrated in Figure 12. Modified from Winker and Howard (1977).

this sequence shows a systematic variation in elevation from north to south (higher to the north and lower to the south) of at least 16.5 feet. The Effington Sequence, on the other hand, *is characterized by broad, gently seaward-sloping strand plains that form prominent capes and cuspate deltas readily distinguished from older and younger sequences* (Figure 12). This sequence shows a systematic variation in elevation from north to south (higher to the north and lower to the south) of around 50 feet (Figure 13). The so-called Trail Ridge Sequence is *a broad sand ridge in Georgia and north Florida (Pirkle and Oho, 1970)* that *can be traced northward to where it becomes the Orangeburg (or Citronelle) scarp* (in South Carolina; Colquhoun, 1974). As best we can tell, the Trail Ridge Formation correlates with the Wicomico Formation further to the north in Georgia. Clearly, then, the Trail Ridge/Wicomico Sequence is the oldest of the Pleistocene sequences on the Lower Coastal Plain of Georgia, a considerable length of which consists of an erosional scarp, rather than a depositional barrier island assemblage. The variation in elevation of this sequence from north to south (higher to the north and lower to the south) is almost 200 feet (Figure 13). A major conclusion of the Winker and Howard paper was that *all three shoreline sequences have been deformed. Warping continued through Pleistocene time, following persistent Cenozoic structural features*, namely the Cape Fear Arch (to the north), the Ocala Arch (to the south), and the Southeast Georgia Embayment (in the middle; see Figure 4). That is, these measures of vertical change are the result of uplifting along the two border arches and relative downwarping along the central Georgia Coast (Murray, 1961; LeGrand, 1961; Henry and Kellam, 1988).

[NOTE: The term cuspate delta refers to a delta that protrudes out into the ocean in a triangular shape, something like the major cuspate forelands Cape Fear and Cape Canaveral. However, these capes are not deltas, but are the result of converging, opposing wave directions. The surface of a cuspate delta is composed of a series of parallel, prograding sandy beach ridges. Another term for this type of delta is wave-dominated delta; therefore, their presence in the older Pleistocene sequences presumably means that either the waves were larger or the tides were smaller at that time than today on the Georgia Coast, where capes and cuspate deltas are absent. Deltas will be discussed later.]

QUATERNARY SEA LEVELS

Although understanding the history of the Quaternary period (Pleistocene and Holocene epochs) is still in its relative infancy, in recent years some progress has been made with regard to comprehending the changes of sea level during this period as the result of new developments in Quaternary dating methods. Two of the primary methods are: a) Measurements of uranium-series disequilibrium; and b) Amino acids.

As noted by Muhs et al. (2004), *the direct dating of emergent marine deposits is possible because U* (uranium) *is dissolved in ocean water but Th* (thorium) *and Pa* (protactinium) *are not. Certain marine organisms, particularly corals, but not mollusks, co-precipitate U directly from seawater during growth.* Because the other two elements are not co-precipitated, if they are present in the coral, they got there by the decay of one of the three naturally occurring isotopes of U incorporated into living corals. Thus, there are *three independent clocks for dating the same fossil coral* (Gallop et al., 2002). This method can date materials as old as 500,000 years. It is an excellent method to use when fossil corals are available. If not, there are two other methods to rely on. Carbon-14 dating is commonly used for preserved formerly living plant or animal remains less than 50,000 years old. For older materials, amino-acid measurements (Wehmiller and Miller, 2000) have provided a complementary method for systems such as the Pleistocene barrier sequences on the coastal plain of the Georgia Bight. Muhs et al. (2004) pointed out that *the basis of amino-acid geochronology is that proteins of living organisms (such as marine mollusks) contain only amino acids of the L configuration. Upon the death of an organism, amino acids of the L configuration convert to amino acids of the D configuration, a process called racemization.* This racemization process continues at a rate that can be measured, hence the time since the organism died can be determined. These methods are best applied to fossil mollusks, which occur on virtually every coast of the world.

With regard to Pleistocene and Holocene sea level histories, the record for both glacial and interglacial episodes and, hence, sea level changes, is reflected in oxygen-isotope composition (i.e., relative amounts of ^{16}O and ^{18}O) of carbonate shells of foraminifera preserved in deep-sea sediments (Emiliani, 1955; Imbrie et al., 1984). The oxygen-isotope compositions of foraminifera *are a function of both water temperature and the oxygen isotope composition of ocean water at the time of shell formation. Foraminifera precipitate shells with more* ^{18}O *in colder water. The oxygen isotope composition of ocean water is a function of the mass of glacier ice on land, because glacier ice is enriched in* ^{16}O*. Thus, relatively heavy (*^{18}O*-enriched) oxygen isotope compositions in foraminifera reflect glacial periods, where light compositions (*^{16}O*-enriched) reflect interglacial periods.*

[NOTE: Foraminifera are small (usually less than 1 millimeter in size), single-celled marine organisms that float freely in the ocean in large numbers. They have a calcareous shell, or test. About 275,000 species are recognized, both living and dead. The first ones appeared about 500 million years ago. Many extinct species existed for geologically short periods of time and both extinct and living species can be limited to a very narrow environmental range (UCal Berkeley, 2004). Consequently, geologists use them for both dating relative ages of rocks and sediments and for interpreting past environments.]

To put this all in perspective, Figure 14A shows a sea level curve based on the changes in deep-sea oxygen isotope records based on measurements on foraminifera tests that are present in deep-sea sediment cores. The peaks in the curve represent times of warming and the lows represent periods of cold that are usually accompanied by widespread glaciation. This record, which goes back 500,000 years, shows five major periods of warming (interglacials?). The last major interglacial showing a low ice volume similar to that of today *started 130,000 years ago, peaked at 125,000 years ago, and terminated 116,000 years ago* (Kukla, 2000; indicated by the P on Figure 14A).

Another way to determine these major climatic cycles is to study cores of ice taken in areas such as the Antarctic ice field (Petit et al., 1999) and from the Greenland Ice Core Project (Kukla, 2000). In January 1998, a collaborative cooperative ice-drilling project among Russian, United States, and French scientists carried out at the Russian Vostok station in East Antarctica resulted in their taking an ice core that reached a depth of nearly 12,000 feet, the deepest ice core ever recovered. This allowed the acquisition of ice up to 450,000 years old. The peaks and valleys in the blue curve in Figure 14B show the past changes in Earth temperatures relative to the present based on measures of either irradiance and/or hydrogen and oxygen isotopes of the ice cores.

The green curve shown in Figure 14B, also based on the Vostok data, is a record of the historic volumes of carbon dioxide (CO_2) in the atmosphere, determined by measuring the CO_2 content of bubbles of air (fluid or gas inclusions) trapped in the ice. Clearly, the graph shows that the amount of CO_2 typically increases with increasing temperature. It appears that even though glacial cycles are thought to be caused by changes in the Earth's orbit, or the Milankovitch cycles, discussed earlier, these changes must also influence the carbon cycle. This is an interesting issue, which goes beyond

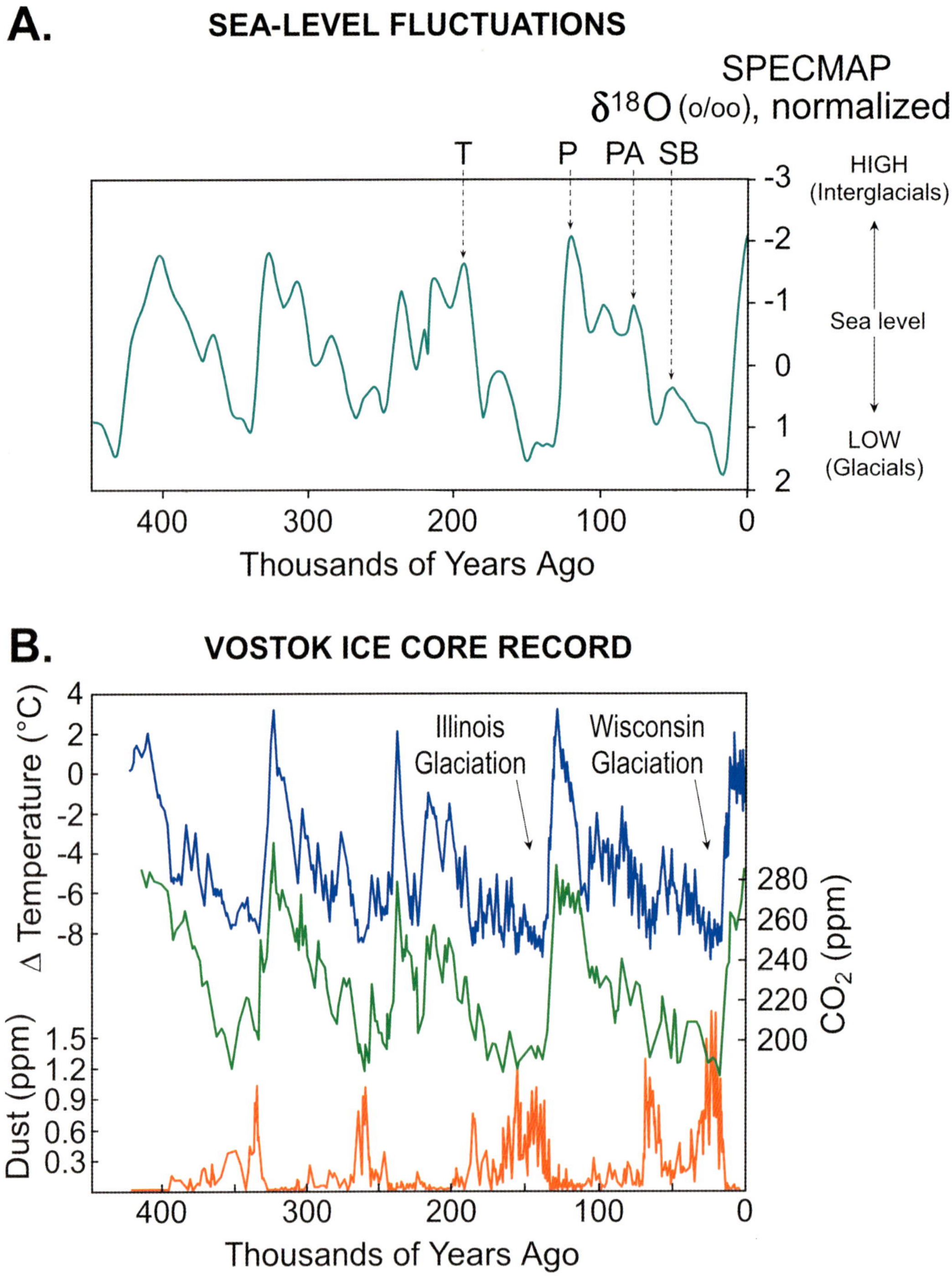

FIGURE 14. Data indicating the occurrence of glacial and interglacial periods extending from the present to 450,000 years ago. (A) Sea level curve based on oxygen-isotope data collected from calcareous foraminifera in deep-sea sediments. The letters signify high points on the curve that might coincide with the highstands during which the Silver Bluff (SB), the Princess Anne (PA), Pamlico (P), and Talbot (T) Pleistocene barrier island sequences (Figures 9 and 10) may have formed. Data derived from the SPECMAP record (the standard chronology obtained from several oxygen isotope records) presented by Imbrie et al. (1984). Slightly modified from Muhs et al. (2004). (B) Graph of temperature change relative to present (blue), CO_2 (green), and dust concentration (red) measured from the Vostok, Antarctica ice core as reported by Petit et al. (1999).

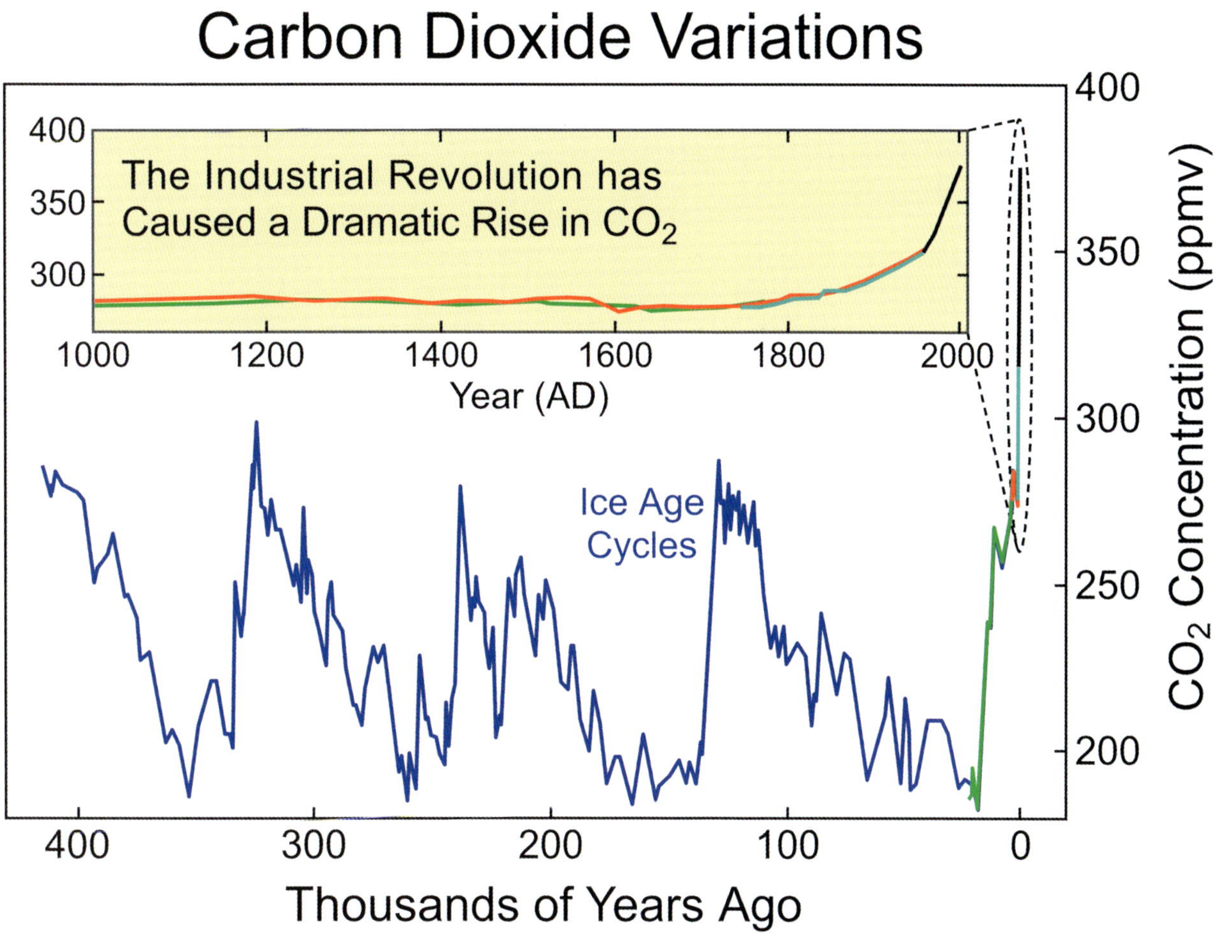

FIGURE 15. The historical level of carbon dioxide which is averaged by diffusion of gases in the Vostok ice core (Antarctica) and sampled once per approximately 1,000 years, compared to the current level of carbon dioxide, which is sampled yearly. An obvious conclusion is that, since the Industrial Revolution started (around 1900), the burning of fossil fuels has caused a dramatic increase of CO_2 in the atmosphere, reaching levels unprecedented in the last 400 thousand years. This figure was prepared by Robert A. Rohde from publicly available data and is incorporated into the Global Warming Art project.

the scope of this book. However, we feel it would be of interest to include the curve shown in Figure 15, which, if the data are sound, indicates that since the industrial revolution started around 1900, the burning of fossil fuels has caused a dramatic increase in CO_2 in the atmosphere (Rhode, 2009), a scary prospect indeed. We don't want to be the prophets of doom or anything, so we will let someone else do it, namely W. S. Broecker in his 2002 book on glaciers in which he said: *With population increasing and with the world's poor becoming less poor, energy consumption is bound to increase. In the absence of a minor miracle, this energy will come largely from burning fossil fuels. I believe that unless some drastic action is taken, by early in the next century* (this book was first published in 1993 and later revised, so we think he is referring to the century we are living in now), *we will have quadrupled the atmosphere's CO_2 content (i.e., from pre-industrial 280 ppm to ~1100 ppm). Compared to the minuscule changes in solar luminosity (a few parts in 12,000), this rise will alter the radiation balance by 15 to 40 parts in 1000. To assume that such a change will not lead to significant impacts on wildlife, agriculture, water availability .... In my estimation is ludicrous.* And, needless to say, a marked rise in sea level should also occur (discussed in more detail later).

[NOTE: In May 2013, the global concentration of CO_2 reached 400 parts per million. This is a stark reminder of the human impact on our environment.

The red curve in Figure 14B shows the volume of dust in the air for the same time interval as the other two curves based on the Vostok core. The periods of highest dust appear to correlate to some extent with the glacial episodes (cold dry periods?).

Based on a variety of literature sources, the highstands of sea level during which the barrier island sequences on the Lower Coastal Plain of Georgia (Figures 9 and 10) were formed occurred at the following times:

1) **Holocene** – Easily the most reliably dated of the seven sequences. Numerous studies place the beginning of this last highstand at 4,500-5,000 years ago (Colquhoun and Brooks, 1986 – see Figure 8; DePratter and Howard, 1980; Moslow, 1980; Ruby, 1981).

2) **Silver Bluff** – Possibly the most debatable of all the timing estimates – 40-50,000 years ago (Henry, 2009; Muhs et al., 2004). Blackwelder, Pilkey, and Howard (1979) sampled intertidal marsh sediments with in-place oyster shells on the continental shelf off the South Carolina Coast (mostly) and concluded that 36,000 years ago, sea level stood at its present elevation, which isn't that different from the other estimates. According to Hails and Hoyt (1972), the present elevation of the Silver Bluff barrier island system is 4.5 feet. The sea level at the time of its formation was probably a little higher than that, because of the continued slow sinking of the land.

3) **Princess Anne** – 85,000 years ago (Willis, 2006).

4) **Pamlico** – 125-130,000 years ago (Willis, 2006; Muhs et al. (2004). This one appears to match the last maximum interglacial period similar to the present one, with respect to the amount of ice on the surface of the Earth during the two different time intervals (Kukla, 2000).

5) **Talbot** – 200-240,000 years ago (Willis, 2006; Muhs et al., 2004).

6) **Penholoway** – 780,000 years ago (Muhs et al., 2004).

7) **Wicomico/Trail Ridge** – 1.0-1.7 million years ago (Winker and Howard, 1977). Said to be 2.2 million years ago by Burdette et al. (2013) using the ESROD method described earlier.

[NOTE: Clark Alexander (pers. comm., 2013) pointed out that John Wehmiller says that *Skidaway is 80 kys old, and my OSL date for Pleistocene Ossabaw was 100 K, so there is no evidence for 40-50 ky ages* (for the Silver Bluff) *as far as I've seen*. However, Turck and Alexander (2013) state that *two OSL samples from the Silver Bluff-age Skidaway Island ... returned ages of 46,440 and 45,740 B.P., respectively and also a sample of the Pleistocene Sapelo Island returned an age of 56,340 ± 9,000 B.P.* And in that same paper, the following statement was made: *In a recent study by Booth and Rich (1999), freshwater peat from a core extracted from the Silver Bluff section of St. Catherines Island was radiocarbon dated to over 40,000 B.P. The peat was most likely deposited during a time of lowered sea level, indicating that Silver Bluff sections of the present-day barrier islands formed over 40,000 B.P. (Linsley, Bishop, and Rollins, 2008). Wehmiller et al. (2004) report that the Slver Bluff islands are approximately 80,000 B.P. based on U/Th dating.*

So, as you can see, for some reason, the dates for the Silver Bluff are all over the place (we could cite several more but you probably get the point). Some of these later dates were arrived at using the new OSL and ESROD methods. Sorry to bring out all this confusion, but that is the way science goes sometimes. One fine day, all of this will be worked out and a more updated sequence of events will be published. Throughout the remaining discussions in this book, we will stick with the 40-50,000 dates that were originally reported by Henry (2009) and Muhs et al. (2004).]

The times proposed for the Silver Bluff, Princess Anne, Pamlico, and Talbot sequences are indicated by arrows on Figure 14A, showing a pretty good correlation of these highstands with warmer temperatures in the deep oceans, although these exact dates are far from certain. The "Silver Bluff peak" indicated on Figure 14A is rather small, perhaps because the warming that caused that highstand was too short-lived to significantly impact the temperature of the water in the deep ocean and, therefore, its oxygen isotope ratio. Also, the curve in Figure 14A does not purport to show the actual sea levels. Those are more accurately determined by on-land studies, such as the ones by Hails and Hoyt (1972; 1974). Any other theories to explain this anomaly between ocean water temperatures and actual sea levels would be welcome at this time.

As already noted, research continues on this topic at a brisk pace, with an unsurprising large amount of disagreement, notably from workers studying different geographic locations around the globe.

3 COMPARISON WITH OTHER COASTS

All coasts have a rather complicated history, such as the one just described for the coast of Georgia. To put the previous discussion in context, a brief look at how other coastlines around the world have formed and how the Georgia Coast compares with them follows.

Based on our experience working along many coastlines around the world, as well as a long-term study of global patterns using charts and satellite images, we now conclude that the following hierarchy of factors usually determines the geomorphological makeup of a coast–in decreasing order of importance: 1) Global tectonic crustal movements; 2) Hydrodynamic regime; 3) Climate (including storms); 4) Sediment supply and sources; and 5) Local geological history and sea-level change.

When Hayes was a graduate student at the University of Texas in the 1960s, very few of the professors in the elite, Ivy League universities "believed in" continental drift, although the idea had been seriously proposed and discussed by scientists since almost the beginning of the 20th century. The problem for the "non believers" was that no credible mechanism had been proposed to explain why the continents had drifted apart. Now, of course, the concept of **plate tectonics** provides the basis for almost all current thinking relative to most topics in geology, including the mechanism for continental drift. Geophysical evidence, such as increasing age of parallel magnetic bands of volcanic rocks on the sea floor away from the mid-ocean ridges, provided indisputable evidence that the continents are indeed drifting apart. The exact mechanism causing this movement is more in doubt. The favored theory at this time is that huge individual slabs of rock (called **plates**, the existence of which is not in doubt) are slowly moved apart by convection cells of hot, softened mantle material located below the rigid plates.

To better understand plate tectonics, we start with the internal structure of the Earth, illustrated in Figure 16, which is composed of three major components:

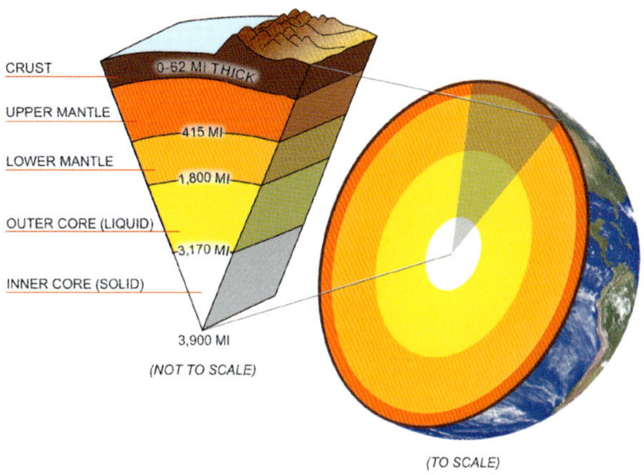

FIGURE 16. Cross-section of the Earth showing its internal structure [modified from a diagram created by Jeremy Kemp, which was based on elements of an illustration by the USGS (2005)].

1) The surface material is called the Earth's **crust**. The crust is composed of a variety of igneous, metamorphic, and sedimentary rocks. The crust under the oceans, which average about 6 miles thick, is composed mostly of rocks that contain an abundance of iron and magnesium-bearing minerals (e.g., basalt), whereas the crust on the continents, which averages 20-30 miles thick, but can be up to 62 miles thick, is composed of rocks made up of lighter minerals (e.g., granite). Because the crust is relatively cold, it is comprised primarily of brittle rocks (plus unconsolidated sediments in some areas). These rocks and sediments are composed predominantly of a group of minerals called silicates, the fundamental building block of the Earth's crust, with the atoms of silicon and oxygen combined making up about 75% of it. Because of their relative compactness, the rocks that make up the crust are susceptible to shearing and breaking, which, of course, causes the earthquakes that occur with some degree of regularity in areas where the tectonic plates are converging (e.g., California, Chile, Haiti, and Japan).

2) The **mantle**, which contains more of the heavier elements than the overlying crust, such as iron (Fe) and magnesium (Mg), is about 1,800 miles thick and makes up approximately 70% of the Earth's total volume (Figure 16). The major component of the mantle is a dark-colored, notably dense rock called peridotite. The most common minerals in this rock are olivine [magnesium iron silicate; $(Mg,Fe)_2SiO_4$], pyroxene (several varieties; e.g., pigeonite, $MgFeSi_2O_6$), as well as some garnet. The uppermost part of the mantle, which is relatively rigid, is combined with the crust to form the **lithosphere**. The lithosphere is underlain by a part of the mantle known as the **asthenosphere**, which is more plastic than the overlying layer as a result of heating by radioactive decay of elements at depth. It is within this more fluid layer that the convection currents mentioned earlier transfer heat to the surface and initiate the process of the "drifting apart" of the individual tectonic plates, which are part of the lithosphere (i.e., if the convection cell theory is correct).

3) The mantle is underlain by the Earth's **core**, which is composed mostly of iron and makes up approximately 30% of the Earth's volume. The core consists of two layers, an outer extremely hot zone, which is molten, and in inner, also very hot, layer that is solid because of the intense pressure at those depths within the Earth.

[NOTE: Ask GeoMan (2012) gave the following approximate values for the **density** of these different units (in grams per cubic centimeter):

Continental Crust: 2.7 to 3.0
Oceanic Crust: 3.0 to 3.3
Mantle (silicates): 3.3 to 5.7 (increasing with depth?)
Outer Core (liquid): 9.9 to 12.2
Inner Core (solid): 12.6 to 13.0]

Based on a study of the world's inner continental shelves that was sponsored by the U.S. Navy through the Defense Research Laboratory at the University of Texas, Hayes (1964; 1965) proposed the following preliminary coastal classification based to a large extent on the concept of plate tectonics:

Class	Tectonism
A. Tectonic coasts	rapid uplift
1. Young mountain range coasts	
2. Glacial rebound coasts	
B. Plateau-shield coasts	stable
C. Depositional coasts	rapid downwarp

Tectonism is defined as the forces involved in or producing deformation of the Earth's crust, such as folding or faulting of the rocks, with plate motion usually being the driving mechanism for this deformation. In this case, the term rapid is used in a geological sense (measured in millimeters or centimeters per year). A similar concept that related coastal types to the plate tectonic postulate was presented by Inman and Nordstrom in 1971. They used the term **collision coasts**, a category into which most young mountain range coasts would fit, the term **trailing-edge coasts**, into which many depositional coasts would fit, and so on.

The cross-section in Figure 17 shows one of the basic concepts of plate tectonics, in which a plate made up of a continental crust (e.g., the North American Plate, which is relatively light because of the abundance of aluminum, silica, sodium, etc. in the rocks) is riding over a sinking oceanic plate (e.g., the Pacific plate, which is relatively heavy because of the abundance of iron, magnesium, etc. in the rocks). If we look at one of the simpler examples, such as the South American plate, a striking contrast can be seen between the western coast of the continent and the eastern coast. Along the western shore, where the continental and oceanic plates collide, or leading edge of the continental plate, a series of young mountain ranges flank the coast. However, on the eastern shore, or the trailing edge of the continental plate, which is moving westward at a rate of around 4 inches/year (Monroe and Wicander, 1998), broad deltaic and coastal plains are abundant. To use this simple example, the western coast of the South American continent, and much of the North American continent for that matter, is dominated by **young mountain range coasts** characterized by: 1) High mountains caused by relatively recent mountain building activity; 2) Bedrock of variable age, but dominated by relatively young sedimentary and volcanic rocks; 3) Active tectonic uplift (e.g., with old beach lines lifted up the sides of the mountains thousands of feet in some places); and 4) Mostly short, high-gradient (steep) rivers emptying into the sea. The eastern sides of the two continents are dominated by **depositional coasts** characterized by: 1) Coastal zone made up chiefly of broad coastal and deltaic plains; 2) Bedrock composed generally of Tertiary and Quaternary sediments (young geologically speaking); 3) Tectonically subsiding areas (e.g., thousands of feet of sediments have accumulated in some areas); and 4) Many large and long rivers emptying into the sea. It is obvious then, that the Georgia Coast

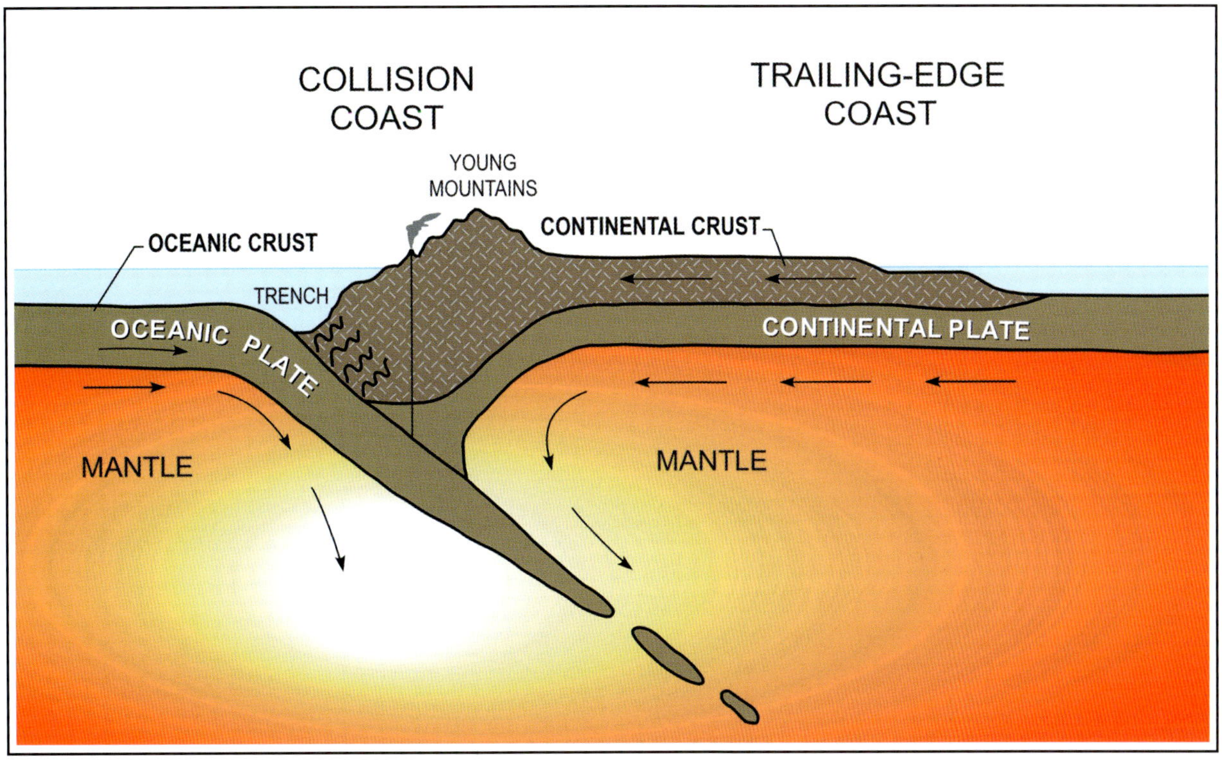

FIGURE 17. Cross-section illustrating the collision of continental and oceanic plates. At the collision point, the continental plate rides over the oceanic plate, creating a high, young mountain range. Shorelines along the collision juncture tend to be dominated by rocky shores, wave-cut cliffs, and relatively short pocket beaches (e.g., coasts of western North and South America). Coastlines on the trailing edge of the continental plate are dominated by coastal plains that contain abundant estuaries and barrier islands (e.g., east coast of the United States). Modified after Davies (1973, Fig. 3) and Inman and Nordstrom (1971).

belongs in the general category of depositional coasts. In fact, it is one of the prime examples, with its wide coastal plain, delta systems, and network of rivers delivering sediments to the sea.

Because the Georgia Coast is classified as a depositional coast, the rest of this discussion on the controls of coastal types will focus only on depositional coasts. Accordingly, the factor that exerts the most control on the geomorphological nature of a depositional coast is the interaction of water in motion with the coastal sediments–typically called the **hydrodynamic regime** (*hydro* = water; *dynamics* = kinetic energy of the water). For purposes of interpreting the morphology of coastal features, the hydrodynamic regime is commonly expressed as the ratio of wave energy to tidal energy (i.e., how big the waves are versus how large the tidal range is). During Hayes' early study of the world's coasts for the U.S. Navy, he observed a striking correlation between tidal range (vertical distance between high and low tide) and the physical characteristics of depositional coasts. At first, he classified depositional coastal types based on this observation, differentiating among **microtidal** (tidal range less than 6 feet), **mesotidal** (tidal range = 6-12 feet), and **macrotidal** (tidal range greater than 12 feet) coasts.

[NOTE: These terms and their limits were originally suggested by Davies (1973), but he did not make the kind of geomorphological correlations that Hayes did. In some cases, major inlets on microtidal, depositional coasts may have some of the characteristics more commonly found on mesotidal coasts (e.g., large shoals offshore of the inlets called ebb-tidal deltas). In that case, **tidal prism** (volume of water entering and leaving the inlets) becomes a factor (Davis and Hayes, 1984).]

Other terms proposed by Hayes (1976; 1979), based partly on publications by the original *guru* of coastal geomorphology, W. Armstrong Price, include **wave-dominated** (common on microtidal depositional coasts

– tidal range = <6 feet), **tide-dominated** (common on macrotidal depositional coasts – tidal range = >12 feet), and **mixed-energy** (common on mesotidal depositional coasts – tidal range = 6-12 feet). These general types are presented in Figure 18. It is important to remember that these terms are most accurately applied to depositional coasts (broad coastal plain and deltaic shore zones). On young mountain range and plateau-shield coasts, tectonic influences and sand supply exert considerable influence on the coastal geomorphology.

Curiously, Hayes (1965) also noted a relatively equal distribution of the three originally defined classes of tidal range around the world's coasts, which was a surprise to him, because at that time, most of his field experience had been limited to the microtidal shores of North Carolina and Texas. Consequently, upon leaving Texas, much of his early research was focused on the mesotidal coasts of New England and South Carolina. Eventually, macrotidal coasts with tidal ranges greater than 30 feet were also studied in Alaska, France, and Chile.

As a generalization, most **wave-dominated depositional coasts** are **microtidal** and characterized by:

- Deltas with smooth outer margins and sand beaches (except in the case of huge delta systems on coasts where the waves are of modest size, like the Mississippi Delta);
- Abundant long barrier islands;
- Sediment grain-size distributions with coarser sediments onshore and finer sediments offshore, because of the role of waves (Breaking waves keep fine-grained sediments in suspension in the nearshore zone.); and
- Multiple nearshore "breakpoint" bars offshore of the barrier island's beaches (Figure 18A).

In North America, wave-dominated depositional coasts occur on much of the Texas Coast, the Outer Banks of North Carolina, and in Kotzebue Sound, Alaska.

On the other hand, most **tide-dominated depositional coasts** are **macrotidal** and characterized by:

- Open mouthed, multi-lobate river deltas;
- No barrier islands;
- Sediment grain-size distribution with coarser sediments offshore and finer sediments onshore, because of the role of tidal currents (In the more offshore areas, tidal currents are strong enough to keep fine-grained sediments in suspension, whereas near the high-tide line, weaker currents allow the fine-

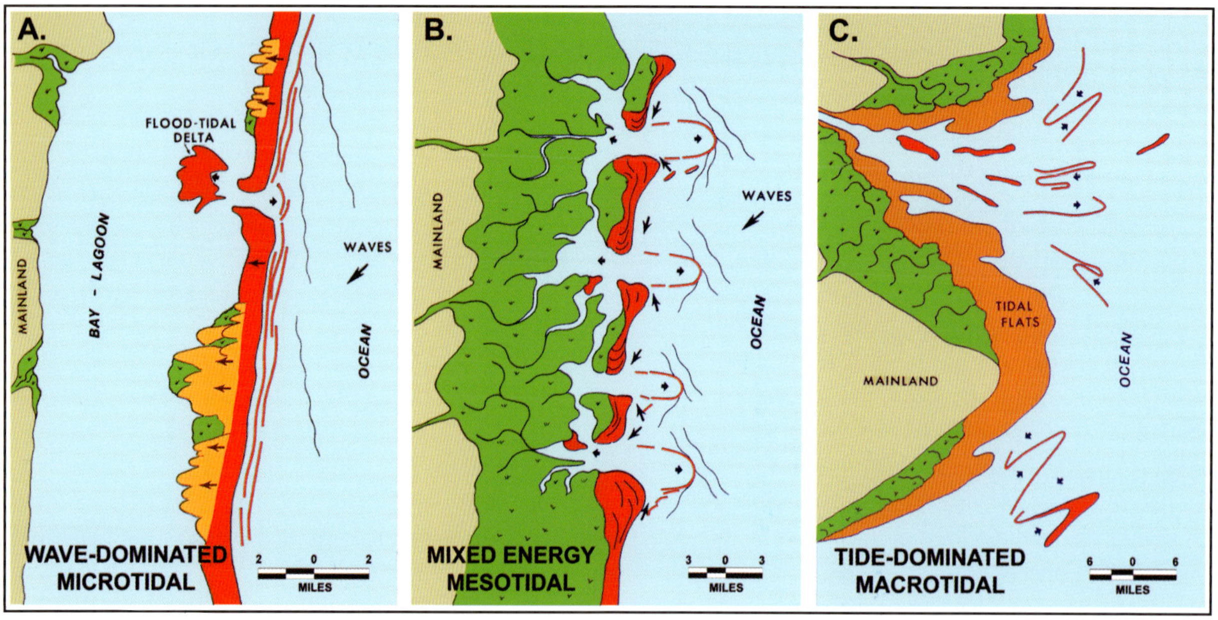

FIGURE 18. Generalized models of the three basic types of depositional coasts. (A) Wave-dominated (usually microtidal; tidal range less than 6 feet), (B) Mixed energy (usually mesotidal; tidal range = 6-12 feet), and (C) Tide-dominated (usually macrotidal; tidal range greater than 12 feet). The red lines in A represent nearshore, subtidal wave-formed sand bars, and those in B and C indicate intertidal sand deposits. The Georgia shoreline is a classic example of a mesotidal (mixed-energy), depositional coast.

grained sediments to settle out. Because of the wide intertidal zones, waves break further offshore, thus playing a more minor role in sediment distribution in the intertidal zone.);
- Extensive tidal flats and coastal wetlands; and
- Offshore linear tidal sand ridges (Figure 18C).

In North America, tide-dominated depositional coasts occur in Bristol Bay and Cook Inlet, Alaska and in the northern reaches of the Gulf of California.

To complicate matters a little bit, most **mesotidal depositional coasts** have intermediate-sized tides and waves, thus another class of coast, **mixed-energy coasts**, has been recognized by us. Most mixed-energy depositional coasts are characterized by:

- Short, drumstick-shaped barrier islands;
- Numerous tidal flats and coastal wetlands; and
- Very complex sediment distribution patterns (Figure 18B).

The Georgia Coast is the classic example of a mixed-energy depositional coast. Other areas in North America where mixed-energy depositional coasts occur are the South Carolina Coast and the Copper River Delta coast of Alaska.

[NOTE: Because there are some exceptions to the general rules just expressed, certain coastal geologists prefer that the terms wave-dominated, mixed-energy, and tide-dominated be used with no strict adherence to the tidal ranges themselves. However, keep in mind that, in this case, we are only referring to depositional coasts. For example, leading-edge mountainous coasts or some plateau-shield coasts do not necessarily conform to the general patterns suggested in Figure 18. Also, these generalizations regarding coastal depositional systems are based on the present conditions of a relatively stable and/or slowly rising sea level. If sea level were to drop markedly and continue to do so, conditions would be different (e.g., valleys would be carved across the continental shelf).]

A second major factor that determines certain key characteristics of depositional coasts is the regional **climate**. Wind patterns along a coast and offshore storms determine the average size of the waves, which has a major input in shaping the coast, as discussed earlier. Knowing whether or not the coast is affected by hurricanes, extratropical cyclones (e.g., northeasters), or monsoons is key to understanding the seasonality of the variability of wave conditions along any given shore. In Georgia, hurricanes generate by far the largest waves that occur on the beaches in the state, but extratropical cyclones are also important annually.

The **type of sediments** on tidal flats and beaches is strongly related to the regional climate. For example, in tropical areas beach sediments are commonly composed of coral and algal fragments, carbonate precipitates, and shells. In temperate regions, quartz sand is usually the dominant sediment type, with rock fragments and feldspar being abundant in sand near river mouths and along coasts with eroding bedrock.

Other obvious climatic influences include:

- The presence or absence of major rivers;
- Effects of glaciers, freeze/thaw, and presence of permafrost in polar regions;
- Types of weathering (chemical or mechanical) that produce the sediments that make up the coastal systems (Hayes, 1967a); and
- Presence of beachrock in the tropics and permafrost in polar regions.

Although tides and waves determine the general framework of a depositional coast, the climate determines the type of vegetation that occurs on the surface of the coastal habitats. Intertidal zones sheltered from waves are commonly populated by mangroves in the humid tropics, by salt marshes in more temperate regions, and by algal mats and evaporites in the arid tropics. The Georgia Coast has a humid subtropical climate, thus the intertidal areas contain abundant salt marshes.

Another control on the nature of depositional coasts that is somewhat overrated in our opinion is the **source and supply of sediments**. Without such a supply, there would be no depositional coasts. On the Georgia Coast, the sediments at the shore are derived from rivers, erosion of the continental shelf, and erosion of areas elsewhere along the coast (most commonly to the northeast).

As shown in Figure 19, there are three major classes of sediments found on shorelines–gravel (particles >2 mm in diameter, which consists of four separate classes, culminating with the largest–boulders), sand (particles between 0.0625 and 2 mm in diameter, ranging from very fine- to very coarse-grained), and mud (particles smaller than 0.0625 mm in diameter and consisting primarily of silt and clay). On the Georgia Coast, most of the sediment along the outer shore is fine-grained sand, whereas mud is by far the most common sediment type in the estuaries.

General Class	Wentworth Scale (Size Description)		Grain Diameter d (mm)
GRAVEL	Boulder		
			256.0
	Cobble		
			64.0
	Pebble		
			4.0
	Granule		
			2.0
SAND	Sand	Very Coarse	
			1.0
		Coarse	
			0.5
		Medium	
			0.25
		Fine	
			0.125
		Very Fine	
			0.0625
MUD	Silt		
			0.00391
	Clay		
			0.00024
	Colloid		

FIGURE 19. Definition of sediment types. From Wentworth (1922).

The last two factors affecting the nature of depositional coasts are **local geological controls** and **sea-level rise**. Examples of geological controls include such factors as glacial effects, faulting, and variations in shoreline bedrock type. Along the Georgia Bight, local geological effects are more subtle, such as the relatively minor uplifting of the Cape Fear Arch to the northeast (Figure 4) and the gentle downwarping of the subsurface rock layers starting near Charleston, South Carolina and continuing on into Georgia. With regard to sea-level rise, as noted earlier, sea level on the Georgia Coast has been relatively constant within the past 4,500 years, though some minor fluctuations have left their mark (Figure 8). A key question at the moment is — will sea level continue to rise as it has been in the past few years, and, if so, how much? This issue will be discussed in more detail later.

4 COASTAL PROCESSES

INTRODUCTION

Fellow coastal geomorphologist Jim Henry gave the following general introduction to the coastal processes on the Georgia Coast (Henry, 2009): *With the exception of wave action associated with seasonal storms (nor'easters) and infrequent hurricanes, the tide's twice-daily ebb and flow is by far the dominant physical process along the Georgia Coast. Because of the concave shape of the shoreline and a broad, shallow continental shelf, wave energy is low with wave heights averaging from two to less than four feet at the breaking point. Prevailing onshore winds are from the northeast and southeast, whereas prevailing offshore winds are from the northwest and southwest. On the eastern coast of the United States, the prevailing alongshore current and sediment transport is from north to south. However, summer wind conditions frequently cause a south to north nearshore sediment transport. (It should be noted that current direction relates to destination, while wind direction relates to origin.)* Waves crossing a wide and shallow continental shelf would be impacted by bottom friction (and hence diminished in height) more so than those waves of similar size crossing a narrow and steep shelf.

TIDES

As shown in Figure 20, when the Sun and Moon are in line (full moon and new moon; syzygy), the gravitational attraction of the Moon and Sun are combined, giving rise to the maximum tides of the month (called **spring tides**). Minimum tides (called **neap tides**) occur during the first and third quarter of the Moon (quadrative), when the two forces work in opposition. During some lunar cycles, the tidal range during spring tides can be almost twice that during neap tides. Because of its location so far from the Earth, the gravitational pull (tide-raising force) of the Sun on the world ocean is only 0.455 that of the Moon. Therefore, the Moon has the greater influence on the tidal cycle, controlling the timing of high and low tides.

In the much-simplified diagram shown in Figure 20, during syzygy two bulges of the high tide occur on opposite sides of the Earth, one facing toward the Moon and another one facing away from it. The bulge facing the Moon is the result of the gravitational pull of the Moon and Sun. The bulge on the opposite side is much more difficult to explain. In actuality, the Earth and the Moon revolve together around the center of mass (barycenter) of the combined Earth-Moon system (Figure 21A). The barycenter is displaced a distance from the center of the Earth toward the Moon, and it is always located on the side of the Earth turned momentarily toward the Moon, along a line connecting the individual centers of mass of the Earth and Moon (Wood, 1982). To put it simply, as this system revolves around the barycenter, centrifugal force comes into play. This force is defined as the apparent force, equal and opposite to the gravitational force, drawing a rotating body away from the center of rotation, caused by the inertia of the body. Thus, the combined Earth-Moon system acts as a lever as the two equal and opposing forces revolve about a fulcrum, the barycenter. Therefore, the gravitational attraction of the Moon (and Sun) creates the bulge facing the Moon and centrifugal force creates a bulge of relatively equal size on the opposite side of the Earth (Figure 21B).

The diagram in Figure 20 shows that, as the Earth rotates about its axis, a given beach location on the Earth's surface will pass by the peak of the two bulges every 24 hours and 53 minutes. The extra 53 minutes is required because that is how long it takes that beach location to "catch up" with the Moon, which revolves all the way around the Earth in about 28 days. Because of the intervening continents, a single bulge cannot pass all the way around the Earth. Instead, individual **amphidromic systems** of different sizes are set up within the world's ocean basins. These systems are separate, more-or-less circular, gyrating tidal systems (standing waves) with a

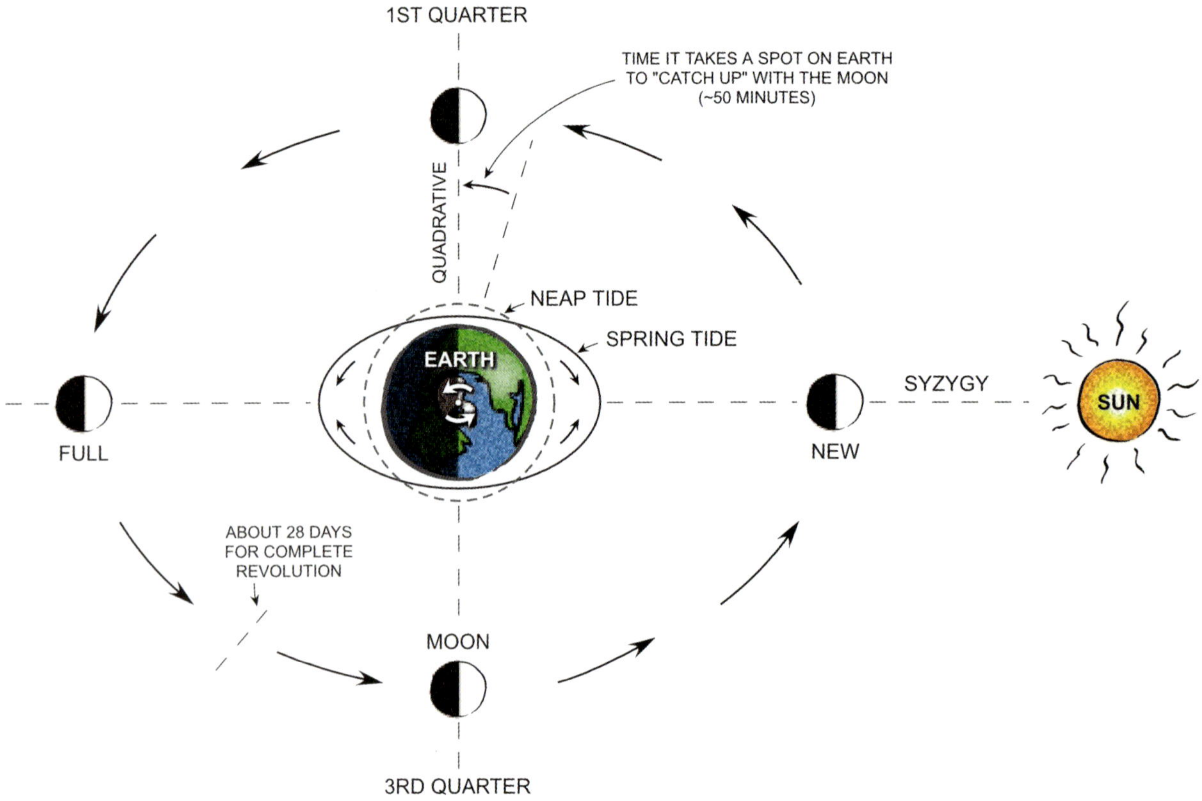

FIGURE 20. Forces involved in the generation of tides.

nodal point in the middle. This complicates the story, but the primary forces just described above that cause the tides and are illustrated in Figure 20 still provide the basic explanation of the process along the Georgia Coast.

There are many other complications to the simple story presented in Figure 20. For example, because the Moon revolves in a slightly elliptical, rather than a circular, orbit around the Earth, *the distance of the Moon from the Earth may vary between extreme limits separated by some 31,278 miles* (Wood, 1982). When the Moon is closest to the Earth, this "close encounter" is called the *perigee,* which is 13% closer than when it is furthest away (called the *apogee*). Thus, perigean tides may be 20% larger than normal mean tides. Furthermore, if the perigean tide corresponds with the syzygy (full and new moons), the so-called *perigean spring tides,* an increase of the tidal range of as much as 40% might be created (Wood, 1982).

The tides on the Georgia Coast are **semi-diurnal** (Figure 22), because two complete tidal cycles occur within 24 hours and 53 minutes, as opposed to some other areas in the world that have only one high and one low tide a day (**diurnal** tides; e.g., the Texas Coast during some parts of the month). Therefore, low and/or high tide is about one hour later each day on the Georgia Coast.

The tidal range, as illustrated in Figure 22, is the vertical distance between high and low water. The spring tidal range in Georgia varies from as much as about 9 feet at the South Carolina border (Figure 22) to about 7 feet at the Florida border. This places the Georgia coastline in the mesotidal class (tidal range 6-12 feet).

Whether the average tidal range in an area is large or small is dependent upon several factors, with the largest tides occurring where: 1) The water body has on open connection with the world ocean; 2) The continental shelf is wide; and, most importantly, 3) Along coastal embayments.

Other factors, such as the hydrodynamic forcing of oceanic currents, position within an amphidromic system, and presence of resonant standing oscillation play a role in some areas, but none of these are as relevant as the first three factors with respect to the tides

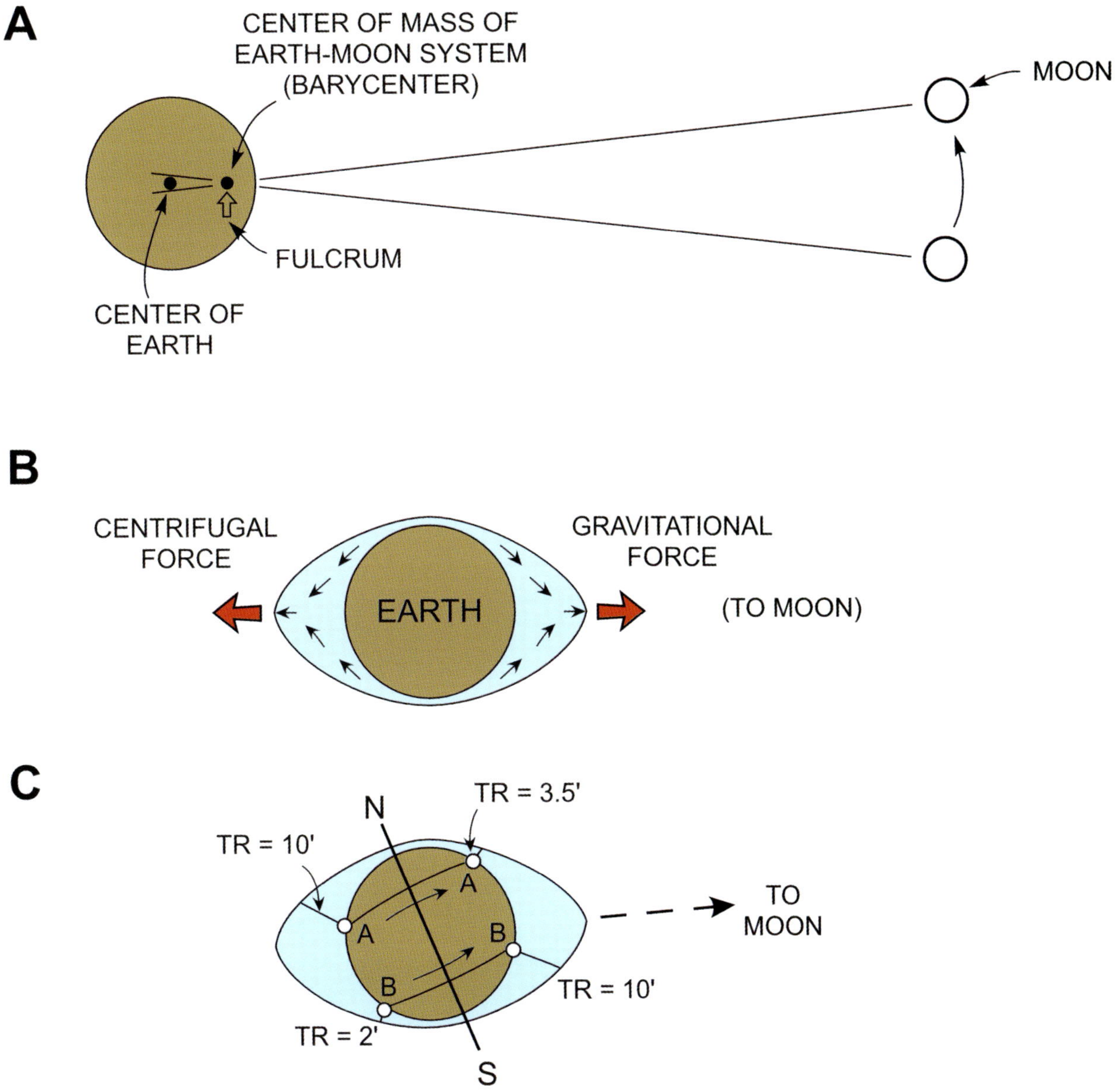

FIGURE 21. Complexities of the Earth's tides. (A) General location of the barycenter and its role as a fulcrum in generating tides. (B) Creation of tidal bulges on opposite sides of the Earth. (C) One of the factors involved in creating the diurnal inequality of the tides during a complete daily tidal cycle is the tilt of the Earth's axis that creates different tidal ranges on opposite sides of the Earth.

on the Georgia Coast. As illustrated in Figure 23, the largest tidal range along the southeastern coast of the United States occurs at the head of the Georgia Bight, a V-shaped embayment, which explains why the tides are so large along the Georgia Coast. Except for the Bay of Fundy, an area where resonance plays a role, most of the larger tidal ranges around the world occur in coastal embayments, such as the Georgia Bight. Komar (1976) gave the following verbal explanation for this – "*as the tidal front approaches a narrowing indention of the coastline, …the enveloping shores constrict its movement and wedge the water together.*"

Currents generated by the tides are very strong at the entrances of the numerous tidal inlets that occur along the Georgia Coast. In some of the major inlets, current velocities may reach 3 or 4 knots [one knot = one nautical mile per hour (nautical mile = 6,076 feet)]. Tidal currents are also strong in the major tidal channels

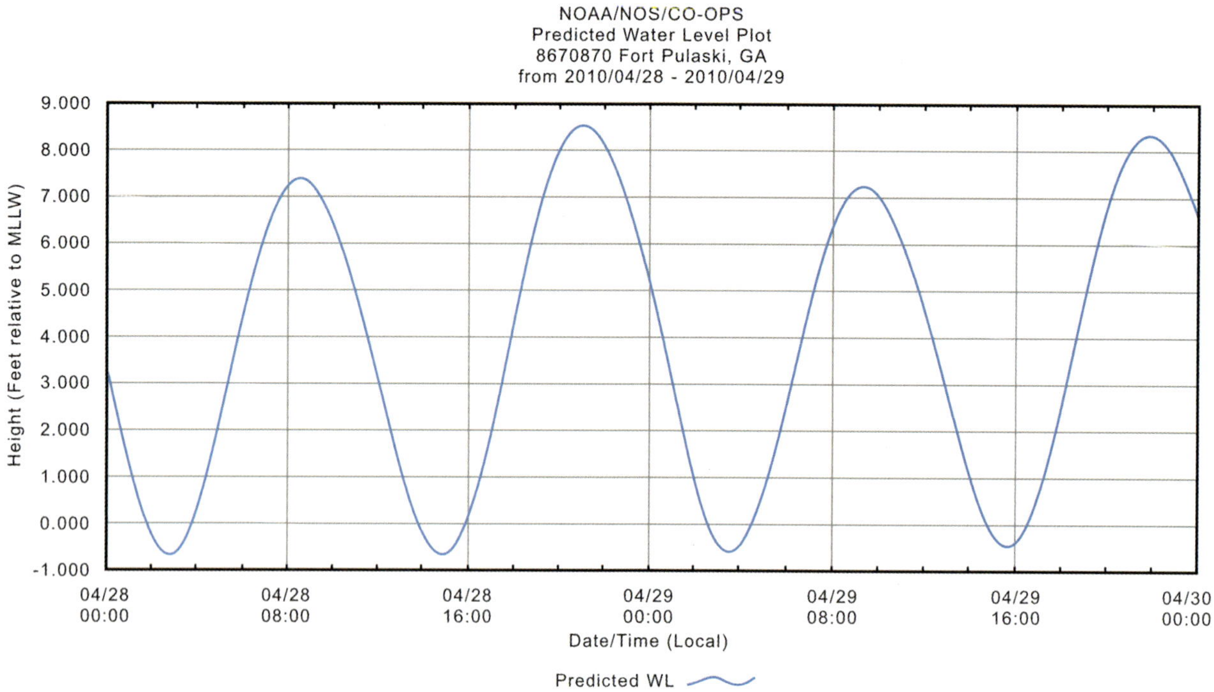

FIGURE 22. Typical tidal curve for Georgia, showing two days. Note that there are two high tides and two low tides per day (plus about an hour), and the elevations of the consecutive highs and lows are somewhat different. Thus, there is a "high" high tide and a "low" high tide each day. This is called a diurnal inequality. One of the causes of this is illustrated in Figure 21C.

scattered throughout the estuaries and other backbarrier regions along the coast.

WAVES

Water waves are generated by the transfer of kinetic energy to the water surface as the wind blows over it. As shown in Figure 24A, a typical wave is defined by its **wave length** and **wave height**. The length of time it takes two succeeding wave crests to pass a single point is called the **wave period**. In Georgia, the average wave height is around 2.5 feet according to Nummedal et al. (1977; see Figure 23), but waves exceed this by several feet during a typical storm.
[NOTE. This number, 2.5 feet, is based on offshore measurements. In South Carolina, the average breaker height along the beach is 2 feet, according to Tim Kana (pers. comm.), which is probably a good estimate for the Georgia Coast as well.]

During hurricanes, waves in the open ocean with heights greater than 50 feet have been observed. The factors that determine the height of waves include:

- The fetch of the wind (length of water surface over which the wind blows to form the wave);
- Wind velocity;
- Duration of the wind; and
- Incident swell.

As indicated in Figure 24B, the waves in the area where the wind that forms them is blowing are referred to as **seas**. Once the waves pass out of the area of formation, they are referred to as **swell.** The differences between seas and swell are evident to even the casual observer, in that the water surface is choppy under sea conditions and smooth in areas where swell conditions predominate. Also, sea conditions are usually indicated by the presence of whitecaps, whereas whitecaps are not typically present on pure swell-type waves. Under sea conditions, the sea surface is irregular and the waves are asymmetrical (two sides not balanced), whereas the surface of swell is regular (relatively smooth) and the waves are symmetrical. An observer standing on a beach may have difficulty distinguishing between the two types of waves, and it is very common for both types of waves to interact to form the resulting waves in the surf zone. To identify sea conditions, note the direction the wind is blowing on that day and look for the presence

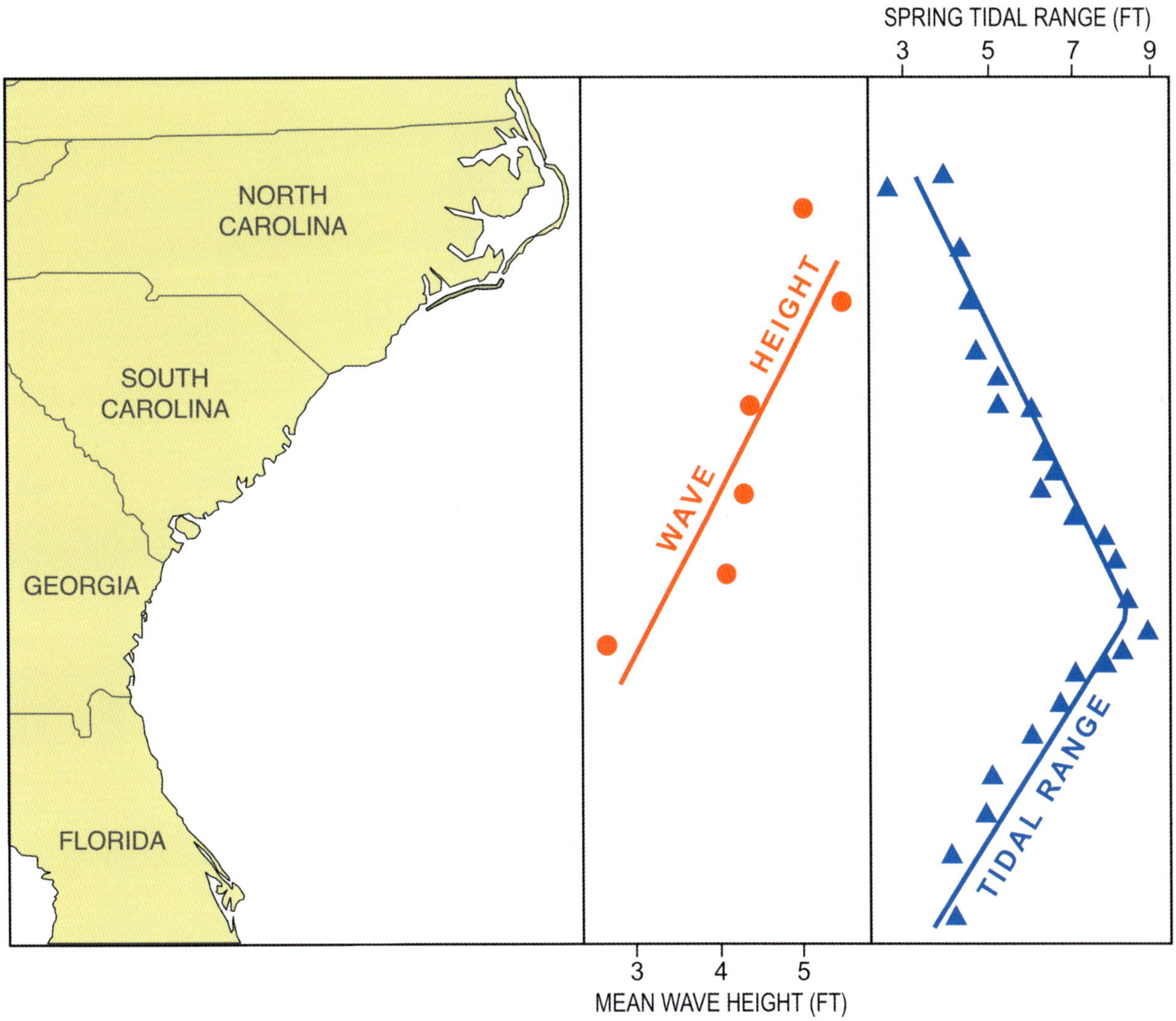

FIGURE 23. Variation of the tidal range and wave height along the shoreline of the Georgia Bight. Tide data from NOAA tide tables and wave data from Nummedal et al. (1977). Note that the largest tides and smallest waves occur at the head of the Georgia Bight (near South Carolina/Georgia border).

of whitecaps. Also, the period (timing between the passage of the wave crests) of swell waves is typically considerably longer than those waves generated by local onshore winds. Therefore, swell waves are usually formed by offshore storms that generate large waves. The waves generated by the storms move away from their zone of formation at a certain velocity with longer waves moving faster than shorter ones. A classic example of this escape of long waves from their zone of formation is the forerunner waves that precede the landfall of hurricanes. Commonly, surfers flock to the shore to take advantage of these huge, long-period "forerunner" waves.

As waves approach the beach from offshore, they increase in steepness and break in different ways to be discussed below. To understand why this happens, consider a cork on the water surface as a wave passes by. The cork makes a circular motion as the wave passes, in the process moving a small distance in the direction the wave is moving. As shown by the diagram in Figure 25A, the water directly beneath the wave also undergoes a circular orbit, with the size of those orbits decreasing exponentially with depth until completely ceasing to orbit at a depth of about one half the wave length.

As the wave moves onshore, it begins to "feel the bottom," that is, the orbital motion of the water under the wave begins to bump up against the bottom sediment

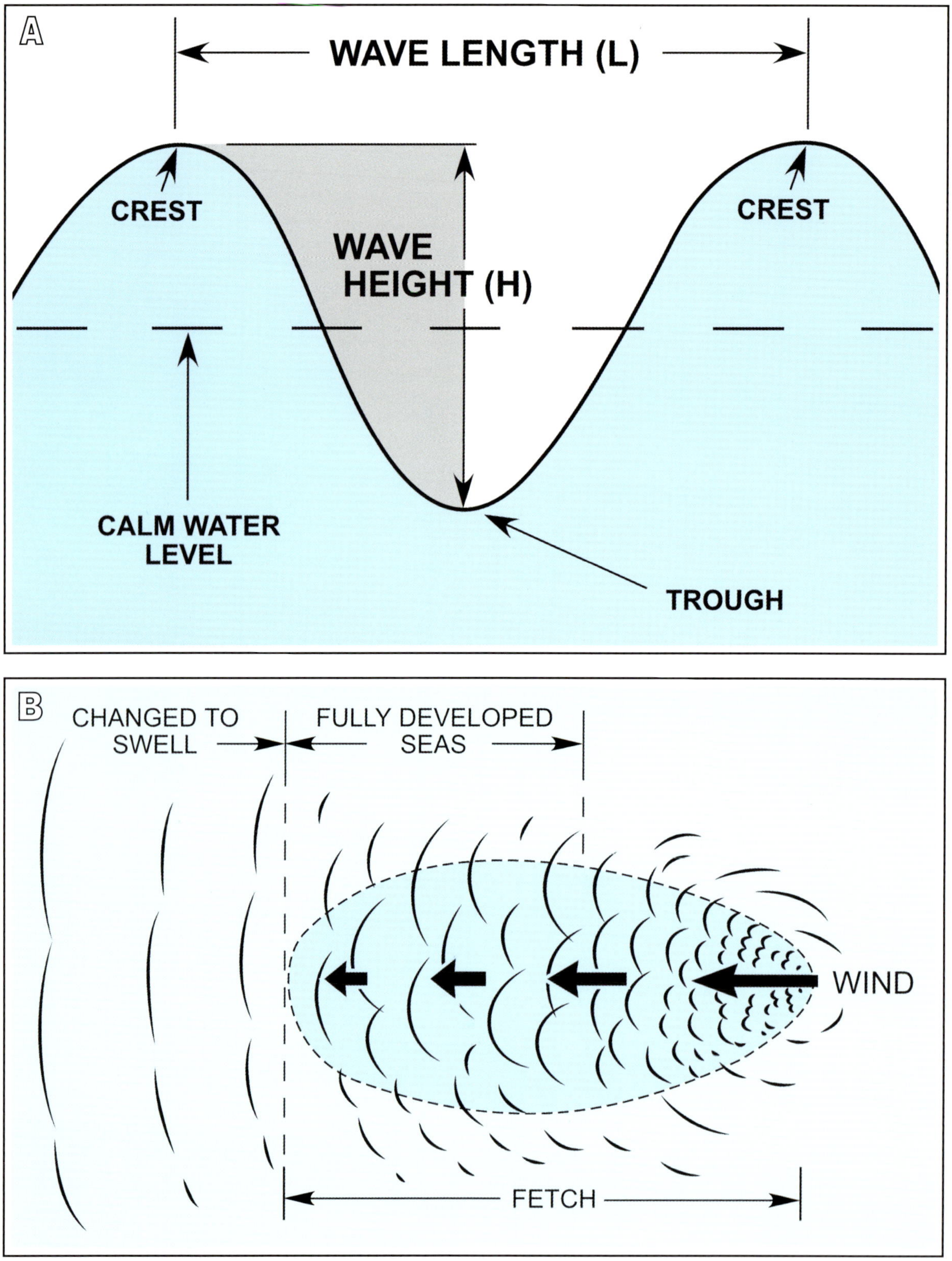

FIGURE 24. Waves. (A) Components of a typical water wave. (B) General pattern for the generation of waves by wind blowing across the water surface.

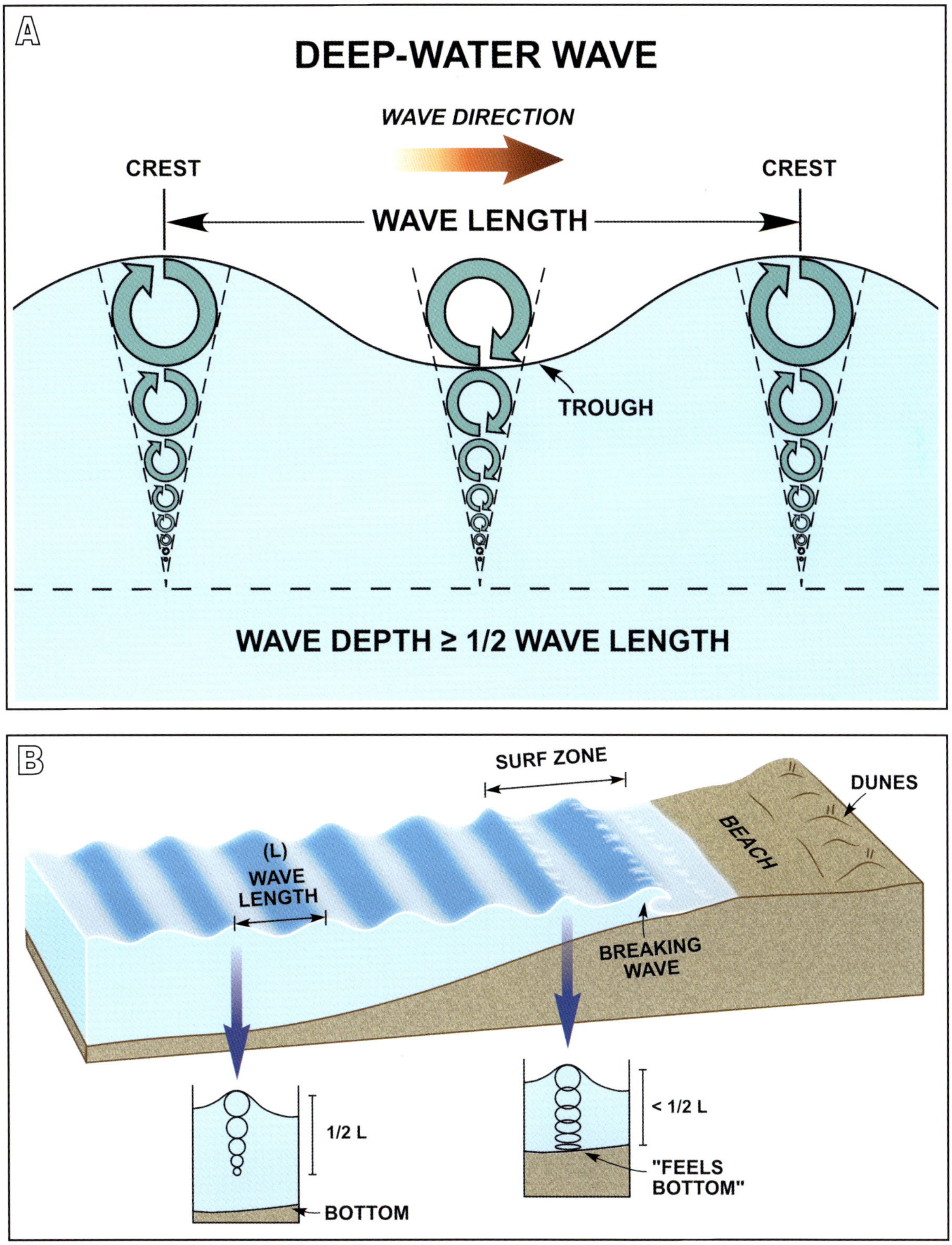

FIGURE 25. Waves approaching shore. (A) Orbital water motion at depth under a passing wave. (B) Characteristics of the surf zone under the influence of wave action. Note that in water shallower than approximately one half the wave length, the orbits of the wave-generated water motion under the waves (illustrated in upper diagram) flatten out as the wave begins to "feel the bottom." This causes the waves to slow down and become steeper and steeper until they eventually break.

where the water becomes shallower than one half the wave length. On the Georgia Coast, most storm waves, with the exception to those generated by major hurricanes, start to "feel the bottom" in maximum water depths of around 25-30 feet. This depth typically defines the base of an offshore environment called the **shoreface**, which, in this area, is considered to be the absolute base (outer limit) of the nearshore portion of the barrier islands. The sediments typically become muddier seaward of the base of the shoreface, because the more common storm waves do not agitate the bottom sediments and, hence cause the fine-grained sediments to go back into suspension once they have settled to the bottom.

Thus, if the typical storm waves on the Georgia Coast "feel the bottom" at a maximum depth of 25-30 feet, they have maximum lengths of up to about 50-60 feet. As a wave approaching shore moves further landward, the orbits flatten out and become unstable, the waves get higher and the crests become sharper because they become compressed like an accordion. The velocity of the water in the top of the wave exceeds that at the bottom to such an extent that the surface of the wave eventually breaks. The zone where the waves are breaking is referred to as the **surf zone** (see Figure 25B). After breaking, the water released by the broken wave washes onto the portion of the beach called the beachface (also referred to as the **swash zone**). The water moving up the relatively steep beachface, known as the wave uprush, eventually loses its momentum and that portion of the water that doesn't percolate into the beachface returns back down the slope under the influence of gravity, defined as the backwash.

Breaking waves are usually of three distinct types– plunging, spilling, and surging. As pictured in Figure 26A, plunging waves take on a cylindrical shape (Hawaii Five-O type) and fall abruptly down with considerable force, usually breaking in the near vicinity of the beachface. The surfaces of spilling waves, on the other hand (shown in Figure 26B), start disintegrating into foaming lines fairly far offshore and continue to foam their way to shore, gradually decreasing in height as they go. Surging waves approach shore and wash directly up the beachface without breaking. As a generalization, for waves of any given height, the slope of the nearshore zone determines the type of breaking wave present. Plunging waves occur more commonly where the nearshore slopes are quite steep and spilling waves where the nearshore slopes are relatively flat. On the Georgia Coast, spilling waves are by far the most common type, with plunging waves mostly occurring during storms or under conditions of exceptionally large swell generated by offshore storms. This is no doubt due to the gentle slopes that characterize the nearshore zone off the Georgia Coast, because of its generally depositional nature. Under some conditions, a single wave may have the characteristics of both plunging and spilling waves. Surging waves are very rare along Georgia's ocean coast.

Wang, Ebersole, and Smith (2002) determined that plunging waves create much greater sediment suspension in the breaking zone than spilling waves, as well as more sediment transport in a wider and more energetic swash zone. This helps explain the authors' general impression that beaches are more apt to accrete (build out) when the waves are spilling than when they are plunging [an observation supported by Tim Kana's Ph.D. research on the South Carolina Coast, which was carried out while he was a Ph.D. student under Hayes at USC; Kana, 1977; 1979)].

When offshore swell waves are viewed from the air, their crests are usually linear and parallel, forming what are known as wave fronts. In many situations, the wave fronts bend as they approach underwater features, this bending being referred to as refraction and diffraction. Wave **refraction** is defined by the U.S. Army Corps of Engineers (2002) as "any change in direction of a wave resulting from the bottom contours." As noted by Woodroffe (2002), *the speed with which the wave travels in shallow water is related to water depth. Those parts of the wave that enter shallow water move forward more slowly than those in deeper water, causing the wave crests to bend toward alignment with the bottom contours.* During the process of refraction, a segment of the wave front may bend around an island or bar, as is illustrated Figures 27A and 27B. In the purest sense, **diffraction** is defined as the lateral spread of wave energy along the wave crest from a point (an example is the wave formed when a pebble is dropped into a pool of still standing water). The U.S. Army Corps of Engineers (2002) defined diffraction as *the phenomenon by which energy is transmitted laterally along a wave crest. When part of a wave in interrupted by a barrier, the effect of diffraction is manifested by propagation of waves into the sheltered region within the barrier's geometric shadow.* During incidents of wave diffraction near the shore, a segment of the wave front may bend into something like a navigational channel or into the lee of an offshore breakwater or reef, relative to

FIGURE 26. Two types of breaking waves. (A) A plunging wave on the Mexican coast near Acapulco. The idiot running into the surf to collect a sediment sample is co-author Hayes (photograph was taken by Robert L. Folk, summer 1961). (B) Small spilling waves to the south of the south groin and offshore breakwater on Sea Island, Georgia (arrow). Photograph taken by Joe Holmes during a low spring tide on 28 April 2010.

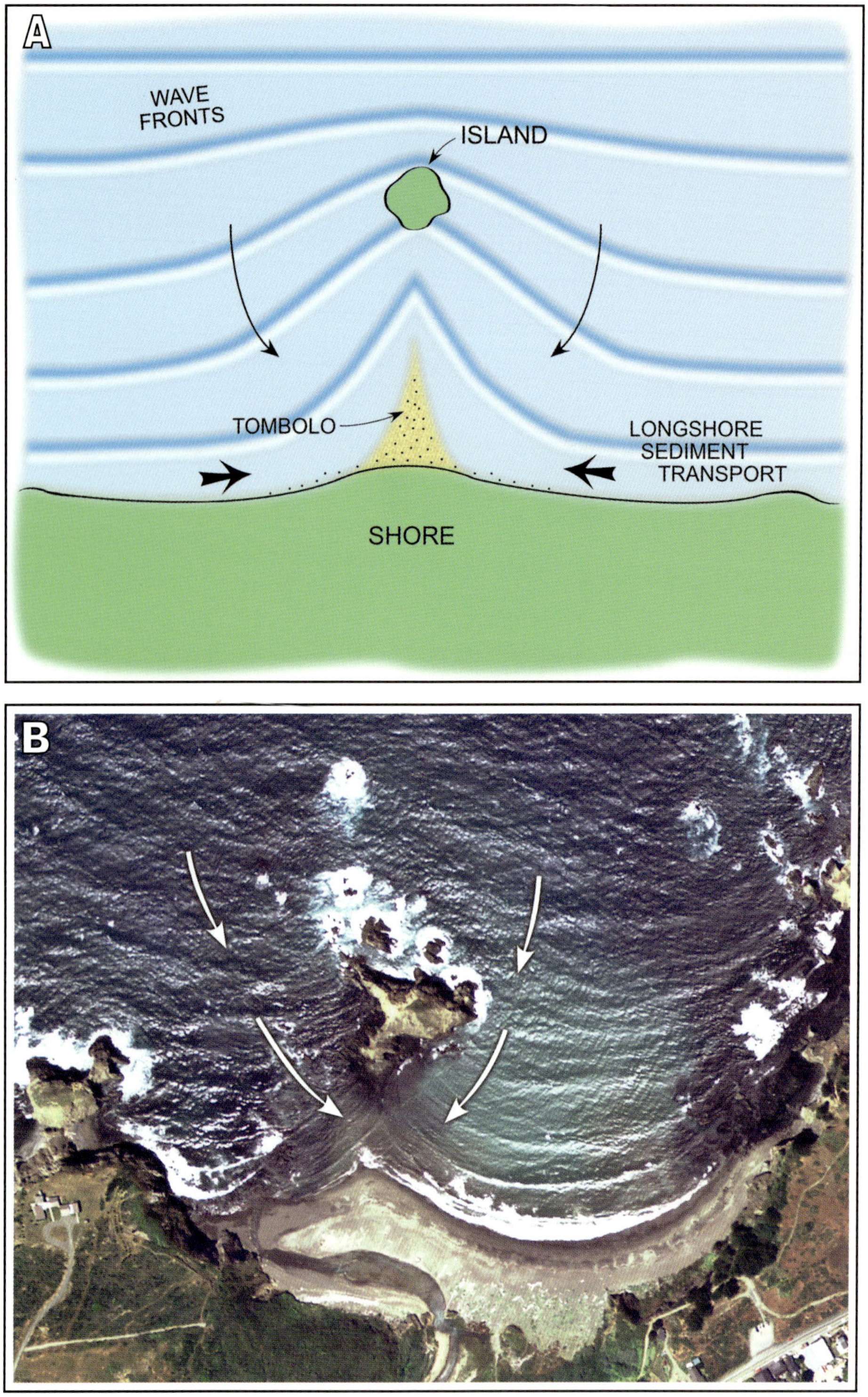

FIGURE 27. Illustrations of wave refraction. (A) Wave fronts bending around an offshore island, generating two opposing directions of sediment transport in the lee of the island. Where the two directions of transport meet, a triangular offshore projection of sand called a tombolo is formed. (B) Wave refraction around Gunderson Rock near Elk, California. The arrows, which indicate the direction of motion of the waves, are oriented perpendicular to the wave crests, or fronts. Digital orthophoto courtesy of the California Spatial Information Library.

the rest of the front. As a result, an arc, or bow, develops in the wave front as it moves into the channel or into the lee of the offshore structure.

When waves approach and break in the surf zone adjacent to the beach, they set up two very distinct, and in some cases very strong, nearshore currents–longshore currents and rip currents. As illustrated in Figure 28 (upper left and lower left), for a wave that approaches a beach at an angle, which many do, two types of water flow are generated that have the capability of moving sediment along shore. This type of longshore sediment transport is a key consideration with regard to the stability of coasts. An illustration of wave-generated longshore sediment transport is given in Figure 27A, which shows waves bending into a sheltered zone in the lee of an offshore island. The result of this process is the creation of a triangle-shaped sand body, called a tombolo, in the lee of the island at the point where two opposing sediment transport directions convergence. An example of this process on the California Coast is shown in Figure 27B.

When waves come ashore at an angle, the wave uprush generated by the breaking wave flows at an angle to the slope of the beachface (Figure 28; lower left),

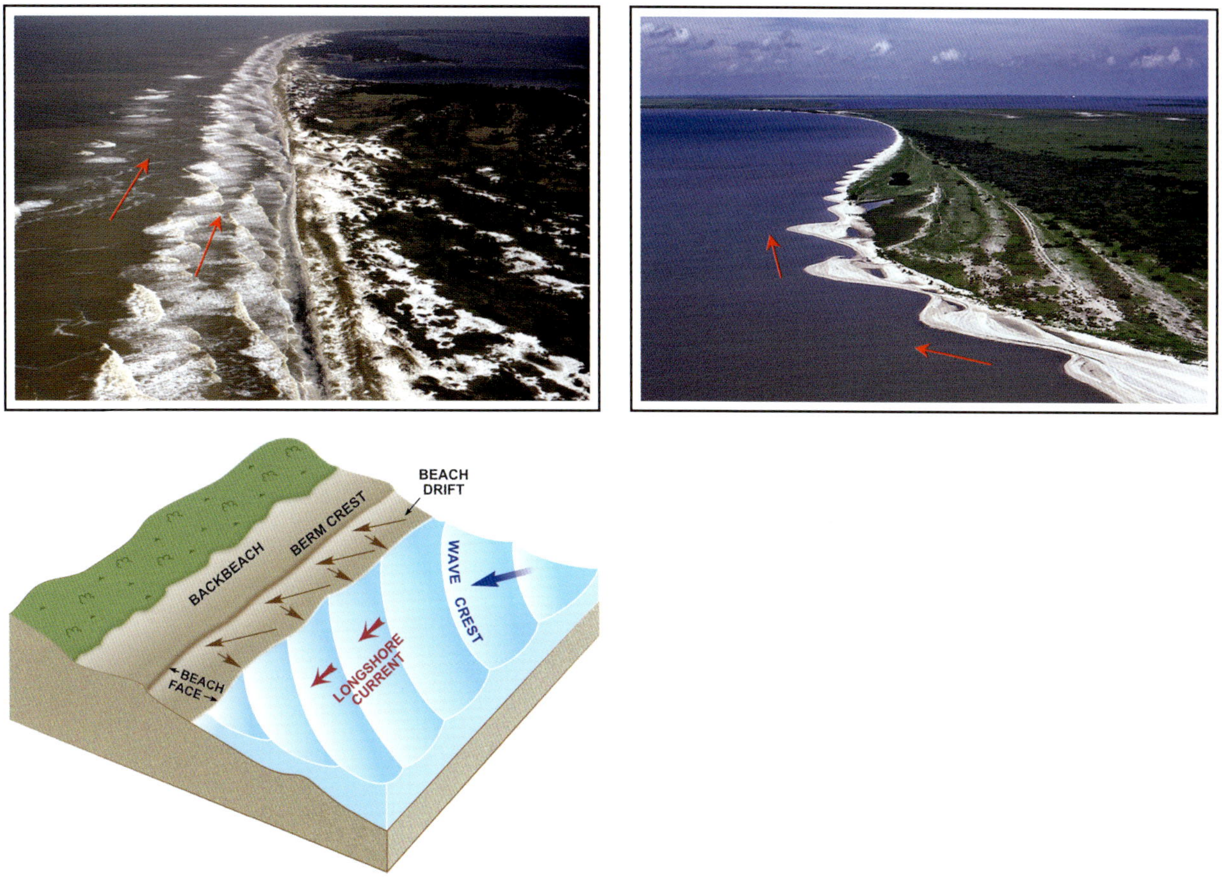

FIGURE 28. Longshore sediment transport. (Lower Left) General model for the generation of *beach drift* and *longshore currents* by waves approaching the beach at an angle. Under these conditions, the sand grains move in a sawtooth pattern along the beachface, and a strong longshore current is developed just off the beachface. (Upper Left) View looking north along St. Joe Spit in northwest Florida showing waves approaching the beach obliquely out of a southerly direction. This high-tide photograph shows the swash of the waves across the beachface. Arrows indicate the direction of the longshore currents generated by the obliquely approaching waves. Note waves breaking on the offshore bar with strong longshore currents generated in the trough on the landward side of the bar. Photograph was taken in April 1973. (Upper Right) Rhythmic topography along the shoreline of Matagorda Bay, Texas. Arrows indicate direction shoreline rhythms were moving through time, also under the influence of southeasterly waves. The water was relatively calm on the day this photograph was taken in August 1979. The white sediments are coarse-grained shell material (mostly oyster shells).

which allows any sand grains moved by the shallow flow of the wave uprush to move obliquely up the beachface. On the other hand, the water in the backwash moves in a perpendicular sense directly down the slope of the beachface under the influence of gravity. Therefore, sand is moved along the shoreline a certain distance with each successive wave, with the length of each advance being dependent upon the size of the wave and its angle of approach (bigger waves and higher angles increase the distance of movement of the sand grains). Thus, the sand grain moves along the shore in a sawtooth pattern called **beach drift** (lower left diagram in Figure 28). This type of sediment transport is restricted to the beachface itself. At the same time, the obliquely approaching wave is piling up water in the surf zone that flows away from the direction of the approaching wave, thus generating a current that runs parallel to the shore called the **longshore current** (upper left diagram in Figure 28). On many coasts, the longshore current is somewhat restricted to a trough that separates the beachface from a nearshore bar. These longshore currents can be quite strong, commonly exceeding a knot or two, and, in concert with the beach drift, they play the major role in transporting sediment along the shore. These currents can also carry swimmers far away from where they entered the water.

Another mode of longshore sediment transport occurs when evenly spaced bulges of sand along the beachface move down the shore like the sinusoidal loops one might throw along a rope. These bulges of sand, called **rhythmic topography** (see photograph in the upper right of Figure 28), are typically spaced 10s to 100s of feet apart. They are created by waves that approach the beach at an oblique angle. Rhythmic topography probably does occur occasionally on the coast of Georgia, but it is typically quite subtle. It is much more prevalent and conspicuous along coasts with deeper offshore areas and bigger waves, such as at Cape Hatteras, North Carolina, Point Reyes, California, and Cape Cod, Massachusetts. When these features were first described, some of the early authors used the term "cusp-type sand waves" in recognition of the fact that they protrude out from the beach somewhat in the same fashion as the very common, smaller perpendicular protrusions off the beach called **beach cusps** (Figure 29). These two features do not form in the same way, thus the more definitive term, rhythmic topography, is used these days for the larger features.

Longshore currents are generated by waves, which are, in turn, generated by the wind. The wind along the Georgia Coast blows from a wide variety of directions. With regard to waves created by the wind and the resulting longshore sediment transport along this shoreline, in late spring and summer the wind is more likely to be blowing from a southerly direction with some resulting

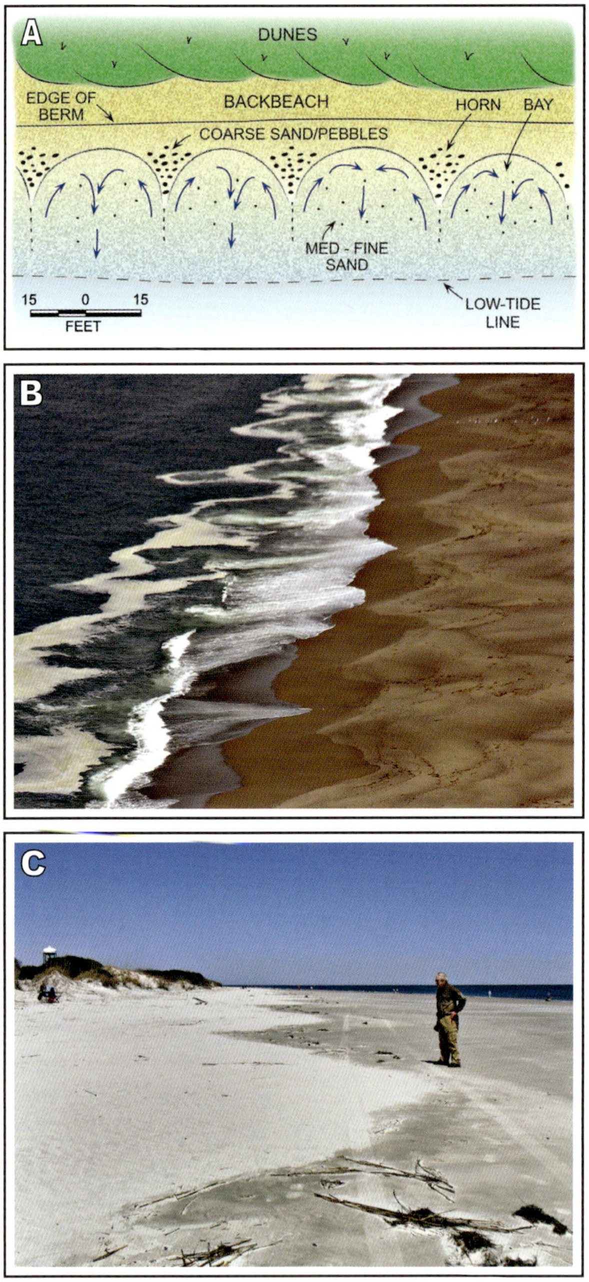

FIGURE 29. Beach cusps. (A) General model. (B) Beach cusps at Point Reyes Beach, California on 18 August 2008. (C) Subtle beach cusps at the high-tide area of the wide sand beach on the south end of Jekyll Island, Georgia (26 March 2010).

longshore sediment transport to the north along some of the beaches (Griffin, 1986; Henry, 2009). These southerly winds are typically generated by clockwise-directed winds blowing around the Bermuda High. In the fall, winter, and early spring, considerably stronger winds associated with extratropical cyclones and the occasional tropical cyclone typically approach the shore from the northeast as they blow in a counterclockwise direction around low pressure centers when the centers of these storms pass offshore of Georgia on the way north. As a generalization, maximum rates of longshore sediment transport on most any shore takes place during major storms. Therefore, the two types of cyclonic storms just mentioned, extratropical cyclones (northeasters) and tropical cyclones (hurricanes), generate by far the strongest and most significant longshore currents (and beach drift; Figure 28) experienced along this shore. Although the details of the cause and effect are a somewhat more complicated story than this simple explanation, the dominant transport direction of beach sand on the state's coast is generally from north to south, except in the vicinity of major tidal inlets. A striking geomorphological indicator of such north-to-south transport along the beaches of the Georgia Coast, a recurved spit on the southwest tip of Blackbeard Island, is shown in Figure 30. Martha Griffin's generalized diagram of this north-northeast to south-southwest trend along the shores of Blackbeard and Sapelo Islands is reproduced in Figure 31. This diagram also shows general directions of tidal currents created by tidal flow into and out of Sapelo and Doboy Sounds.

As pointed out by Oertel (1959b; 1979), the large ebb-tidal deltas on the Georgia Coast have a big influence on longshore sediment transport directions in local areas, with two directions being common. In the 1979 paper, he summarized his ideas as follows: *Prevailing longshore currents do not exist in this section of the Coastal Plain Physiographic Province, although seasonal northeast storms apparently produce a residual southerly flow on an annual basis. Reversals of sediment transport by wave refraction are also caused by inlet shoals.*

Determinations of the volume of sediments that move along the shore in the longshore sediment transport system can be done in several ways, including: 1) Measuring the volume of sand impounded by breakwaters or jetties; 2) Determining the volume of sand dredged from harbors or other collection sites; and 3) Calculations using formulas based on wave and sediment data. Research on the longshore sediment transport rates in South Carolina show that an average of around 150,000 cubic yards of sand move from northeast to southwest along the sand beaches in the state in any given year (Kana, 1976; 1979). We know of no specific data for the Georgia beaches, but we would assume a somewhat smaller number because of the smaller waves. According to R. A. Davis, Jr. (pers. comm.), southerly sediment transport of 500,000 cubic yards of sand per year has been estimated for the northeast coast of Florida, an area that experiences larger waves than those on the Georgia Coast (Beck and Legault, 2012), presumably because of its exposure to the longer fetch of the northeastern winds.

Because the South Carolina number (150,000 cubic yards of sand) is a little difficult to grasp, we once calculated that this volume of sand would go a long way toward filling Williams Bryce Stadium, where the University of South Carolina football games are played. One hundred and fifty thousand cubic yards does seem like a large number, but in fact, it is rather puny by global standards. Places on the California coast have rates at least three times that much, and the Ghana coast in West Africa has rates that exceed one million cubic yards of longshore sediment transport in one year, as a result of the large waves on these exposed, open ocean coasts (Griggs, 1986; Nairn and Hayes, 1997).

Rip currents (Figure 32) form most commonly along beaches where waves come straight on shore (not at an angle usually, but there are some exceptions to this general rule). Also, there is typically some degree of reflection of the waves off the beachface. These currents, which typically flow in a perpendicular direction straight off the beach, tend to be very regularly spaced. Based on extensive research, most rip currents owe their formation to the occurrence of edge waves–standing waves with their crests normal to the beach, according to Bowen and Inman (1969) and Komar (1976). Therefore, the direction of motion of the water within these waves is parallel to the shore and their crests are perpendicular to the shore. Furthermore, they are "trapped" against the beach, decreasing in amplitude exponentially in the offshore direction. If, indeed, they are formed as a result of the occurrence of standing waves, it would follow that the rip currents would be positioned at equally spaced distances where the edge waves have their maximum amplitude, not at the nodal points. Oh well, it would take a Ph.D. in physical oceanography and at least a B.S. in

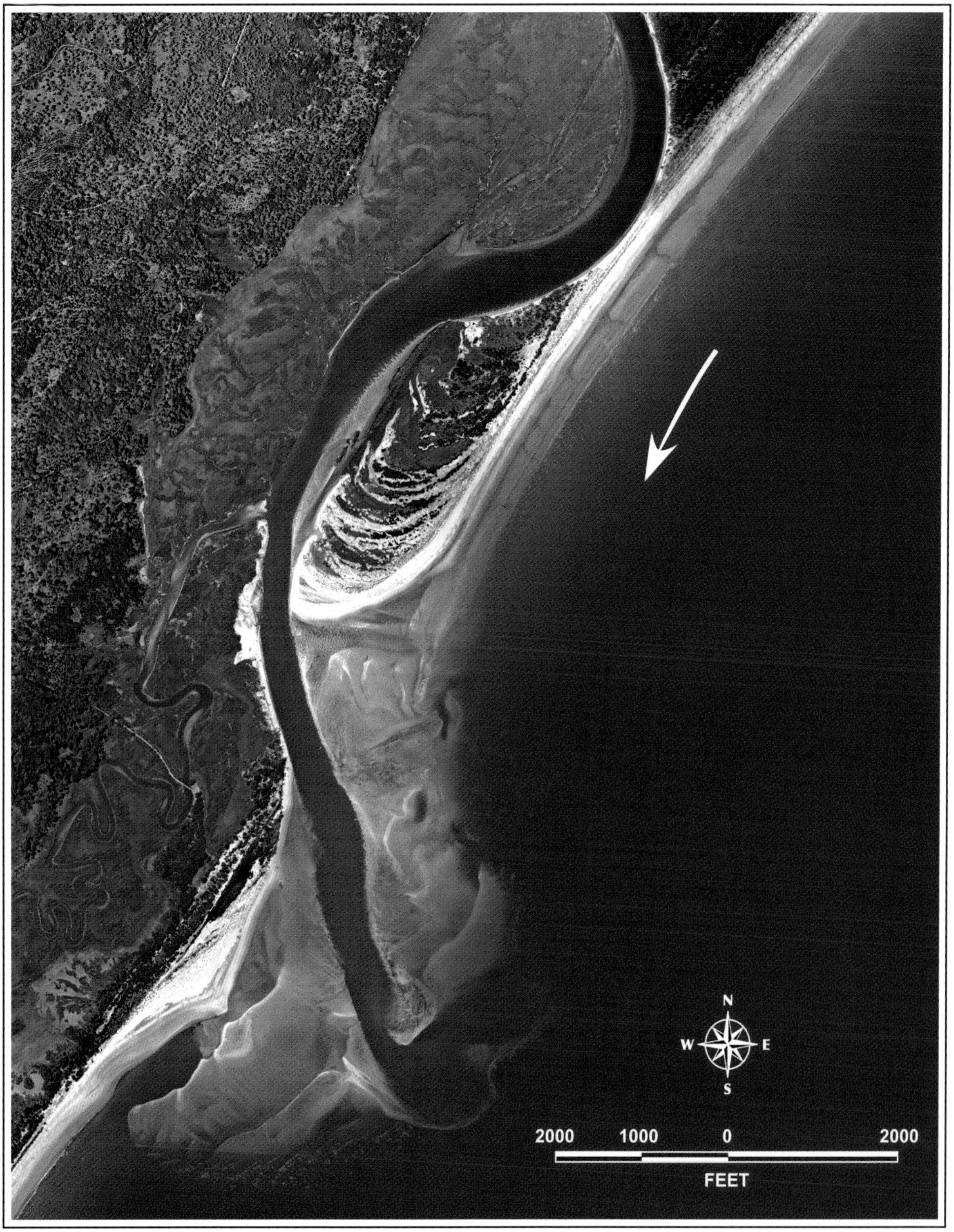

FIGURE 30. The recurved spit at the southern end of Blackbeard Island, Georgia, which indicates a pattern of consistent movement of sand along the shore from northeast to southwest (see arrow). Low tide digital orthophoto acquired in June 2008, courtesy of USGS.

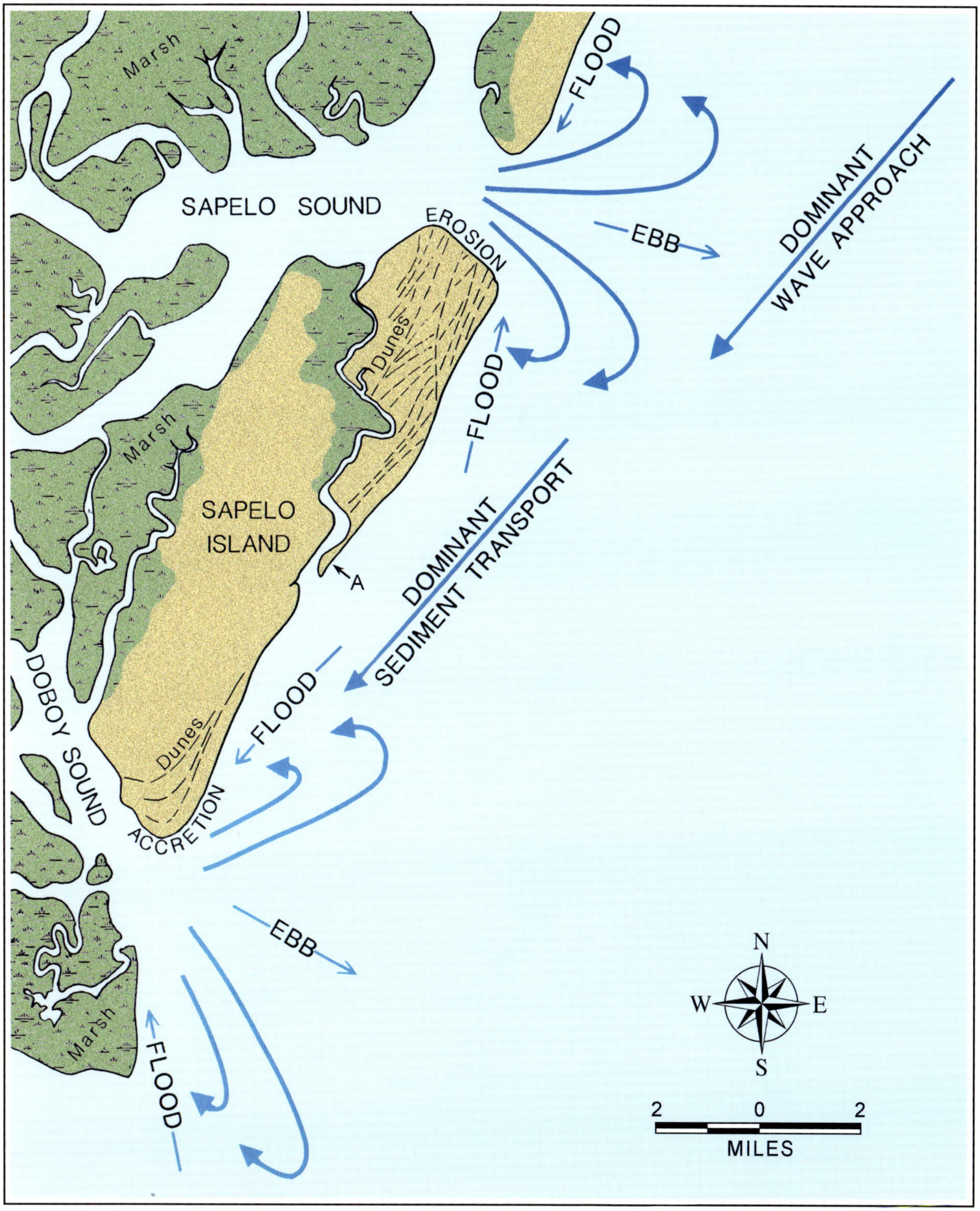

FIGURE 31. Physical coastal processes in the Sapelo Island, Georgia area. Dominant wave approach direction (mostly generated by extratropical and tropical cyclones) and dominant directions of tidal currents are shown. The dominant direction of longshore sediment transport along these beaches is from northeast to southwest. Arrow A points to the recurved spit shown in Figure 30. Modified from Griffin (1986).

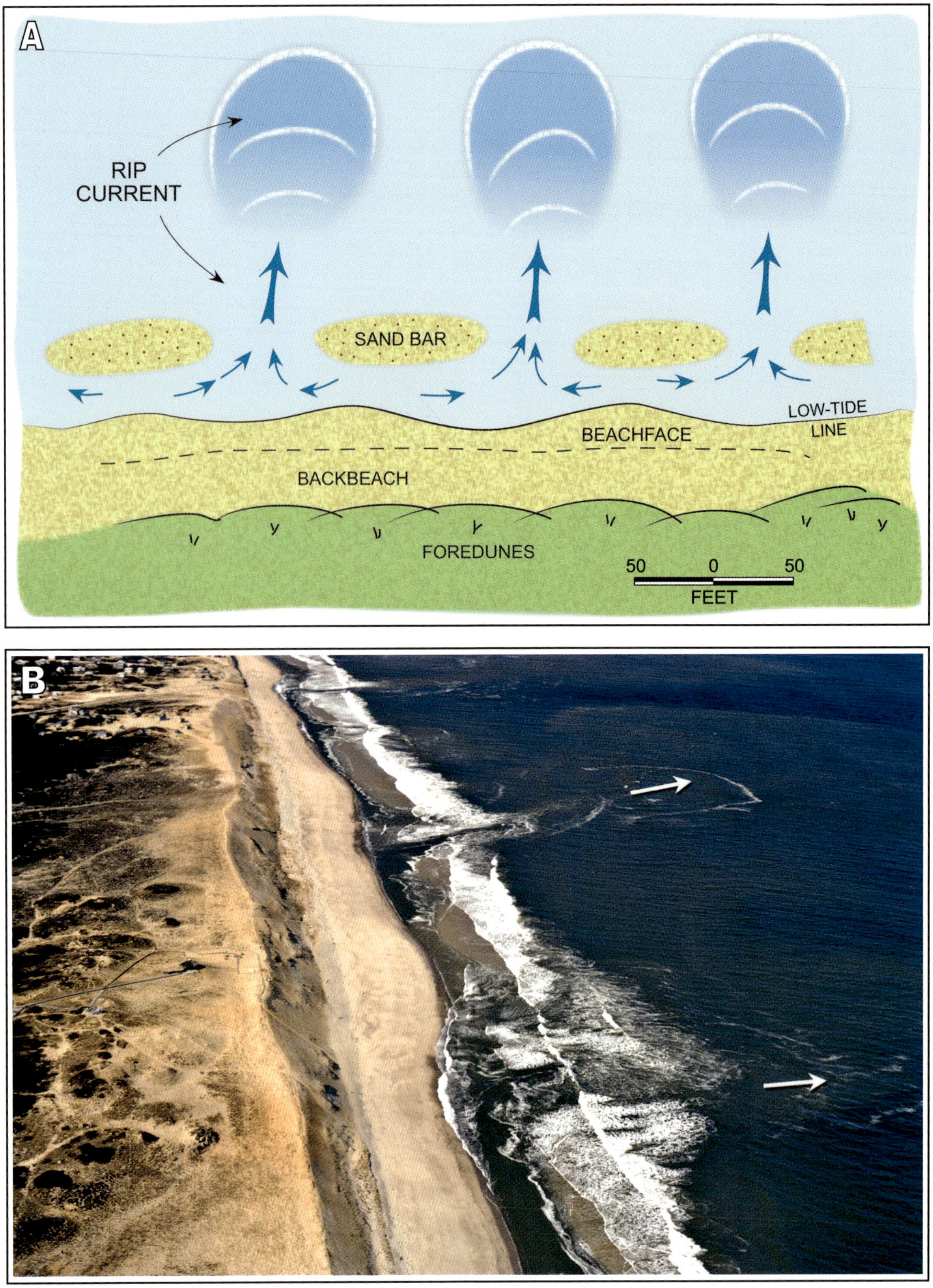

FIGURE 32. Characteristics of rip currents. (A) General sketch. (B) Rip currents along the shore of Cape Cod in April 1972. Note the sand bars coming ashore in the spaces between the rip currents.

physics to completely understand edge waves.

Rip currents may also be associated with some types of nearshore bars and off certain man-made structures (e.g., groins, which are discussed later), but the most common ones appear to be related to wave reflection and edge waves. At any rate, Hayes swears he has seen edge waves with his own eyes off a steep beach in Ghana, but Michel is a non-believer in edge waves, because she says "I have never seen one."

Beach cusps, illustrated in Figure 29, are relatively small-scale features (a few tens of feet maximum) that develop on the scalloped seaward margin of a mound of sand deposited on the intertidal beach called a berm (described in detail later). They consist of a regularly spaced sequence of indented bays and protruding horns. An example of subtle cusps at the high tide line on Jekyll Island is shown in Figure 29C.

Beach cusps are thought to be primarily depositional features (i.e., on a prograding beach) that form as a result of the interaction of shore-normal standing waves (edge waves) with shore-parallel, reflective waves, the same general process that produces rip currents (but at a smaller scale). This mode of origin was originally proposed by Guza and Inman (1975). There are other theories that challenge the edge-wave solution, believe it or not.

The only rip currents (not edge waves!!) the authors have actually seen on the South Carolina or Georgia Coasts were generated by waves reflecting off the steep beachface at the southern end of Edisto Beach State Park, South Carolina. Tim Kana reported that these types of currents are "fairly common" in the Grand Strand region at the northern end of the South Carolina Coast. Rip currents no doubt also occur during the passage of major storms, and they can be very dangerous. If by chance you ever get caught in one while swimming off the beach, swim parallel to the beach so as to escape the current, which has the potential of carrying you far offshore.

HURRICANES

In terms of the scale of dynamic coastal processes, nothing can match those generated during the passage of a large hurricane. As it turns out, these types of storms have played a major role in the scientific careers of the authors of this book. For example, when Hurricane *Carla* crossed the South Texas Coast on 11 September 1961, Hayes, a graduate student at the University of Texas at the time, was well into his dissertation study of the coastal environments of that area. In addition to doing extensive damage to the University of Texas field station at Port Aransas, a considerable amount of the data Hayes had collected up to that time was lost during the storm. Furthermore, the coastal environments themselves had been drastically changed by the hurricane. Ever the opportunist, Hayes turned his dissertation into a study of the geomorphological impacts of a major hurricane. This study became the first detailed geomorphological investigation of such impacts on the coastline of the United States. Another hurricane, *Cindy* (1963), also provided input to that study (Hayes, 1965; 1967).

In 1979, we were investigating the pollution of the South Texas Coast resulting from the *Ixtoc I* blowout in Mexico, the third largest oil spill in history (formerly second largest, but recently surpassed by the *Deepwater Horizon* spill in the Gulf of Mexico in 2010). As the cleanup from that spill continued into the summer of 1980, Hurricane *Allen* crossed the coast just north of Brownsville, Texas in early August. Happily, the storm cleaned the beaches of most of the remaining oil, thus we were able to bring that project to an end.

In September 1989, Hurricane *Hugo* crossed the South Carolina Coast just north of Charleston, doing a great deal of damage to the coast. We participated in some scientific studies of that hurricane (e.g., Sexton and Hayes, 1991), which became a kind of time line for references to the characteristics of the South Carolina Coast that we referred to as we led our field seminars there.

In 2004, we moved to New Orleans to set up a companion business for our home office in Columbia, South Carolina. We had been in New Orleans less than a year when hurricanes *Katrina* and *Rita* struck Louisiana. Our house in Uptown New Orleans suffered only very minor damage (it was above sea level), but the building that housed our 5th floor business office on the University of New Orleans campus was declared unlivable, because of an intense attack of mold. Therefore, we had to move the office to Baton Rouge, a 1.5-hour drive from our house. Meanwhile, Michel was working endlessly for NOAA and the U.S. Coast Guard on the over 135 oil spills in south Louisiana generated by the two hurricanes. After giving this living arrangement some consideration for a few months, we moved that business operation back to the main office in Columbia. At this point, we would be happy not to be involved with hurricanes any more for a while, big oil spills being all the excitement we need for the time being.

Technically, a **hurricane** is a storm of tropical origin with a cyclonic wind circulation (counterclockwise in the Northern Hemisphere) of 74 miles per hour (mph) or higher. The "hurricane" is the North Atlantic member of the **tropical cyclone family** that includes the "typhoon" of the western North Pacific, the "cyclone" of the Indian Ocean, and the "willy-willy" of Australia. Tropical cyclones are the most powerful and destructive of all storms. Tornados have higher wind velocities (central axis velocities may attain 400-500 mph), but tropical cyclones are also intense (winds may reach 150-200 mph) and cover much larger areas. Extratropical cyclones (e.g., northeasters) often cover even larger areas than tropical cyclones, but do not achieve their intensity. Tropical cyclones exceed all other natural catastrophes in total damage to human life and property, and no other weather phenomena receive such widespread and intense attention from the public these days, particularly after Hurricane *Katrina* (2005) resulted in the flooding of New Orleans and Hurricane *Sandy* (2012) devastated large areas of coastal New Jersey and New York.

Over the period of 1851-2010, hurricanes have occurred in the North Atlantic with a frequency of about 5 per year; hurricanes have struck the mainland United States at a frequency of about 1.8 per year (Blake and Gibney, 2011). In recent years, however, there has been a clear increase in the number and intensity of tropical storms and major hurricanes. Since 1995, the number of storms has risen dramatically, with an average at about 15 tropical storms, including about 8 hurricanes, per year between 1995-2010. More significantly, landfall activity during the 2000s has picked up and is now near the frequency seen in the very active 1950s. The 2011 Atlantic hurricane season produced 20 tropical cyclones, 19 tropical storms, 7 hurricanes, and 4 major hurricanes. Like many scientific issues, the cause of this increase appears to be a matter of debate. Some prognosticators relate this increase to global warming, but others, most notably Professor William Gray of Colorado State University, a long-time predictor of hurricane frequency, believe *these multi-decadal variations are mostly due to changes in large-scale ocean circulations referred to as the Atlantic Ocean thermohaline circulation.*

Not in doubt, however, is the fact that tropical cyclones occur most frequently during the months of August, September, and October. These months of maximum occurrence of hurricanes correspond roughly to the time when the Intertropical Convergence Zone, or equatorial trough, has its maximum divergence from the equator. Dunn and Miller (1960) pointed out the importance of this divergence to hurricane formation in the following observation:

> *When the Intertropical Convergence Zone is located within a few degrees of the equator, only small vortices form along its course, but when it migrates northward, the influence of the rotating globe becomes great enough to transfer sufficient spin to converging currents to permit tropical cyclones to develop. In the Atlantic area, this occurs principally in the Cape Verde region and in the western Caribbean just north of the Isthmus of Panama.*

The characteristics of hurricanes that produce the greatest geomorphological impact on shorelines are the storm surge and waves. Strong nearshore currents are also generated during the passage of a hurricane. Not only is the **storm surge** the primary cause of death and property damage during a hurricane, but it is also the quality of hurricanes most responsible for making them such important geomorphological agents. The rise in water level brought about by hurricanes inundates vast areas of low-lying coastal regions, permitting widespread erosion and deposition of nearshore sediments, in addition to catastrophic loss of man-made structures.

The two most important factors in the generation of storm surges are the stress of wind on the sea surface referred to as wind set-up and the reduction of atmospheric pressure, or "inverted barometer effect." The theoretical relation of decrease in atmospheric pressure to water level rise is a 13-inch elevation of water level for each inch of mercury in pressure loss. Winds of North Atlantic hurricanes have frequently reached velocities of 100 to 135 mph and a few have attained 200 mph. The width of the path of a single storm may sometimes extend hundreds of miles. These factors, combined with the fact that a storm may last for several days, give hurricanes the ability to pile up tremendous quantities of water against the coastline.

Other factors that accentuate, or modify, storm surge are:

1) *Shoreline configuration* – Just as with astronomical tides, the funneling of surge waters into coastal embayments, bays/lagoons, and estuaries will raise water levels considerably higher in their upper reaches compared to open coasts.

2) *Shape and slope of the continental shelf* – Greatest amplification occurs on a gently sloping shelf, like those

off the Gulf Coast states and the Georgia Bight.

3) *The angle the hurricane track makes with the coastline* – The counterclockwise circulation of the winds in a hurricane is such that, as the storm approaches the shore, only the winds on the right-hand side of the circular flow are moving in the same direction as the storm. Consequently, there is an additive effect of the wind set-up as the storm moves along, only on the right side of the storm. Therefore, the peak surge usually occurs many miles "to the right" of where the storm's eye crosses the shore. If the storm strikes the shore at an acute angle between the right-hand side of the storm track and the shore, the surge will be amplified even further. An example of this effect was the disastrous Galveston hurricane of 1900, in which more than 6,000 people died. That hurricane approached the northeast-southwest trending Texas coastline from the east-southeast, creating a maximum surge of 14.5 feet that washed over the low-lying barrier island. For more discussion of this storm see Erik Larson's book *Isaac's Storm: A Man, a Time, and the Deadliest Hurricane in History*.

4) *Stage of the astronomical tide (maximum effect at spring high tide)* – This is more important on coasts with large tidal ranges, such as the Georgia Coast.

5) *Rainfall effect* – Hurricanes may produce more rainfall in an area within a span of a few hours than is normally experienced in a whole year. This excess water may add considerably to flooding of coastal regions, especially if stream outlets are blocked by the storm surge.

Probably the most spectacular geological effects produced by hurricanes are the result of erosion by **breaking waves**. The most impressive hurricane effect ever witnessed by the authors, who have studied the impacts of numerous hurricanes, was the absolute destruction of a coastal town called Holly Beach, Louisiana (all houses ripped completely off their foundations) by the destructive waves of Hurricane *Rita* (2005). Hurricane *Ike* (2008) did the same thing to parts of the Bolivar Peninsula and Galveston Island, Texas, with considerably more houses being destroyed by *Ike* than were destroyed by *Rita* in Louisiana.

A cubic yard of water weighs about three-fourths of a ton, and a breaking wave may move forward at speeds of 50-60 mph. Their erosive effects are greatly increased when they ride the crest of a large storm surge, because of the extensive land areas and human developments that are exposed to erosion. Sometimes hurricane waves attain tremendous heights. It is common for wave heights of 40 to 50 feet to be observed offshore during major hurricanes.

Hurricane-generated waves that strike the shoreline obliquely no doubt generate strong longshore currents, although they are almost never actually measured by scientists (wonder why?). Timbers and pilings from Bob Hall fishing pier on northern Padre Island, Texas, which was completely destroyed by Hurricane *Carla* (1961), were found by Hayes for many miles south along the beach after the storm. Exceptionally strong currents also flow through tidal inlets, as well as through new channels cut through barrier islands, during all major hurricanes.

Hurricanes cross the Georgia and South Carolina Coasts much less frequently that they do the coasts of Florida, North Carolina, and Texas, where, on the average, crossings happen more than once a year. This rarity of storm crossings of these two coasts appears to be due to: (1) The shoreline at the head of the Georgia Bight is located so far inland of the general trend of the shoreline of the southeastern United States; and (2) Hurricanes approaching Georgia tend in many cases to swing offshore in a more northeasterly direction, presumably because the global westerly winds "push them" in that direction, where they commonly clip the barrier islands of North Carolina that project far out into the ocean away from the general trend of the shoreline.

HURRICANES THAT HAVE CROSSED THE GEORGIA COAST

The tracks of the 14 hurricanes that have either crossed the Georgia Coast or passed close by between 1881 and 2011 are illustrated in Figure 33. However, with regard to major hurricanes, Lenz (2002a) noted that the *Georgia Coast experienced five major hurricanes in the 1800s, including one that put Tybee Island completely under water, but has been mostly spared in the 1900s.* A succinct summary of the effect of hurricanes on the Georgia Coast is given by Clayton et al. (1992) in their book titled: *Living with the Georgia Shore.* We highly recommend that you buy this book if you either own property along the Georgia Coast or vacation there regularly. With regard to the occurrence of hurricanes on the Georgia Coast, they pointed out that *if the years from 1886 to 1968 can be considered typical, Georgia may*

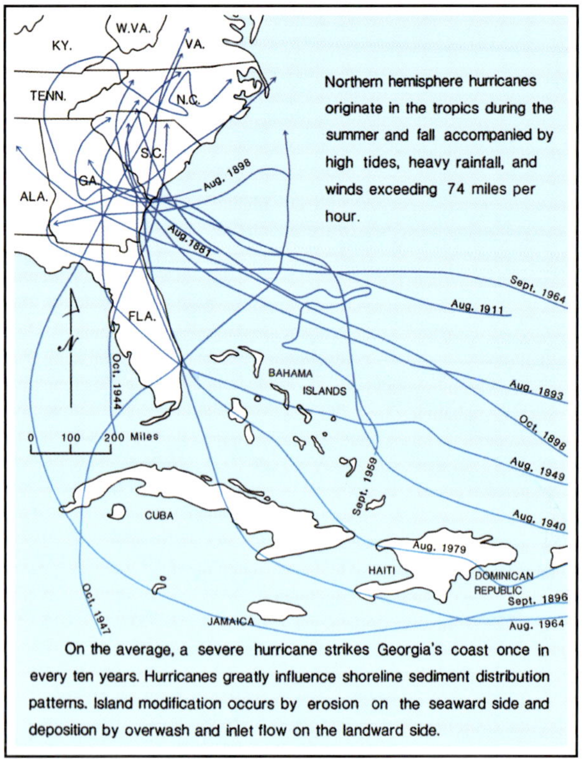

FIGURE 33. Tracks of hurricanes that have impacted the Georgia coast between 1881 and 2011. Modified from Griffin (1986).

expect to receive about one full-fledged hurricane every 10 to 15 years. They also presented a table that gives details on the twenty most significant hurricanes that have impacted the Georgia Coast since September 1752. Three of the most damaging storms were:

1) A major hurricane in August 1881. Winds of 80 mph and a 16-foot storm surge resulted in extensive property damage and 335 deaths in Savannah.

2) Extreme/Great hurricane in August 1893. Winds of 72 mph and a 17-20-foot storm surge damaged nearly all the buildings on Tybee Island. An estimated 1,000-2,000 deaths resulted from this storm, but most of them were in South Carolina, because the storm track crossed over Tybee Island and the biggest storm surge always occurs to the right of the storm track. This storm essentially wiped out the town of Edingsville Beach, South Carolina, destroying 60 homes (Orrin Pilkey, pers. comm.).

3) Extreme hurricane of October 1898. A storm surge of 12 feet and winds of unknown strength created severe damages in the Brunswick/Darien area, with at least 180 fatalities.

The most recent hurricane to cross the Georgia Coast was *David* in 1979. According to Hebert (1980), *hurricane David made landfall in Georgia as a quickly weakening minimal hurricane, bringing a three–five foot storm surge and heavy surf. Its inner core remained away from major cities, though Savannah recorded sustained winds of 58 mph and wind gusts of 68 mph.* No major damage occurred in Savannah.

The Saffir-Simpson Scale, is used by the National Weather Service to rate a hurricane's strength. This scale, which ranks storms based on wind velocity, storm surge, and barometric pressure, is reproduced here as Table 2. Using this scale allows the National Weather Service to communicate the nature of the storm, informing the potentially threatened citizens about what to expect with regard to how strong the wind might be, the size of the potential waves to expect, and potential flooding. Clearly, the higher the category, the greater the possible danger from the storm; therefore, categories 3-5 storms should be given a wide berth. *Surge amplitudes are specifically calculated for the Georgia Coast. These values are higher than those listed in the standard Saffir/Simpson scale because the shallow waters of the Georgia Coast enhance the height of storm surge* (Clayton et al., 1992).

TABLE 2. Saffir-Simpson Hurricane Scale.

Category	Central Pressure (inches mercury)	Wind (mph)	Surge (feet)	Damage	Example
1	>28.94	74-95	4-5	minimal	*Agnes* (1972)
2	28.50-28.93	96-110	6-8	moderate	*David* (1979)
3	27.89-28.49	111-130	9-12	extensive	*Connie* (1955)
4	27.18-27.88	131-155	13-18	extreme	*Hugo* (1989)
5	<27.17	>155	>18	catastrophic	*Camille* (1969)

EARTHQUAKES

Although located in the Georgia Bight, an area not normally associated with such happenings, the earthquake that hit the Charleston, South Carolina area on 31 August 1886 was *the most destructive ever recorded in the eastern half of North America* (Murphy, 1995). The New Madrid, Missouri earthquake of 1811 was more intense, but not nearly as damaging to human-habitation structures because of the low density of population in that part of the country at that time. By the way, according to legend, the Madrid earthquake was caused by the great Shawnee warrior, *Tecumseh*, stomping his foot. Should you be interested in more details on that subject, we recommend the intriguing book, *A Sorrow in Our Hearts, The Life of Tecumseh*, by Allan W. Eckert.

Two books that give fascinating details on the Charleston earthquake are:

1) Chapter 9 in *Carolina Rocks!* by C. H. Murphey (1995), that gives some of the geological background; and

2) *City of Heroes* by R. N. Cote (2006), that treats in great detail the human impacts of the calamity.

Earthquakes in Georgia are rare. However, the state is occasionally subject to tremors radiating from earthquakes in neighboring states, such as the Charleston earthquake of 1886. As noted by UGA/Geology (2009), the relative lack of earthquakes in Georgia results from *Georgia's location on a rifted passive margin of a continent, where there is a stable transition from the continental crust of North America to the ocean crust of the Atlantic,* unlike in areas such as Chile and parts of California and Alaska, where the continental North American Plate collides with and overrides the oceanic Pacific Plate. Although rare, smaller earthquakes do sometimes occur in Georgia, such as on 7 December 2009 when the U.S. Geological Survey reported a small earthquake that registered a magnitude of 3.2 centered in the Deepstep, Georgia area, a small community approximately 90 miles south of Atlanta (Cornelius, 2009).

[NOTE: The Richter scale and the Moment-Magnitude scale are the two scales most commonly used to describe the magnitude of earthquakes. The Richter scale, invented by Charles F. Richter in 1934, is based on the calculation of the maximum magnitudes of the shear waves generated by the quake (as read on a seismograph instrument, which contains a pendulum or a magnet, that, because of inertia, remains still as the Earth moves beneath it). Both the Richter and the Moment-Magnitude scales are logarithmic (base 10). For example, a magnitude 5 earthquake on the Richter scale would result in ten times the level of ground shaking as a magnitude 4 earthquake (and 32 times as much energy would be released) (USGS, 2012). Quakes with Richter scale values of 3.5 to 5.4 are often felt, but rarely cause damage. Those above 7.0 on the scale are classified as major earthquakes and can cause widespread damage. The Moment-Magnitude scale (expressed as M), defined as the amount of energy released on the fault, is based on several factors, including: a) How far the fault slipped (i.e., how far a fence line, railroad line, and so on was offset); b) Length of the fault line that was affected; c) Depth at which movement was initiated; and d) The "stickiness" of the surfaces being faulted (rock rigidity). Clearly, this measure is more difficult to calculate, but it is the one most commonly used in reporting earthquake magnitudes at this time.]

5 MAJOR LANDFORMS OF THE COAST

GEORGIA BIGHT

Introduction

A nearly continuous chain of barrier islands lies along a broad coastal arc bordering the southeastern U.S. states of North Carolina, South Carolina, Georgia, and Florida. This coastal area, referred to here as the Georgia Bight (called the South Atlantic Bight by others), is, in fact, the centerpiece for the longest single stretch of barrier islands in the world, the east and Gulf coasts of the North American continent–Cape Cod, Massachusetts to the Yucatan Peninsula, Mexico.

The shoreline of the Georgia Bight, which extends from Cape Hatteras, North Carolina to Cape Canaveral, Florida (Figures 4 and 23), is over 740 miles long. The Bight flanks the coastal plain on the trailing edge of the North American Plate that is generally tectonically stable to slightly downwarping. Two northwest-southeast oriented tectonic highs (regions that historically are rising rather than sinking), the Cape Fear Arch on the north and the Ocala Uplift on the south, border the Bight (Figure 4). A corollary of the rise of the two arches is the downwarping of the underlying crust in the center of the Bight, in the area know as the "Low Country" in the southern reaches of South Carolina and on into Georgia. This tectonic regime has apparently been in existence since late Paleocene (about 60 million years ago; Colquhoun et al., 1983). The bulk of the older sediments and sedimentary rocks that underlie the coastal plain of the Bight, that are Cretaceous to Holocene in age (see Table 1; a time span of approximately 100 million years), is only 900 feet thick at the North Carolina/South Carolina border, but thickens to 3,000 feet at the South Carolina/Georgia border. This wedge of older sediments thins again further to the south.

As discussed earlier, evidence published by Winker and Howard (1977) indicates that a zone of Pleistocene sediments deposited up to several tens of miles landward of the present shoreline between 130,000 and 40,000 years ago, which they call the Chatham Sequence, has been raised 16.5 feet in North Carolina relative to the center of the Georgia Bight during that period of time. Thus, we can assume that the tectonic regime that originally formed the Georgia Bight (i.e., elevation of the two arches to the north and south and sinking of the crustal material underlying the center of the Bight), however mild, remains active even to this day. This tectonic sinking should not be confused with global sea level rise, which is a different issue altogether.

The Holocene coastline of the Georgia Bight is, for the most part, depositional, with barrier-island, estuarine/bay/lagoonal, and minor deltaic and river deposits blanketing an irregular topography eroded during the low stands of the sea during the Pleistocene epoch (Ice Age).

Previous Work and Database

The Georgia Bight is one of the most intensely researched areas in the world with respect to its coastal geology. John Fisher's graduate research at the University of North Carolina was one of the first studies to call attention to the geology of North Carolina's barrier islands (Fisher 1962; 1967). This work was soon followed by Pierce and Colquhoun's (1970) excellent discussion of the Holocene evolution of the barrier islands of the same area. As noted in the dedication of this book, the 1960s also saw an abundance of papers published by workers at the University of Georgia Marine Institute (e.g., Hoyt and Henry, 1963; Pilkey and Richter, 1965; Hoyt, 1967). Papers such as those by Weimer and Hoyt (1964), Hoyt, Weimer, and Henry (1964), and Hoyt and Henry (1965) highlighted the Georgia Coast as a modern analog for use during exploration for oil-and-gas reservoirs in ancient sediments. This tradition, called comparative sedimentology studies, was carried on into the 1980s, with the publication of papers such as Frey and Howard

(1969), Howard and Reineck (1972), and Howard and Frey (1980). Oertel's work on inlet dynamics and sedimentation (Oertel, 1972; 1977) and Henry's shelf work (Henry and Hoyt, 1968; Henry and Kellam, 1988; Henry and Idris, 1992) complemented those efforts.

Studies of the continental shelf of the Georgia Bight, initiated primarily by Orrin Pilkey's early work (Pilkey, 1963), have added many insights regarding the topography and geology of the shelf, as well as insights into the origin of the barrier islands of the Georgia Bight. Papers by Swift (1975), Hine and Snyder (1985), Hine et al. (1986), Meisburger and Field (1974), and Sexton (1987) are particularly informative in that regard.

The authors of this book have been carrying out research in the Georgia Bight since the mid 1970s that has included:

1) Numerous aerial surveys of the entire Bight, resulting, among other products, in the classification of the area into six geomorphologically distinct compartments. These compartments were subdivided into subunits and described in detail.

2) Numerous M.S. theses and Ph.D. dissertations on the coastal geology of the Georgia Bight were carried out under Hayes' supervision at the Geology Department of the University of South Carolina between 1972 and 1985. During these projects, over 500 vibracores of the Holocene sediments were collected, numerous wave measurements were made, hundreds of sediment samples were analyzed, and so on.

3) The authors also supervised the compilation of detailed maps of the coastal habitats of the entire Bight using an Environmental Sensitivity Index (ESI) they created in 1976 (Hayes, Michel, and Brown, 1977), which has a geomorphic framework. One of the uses of these maps is to determine the relative sensitivity of the coastal environments throughout the Bight to oil pollution (NOAA/RPI, 1996a; 1996b; 1997; FDEP/RPI, 1996), but they also have broader environmental applications. The maps also contain detailed biological information, with some of that for the Georgia Coast being discussed in Section II.

BARRIER ISLANDS

Introduction

Barrier islands are elongate, shore-parallel accumulations of unconsolidated sediment (primarily sand), some parts of which are supratidal (above the high-tide line most of the time, except during major storms). They are separated from the mainland by bays, lagoons, estuaries, or wetland complexes. According to Glaeser (1978), 76% of the barrier islands of the world occur along the coastlines of the trailing edges of continental plates and of large inland seas and lakes (e.g., Caspian and Black Seas). In a recent study of satellite imagery of barrier islands along the world's coasts, Matthew Stutz and Orrin Pilkey of Duke University reported that 63% of the barrier islands occur on trailing edge coasts, 16% on leading edge coasts, and 21% on marginal seas, with their 84% being a little higher than Glaeser's 76%. Barrier islands do not occur on coasts with tidal ranges greater than about 12 feet, because their primary mechanism of formation, wave action, is not focused long enough at a single level during the tidal cycle to form the island, and the strong tidal currents associated with such large tides transport the available sand to the offshore regions. Some barrier islands do occur, primarily as spit forms, on leading edge and glaciated coasts, but they are a relatively minor coastline type in those areas.

Types of Barrier Islands

Depositional, coastal-plain coasts, such as the Georgia and South Carolina Coasts, typically have barrier islands located between major river deltas and estuaries. Two major types of barrier islands are present along the coast of the Georgia Bight, those that consistently migrate landward (**transgressive**) and those that build seaward (**prograding**). The Georgia Coast has these two types of barrier islands, with a third, somewhat more complex type, to be discussed later. Although all barrier islands tend to change to some extent over time, the rates of landward migration and/or seaward growth are different from place to place. In some parts of the South Carolina Coast, for example, the landward-migrating barrier islands may move landward tens of feet per year. However, those that build seaward usually grow more slowly, at rates of less than 10 feet per decade. Which of these two types of islands is present at a particular location along any coast depends upon the rate of sea-level change relative to sand supply at that location (Curray, 1964). Diminished, or low, sand supply and/or rapid sea-level rise promotes the development of landward-migrating islands, and *vice versa* for those that build seaward (Hayes, 1979; 2009). On the Georgia Coast, there are a number of locations along originally prograding barrier islands where the beach is eroding at the present time because of diminished sand supply or other reasons, but the part of the island that remains was formed during the progradational phase of island development. Also, there may be short-term episodes of beach buildup on landward-migrating barrier islands where an unusually large amount of sand comes on shore. However, these are exceptions to the general rule.

As illustrated in Figures 34A, 34B, and 35 (upper half), landward-migrating barrier islands are composed of coalescing **washover fans** or a **washover terrace**, that is overtopped at high tides, usually several times a year. In the process of migration, the entire washover terrace complex moves landward, leaving an eroded nearshore zone in its wake. As a result of this type of migration, in three dimensions the entire complex consists of a relatively thin (<6-12 feet) wedge of sand and shell of the washover terrace that overlies muddy sediment originally deposited in the estuaries or wetlands behind the islands (see cross-section in upper half of Figure 35). Generally speaking, if you are walking on the beach of a barrier island at low tide and you find marsh sediments exposed in the intertidal zone, such as that shown in Figures 34A and 34B, that particular part of the barrier island is migrating landward. The landward-migrating barrier islands in the Georgia Bight are relatively short, 1-5 miles on the average, because new inlets are created where the migrating islands intersect tidal channels.

Prograding (seaward-building) barrier islands (Figures 34C, 35-lower half, and 36) are typically composed of multiple, relatively parallel linear ridges of sand topped by vegetated sand dunes that originally formed as front-line dunes on the backbeach (called **foredunes**). The most notable changes on these types of islands occur where adjacent tidal inlets migrate into them or when the inlets expand dramatically during hurricane storm surges, as shown by the general model in Figure 37. As a result of this type of seaward growth, in three dimensions they consist of a wedge of sand 25-30 feet thick that has built seaward over offshore mud (see cross-section in the lower half of Figure 35).

The morphology of the prograding barrier islands in some cases takes on a characteristic drumstick appearance, as shown in Figure 36. Two factors that enhance the development of the drumstick shape on the coast of the Georgia Bight are:

1) Masses of sand in the form of large intertidal sand bars (**swash bars**) that pass around the outer margin of a

FIGURE 34. Georgia barrier islands. (A) Landward-migrating (transgressive) barrier island at the southern end of St. Catherines Island. Arrows labeled A point to marsh sediments exposed in the intertidal zone and those labeled B point to washover fans. (B). Landward-migrating (transgressive) barrier island in the form of a series of coalescing washover fans on Wolf Island. Arrows point to former tidal channels in the marsh. (C) Prograding barrier island at Bradley Point on the north end of Ossabaw Island. Note the parallelism of these seaward prograding beach ridges of Holocene age (<2,000 years in age). Arrow points to well-developed intertidal sand bar. All three photographs were taken during a low spring tide on 28 April 2010.

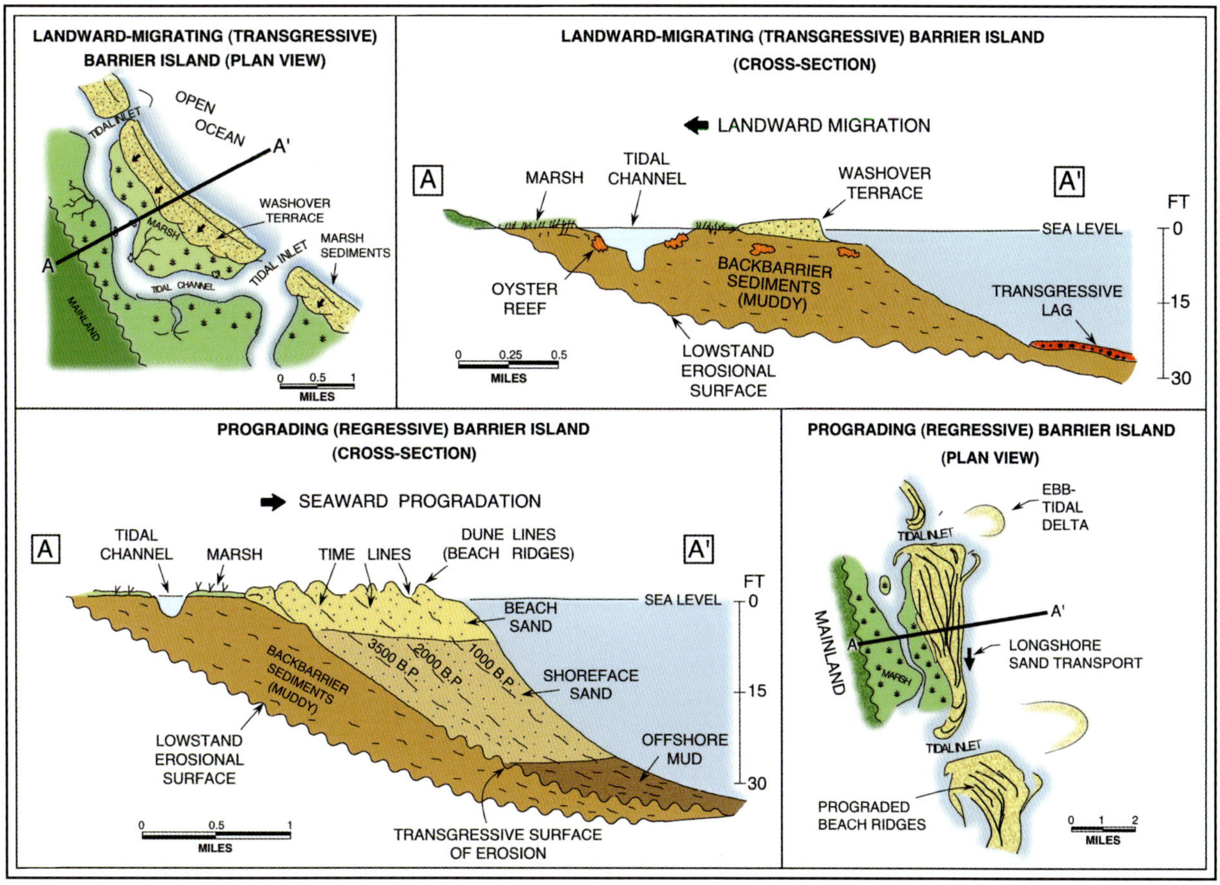

FIGURE 35. Morphology and subsurface three-dimensional configuration (stratigraphy) of landward-migrating (transgressive) barrier islands in the top row, and prograding (regressive) barrier islands in the bottom row. The cross-sections (A-A') are located in each of the plan view maps.

large sand shoal (**ebb-tidal delta**) associated with the tidal inlet on the northeast end of the island. These huge swash bars eventually move toward shore and attach to the beach (Figure 36A), building out the northeast end of the island and accentuating the drumstick shape (see photograph in Figure 36B). This process is clearly illustrated in Figure 36C; and

2) Bending of the dominant northeasterly waves around the ebb-tidal delta, a process that creates sediment transport reversal to the northeast in the area in the shelter of the ebb-tidal delta. This relatively minor reversal from the normal northeast-to-southwest sediment transport direction that occurs along the rest of the island (Figure 36A) allows some sand to remain in the inlet area, which accumulates on the northeast end of the island.

These two processes also commonly combine to create major **downdrift offsets** at many of the tidal inlets in the Georgia Bight. **Drift** is a term engineers use for the volume of sand that typically moves along a beach during some specific interval of time. The term updrift refers to the direction from which the sand comes and downdrift is the direction in which it travels. Normally, one would expect the two barrier islands on either side of a tidal inlet to be oriented along the same general trend of the shoreline. When this is not the case, the adjacent islands are said to be offset at the inlet. Because of the two processes described for the formation of the drumstick shape of many of the prograding barrier islands in the Georgia Bight (Figure 36A), the islands to the southwest of most of the inlets are offset in a seaward direction, including a few on the Georgia Coast.

A very interesting combination of the two barrier-island types, transgressive (eroding) and prograding, occurs on either side of Wassaw Sound. Williamson Island on the southern tip of Little Tybee Island, located

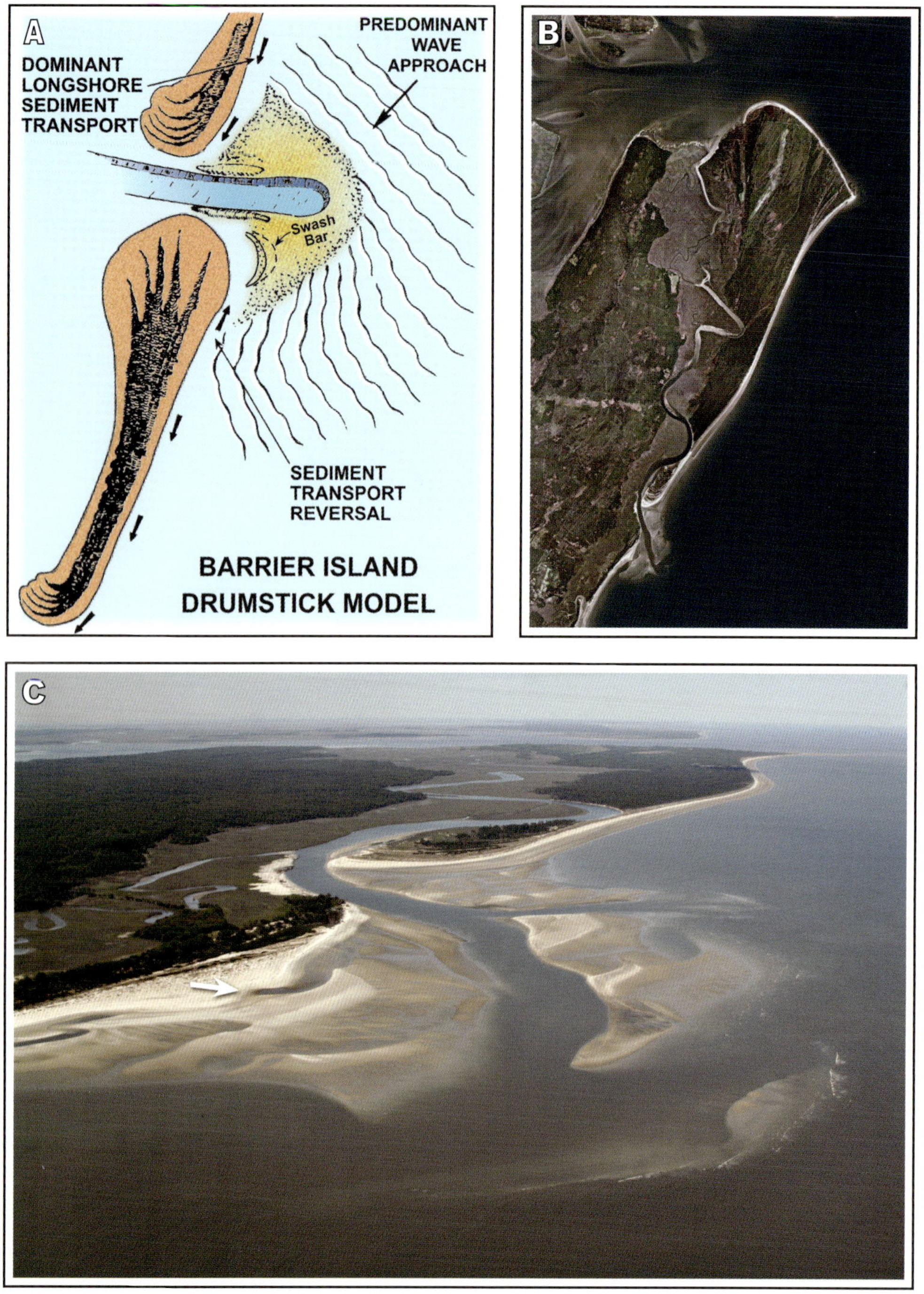

FIGURE 36. Typical morphology of prograding barrier islands in the Georgia Bight. (A) Barrier island drumstick model, primarily the result of transport of masses of sand derived from the updrift ebb-tidal delta in the form of large swash bars. (B) Blackbeard Island, Georgia, a seven-mile-long prograding barrier island that reflects the drumstick configuration. The updrift end of the barrier island is at top of the photograph. Digital orthophoto acquired in June 2008, courtesy of USGS. (C) Low spring-tide photograph (on 28 April 2010) of the ebb-tidal delta at the outlet to Blackbeard Creek (Cabretta Inlet). Arrow points to zone where swash bars are attaching to the beach on the downdrift side of the inlet as is suggested in A.

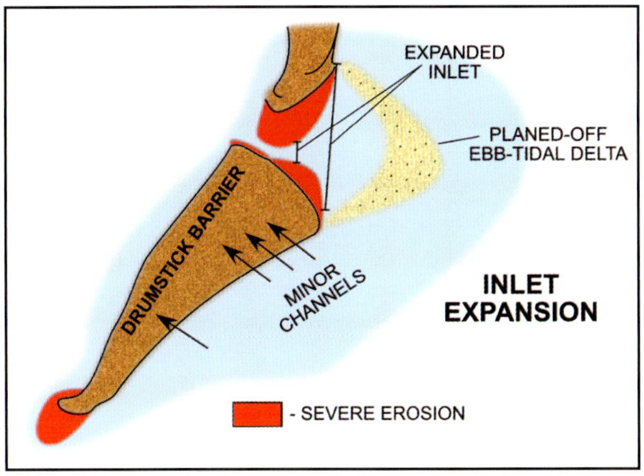

FIGURE 37. Generalized model for the expansion of tidal inlets observed on the South Carolina Coast as a result of hurricane *Hugo* (1989).

on the northeast side of the sound (arrow A on Figure 38), is clearly of the transgressive type. On the other hand, the island on the southwest side of the sound, Wassaw Island (arrow B on Figure 38), is composed of a series of prograding beach ridges, a classic example of a prograding barrier island. Inasmuch as the islands on both sides of the inlet have been subject to similar recent rates of sea-level rise, the difference between them must be explained by a larger supply of sand being available to Wassaw Island in recent times. Why? This is probably related in some way to the behavior of the large sand shoals located off the entrance to the sound, as well as to a diminished sand supply by the Savannah River in recent years because of the construction of dams on the river. Clark Alexander (pers. comm.) pointed out that *Williamson Island did not exist prior to the first renourishment of Tybee Island in 1976, and has been prominent after renourishments (1985, 1995, 2009), and eroded between them.*

Wassaw Island, composed entirely of Holocene sediments, is somewhat of an exception to the general rule on the Georgia Coast, where many of the major barrier islands have either a core, or are closely associated with, a massive body of Pleistocene sediments that was deposited during a highstand of sea level around 40-50,000 years ago (Silver Bluff; an age that is apparently somewhat controversial as discussed earlier). Figure 39A illustrates this point. The Holocene components of the barrier islands are best developed in the near vicinity of the two primary sources of sand to the shore at the present time, the Savannah and Altamaha Rivers. As illustrated in Figure 39B, these two rivers have provided enough sediment in the last 4,500 years to form modest-sized river deltas (discussed in more detail later).

Origin of Barrier Islands

The information we have, based on many years of study of the Georgia Bight coast, indicates that the barrier islands in this area originated in one of four possible ways, but we will cover all of the possibilities that we know of in the following discussion:

1) Elongation of sand spits away from a headland with permanent segmentation of the spits as they grow as a result of the formation of permanent tidal inlets through them during storms, creating individual islands, as illustrated in Figure 40. One of the major proponents of the **spit-elongation hypothesis**, John Fisher, cited the northern part of the Outer Banks of North Carolina as one of his examples of barrier islands that have been formed by that mechanism. Similar ideas were proposed by de Beaumont (1845) and Gilbert (1889) for other areas. The fact that some of the coastal sand bodies that build away from the uplifting Cape Fear and Ocala Arches have a spit-like form is evidence that a spit elongation mode of origin could have accounted for some of the barrier islands in these tectonically uplifting areas as well.

2) Formed at lower sea level and migrated to their present position. In a paper written in 1975, the noted marine geologist Don Swift stated that barrier islands originated at a lower stand of sea level and migrated over the drowning coastal plain as sea level rose, as indicated in Figure 41A and explained further in Figure 41B. This mechanism is probably the case in several localities in the Bight where the islands continue their landward migration even to the present day (e.g., Core Banks, North Carolina and just north of Cape Canaveral, Florida). These islands presently fit the transgressive barrier island type (Figure 35-upper layer). He didn't say exactly how the islands originated at that lower level. It is important to remember that all these barrier islands formed during the relatively recent Holocene transgression of the sea. As pointed out by Oertel et al. (1992), many of the world's coastal barrier lagoons have transgressed landward a sufficient distance to be partially or totally affected by landscape topography. The end result is that the lagoon or backbarrier areas between the island and the mainland tends to be complex from place to place, defying too much generalization. However, on

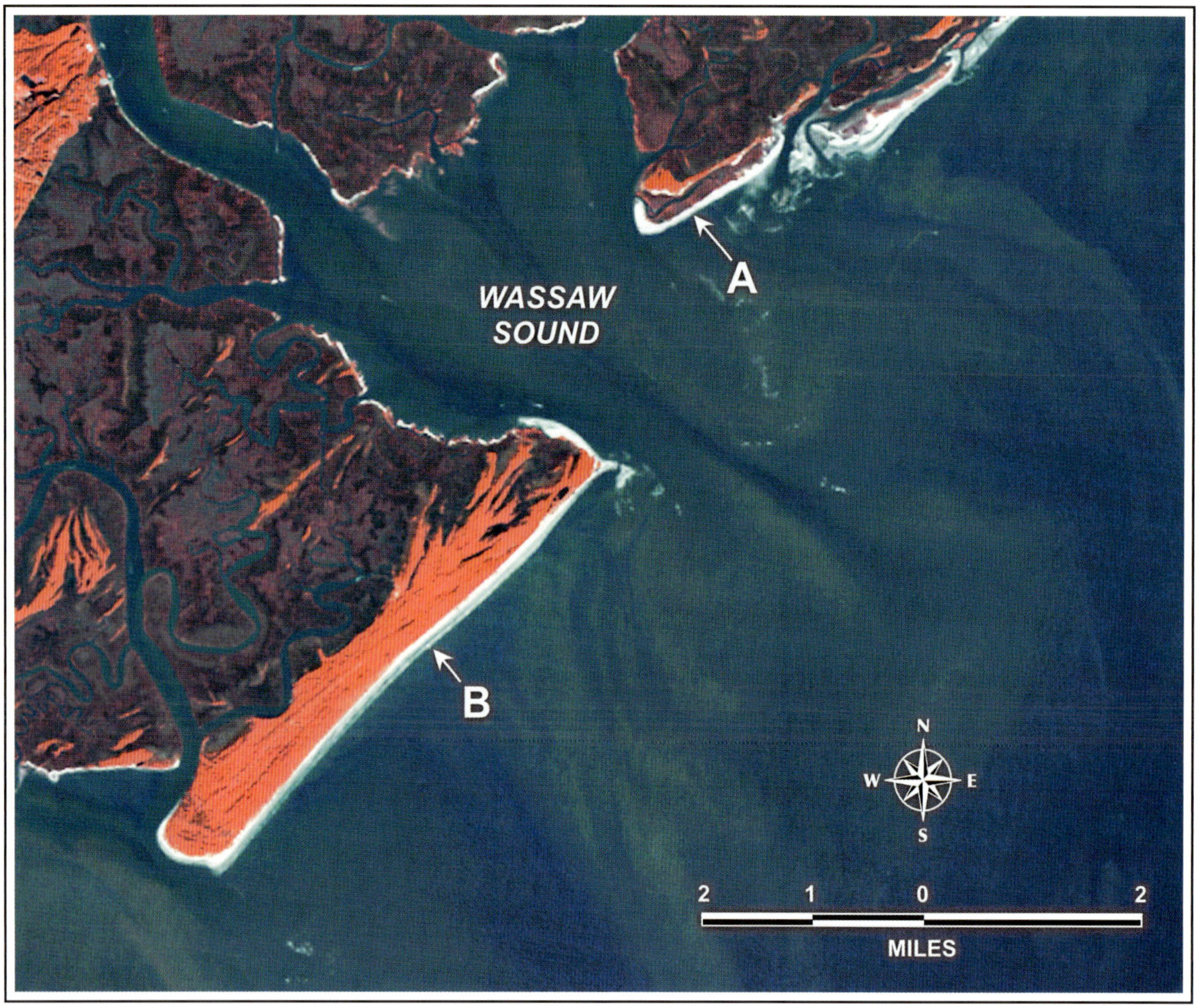

FIGURE 38. Vertical infrared image of the northeastern end of the Georgia Coast. This outbuilding, deltaic shoreline (Figure 39B) is the result of a significant sediment supply historically provided by the Savannah River. Arrow A points to an eroding barrier island on the southwest end of Little Tybee Island, and arrow B points to a prograding barrier island (Wassaw Island). Landsat image acquired in October 2005, courtesy of USGS.

coasts with large waves and small tides the backbarrier is commonly open water bays and lagoons (e.g., the North Carolina Coast). Where the tides are large and the waves are relatively small, it is common for the backbarrier to be wide salt marshes (e.g., Georgia Coast).

3) Evolving from the drowning of coastal sand ridges, a theory proposed by local hero John Hoyt (1967), which is illustrated in Figure 42B. W. Armstrong Price (pers. comm.) proposed a similar idea; however, he attributed their formation to the creation of large offshore bars during the high waters (storm surges) of hurricanes (Figure 42A). These bars would have become emergent as the water level created by the storm surge dropped, with some of them evolving into barrier islands. These are interesting ideas, but we are not convinced that either process is necessary for the formation of the barrier islands in the Georgia Bight.

4) A process we call the transgressive-regressive interfluve hypothesis, which has been well documented by others (Pierce and Colquhoun, 1970; Moslow, 1980). The mode of formation for most of the larger prograding barrier islands on the South Carolina Coast, such as Kiawah Island, is clear and well documented. Four major steps take place in this mode of formation according to the **transgressive-regressive interfluve hypothesis**, which is illustrated in Figure 43.

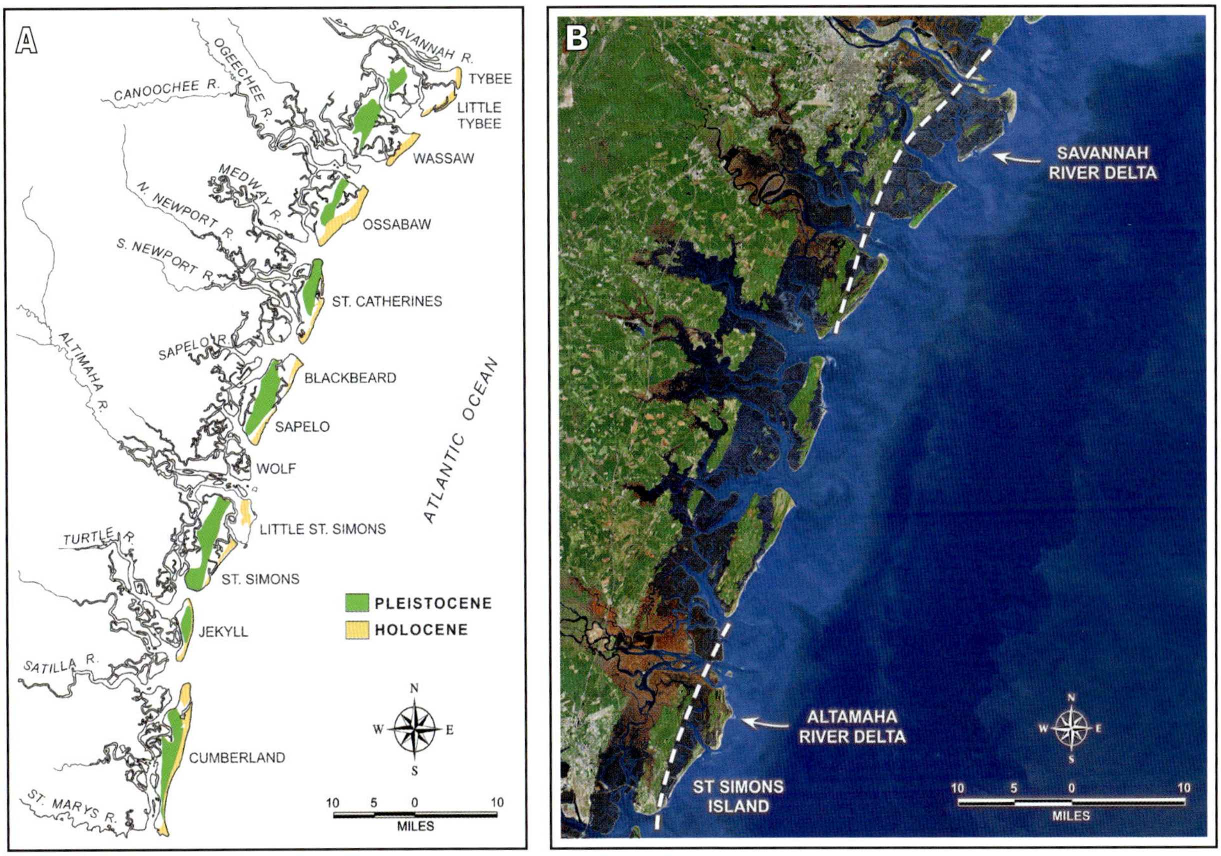

FIGURE 39. Barrier islands of the Georgia Coast. (A) Note how some of the Holocene components are attached to remnants of Pleistocene barrier islands. Slightly modified after Clayton et al. (1992; Figure 3.6). (B) The location of the two significant river deltas that occur along this coast, the Savannah River and Altamaha River Deltas. Dashed line is approximate location of the high-tide line 4,500 years ago. Landsat image mosaic acquired in 2000, courtesy of the NASA Applied Science & Technology Project Office.

Step One As sea level was rising after the end of the last major glaciation (about 12-15,000 years ago), a narrow, landward migrating barrier island, such as the one illustrated in the top diagrams of Figure 35, moved "rapidly" across what is now the continental shelf, leaving behind a thin lag of coarse material on top of an erosion surface scoured across the shelf during this process, called the **transgressive surface of erosion**. There is some question as to whether this was a continuous process. If the sea level was rising too rapidly, the advancing island could have been eroded away at times, having to reform again later.

Step Two The topography over which the shoreline advanced was irregular, and estuarine waters flooded the numerous river valleys carved into the land surface when the shoreline was further offshore during the peak of glaciation. Isolated, primary landward-migrating (transgressive) barrier islands, consisting of washover terraces composed of coarse-grained sand and shell, continued to develop on the exposed interfluves or headlands between the drowned valleys (top diagram in Figure 43);

Step Three When sea level stopped rising and a relative stillstand occurred about 4,500 years ago or somewhat later, shoals developed at the entrances of the estuaries created by the drowning of the valleys and a longshore sediment transport system was initiated along the face of the stranded barrier islands. Over time, beach ridges began to develop, and downdrift migrating spits formed, eventually impinging upon the adjacent estuary entrances (middle diagram in Figure 43). As a well-defined inlet throat evolved, a shoal off the entrance (ebb-tidal delta) formed around which sediment was bypassed, augmenting beach-ridge growth downdrift of

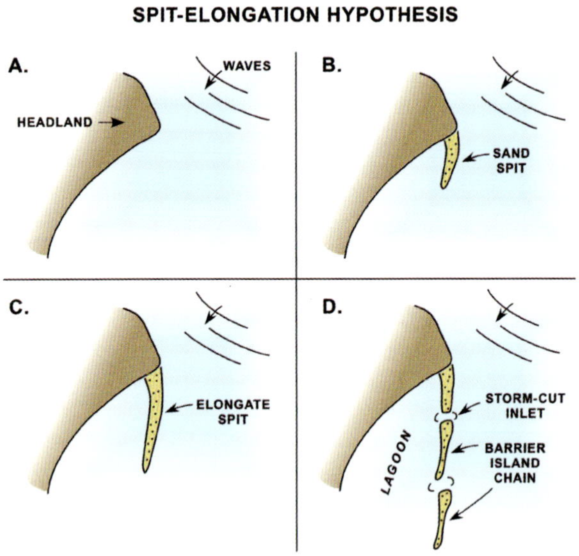

FIGURE 40. Spit-elongation hypothesis for the origin of barrier islands based on Gilbert (1889) and Fisher (1967).

the inlet (bottom diagram in Figure 43).

Step Four As the barrier island matured, and minor fluctuations of sea level occurred, parts of some of the originally prograding beach ridges were eroded as a result of tidal-creek and tidal-inlet migration.

The end result of all this was a prograding, drumstick-shaped barrier island (Figure 36). Perhaps a distinction should be made here between origin and evolution of barrier islands. Figures 40-42 define modes of origin, but the concluding phases of the process illustrated by Figure 43 illustrate the later evolution of the island into its final drumstick shape.

5) And now, here's another idea for a general model of barrier formation. Over the years, we have observed the formation of features we called *micro-barriers* on tidal flats in Iceland, Alaska, and Cape Cod (Figure 44). Although exceedingly small (measured in a few inches to a couple of feet), these features contain many of the elements found on typical barrier islands, such as recurved spits,

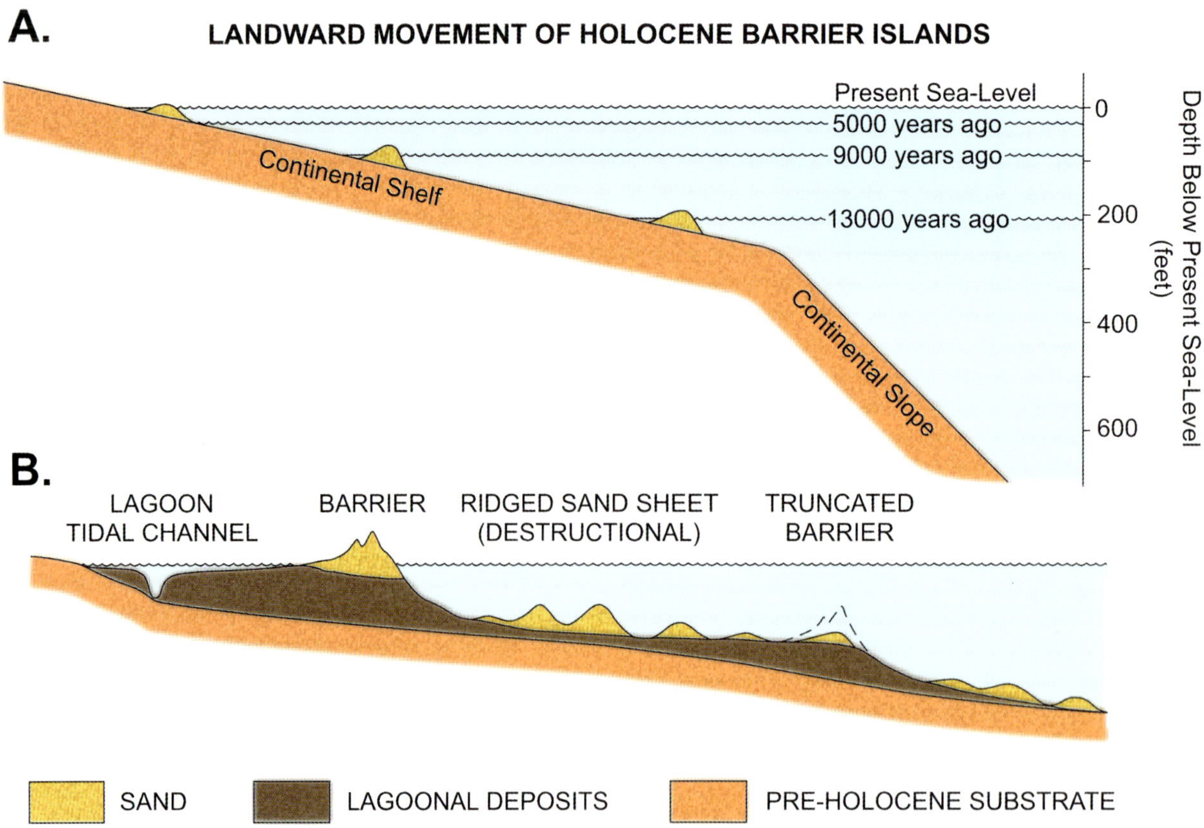

FIGURE 41. Swift's (1975) theory for barrier island origin and evolution. Proposed landward movement of a Holocene barrier island across the continental shelf after sea level rose following the last major glaciation (started moving landward approximately 16-12,000 years ago). Modified slightly after Clayton et al. (1992). (B) The resulting stratigraphy. Modified slightly after Swift (1975, Figure 9).

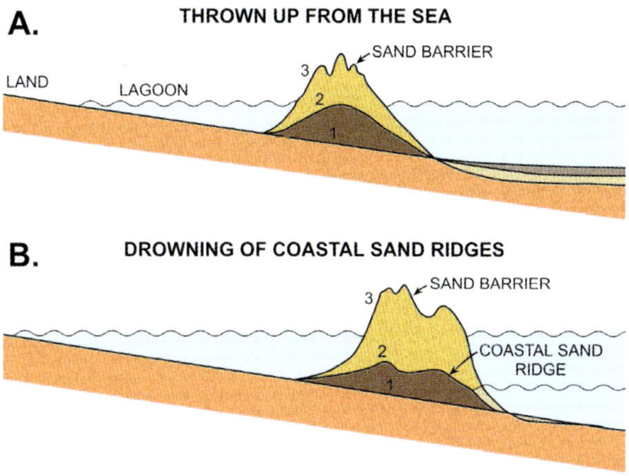

FIGURE 42. Illustration of two theories for barrier island formation. (A) Evolving from an offshore bar (De Beaumont, 1845; W. A. Price, pers. comm.); and (B) From the drowning of coastal sand ridges (Hoyt, 1967b). The numbers represent the time sequence, with 1 representing the earliest (oldest) condition and 3 the latest (youngest).

tidal inlets, lagoonal areas on their landward sides, and so on. In all cases, these features formed within minutes to a few hours. The requirements for their formation was a nearly flat, gently sloping surface suddenly flooded by shallow water with a strong enough wind to form small breaking waves. If you accept the obvious problem of the differences in scale, this illustrates that features like barrier islands can form on relative flat, flooded surfaces with just the right wave conditions. No doubt these conditions exist on depositional coasts in many parts of the world. The only remaining question is exactly what depth, slope, and wind/wave conditions does it take to form the island? How much sand is available is another issue. Once that is figured out, it should be no surprise that long chains of barrier islands occur offshore of lagoons and estuaries, regardless of some of the other theories proposed for the origin of barrier islands. [NOTE: Richard Davis (pers. comm.) pointed out that there are lots of places where islands of "real size" have formed within the past 100 years (e.g., Florida Peninsula, Danish Baltic, and Australia).]

6) **The complex Georgia type.** Jim Henry (2009) gave the following succinct description of the formation of six of the eight largest barrier islands on the Georgia Coast:

> *They are composite barriers consisting of a core of beach and dune deposits formed during the previous and slightly higher, worldwide sea level of the Pleistocene Silver Bluff, approximately 40,000 to 50,000 years ago. Most of the islands are closely fronted by similar deposits formed during the present or Holocene, sea-level rise that began about 15,000 years ago. At that time, the shoreline was located along the edge of the continental shelf some seventy-five miles east of Brunswick, and the exposed sea*

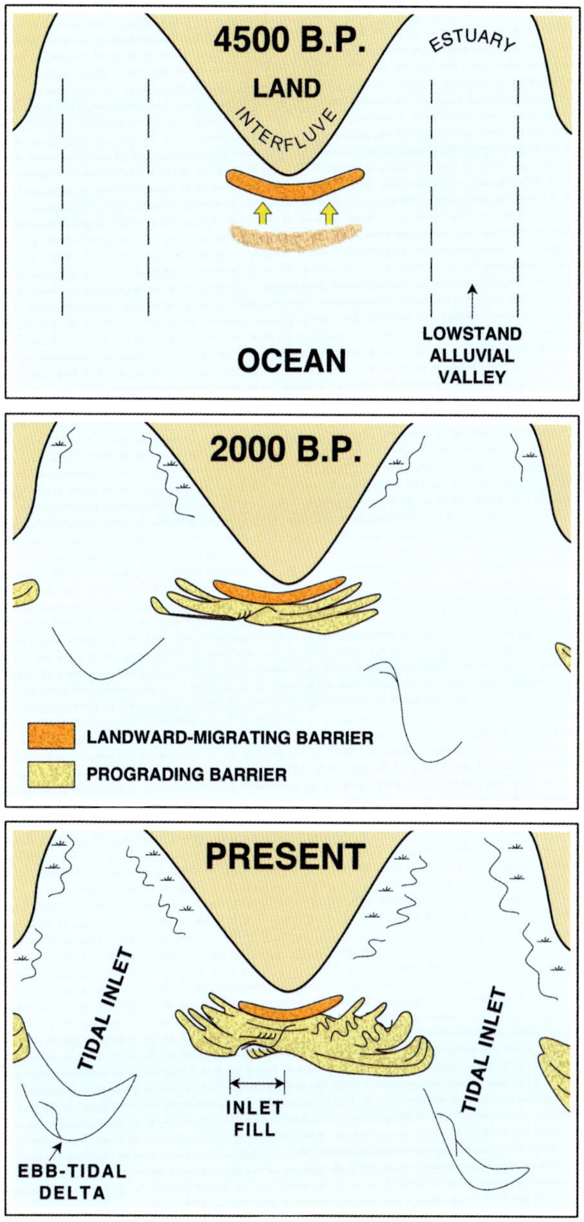

FIGURE 43. Model for the origin of the prograding barrier islands on the coast of the Georgia Bight – *transgressive-regressive interfluve hypothesis*. Based on Pierce and Colquhoun (1970) and Moslow (1980).

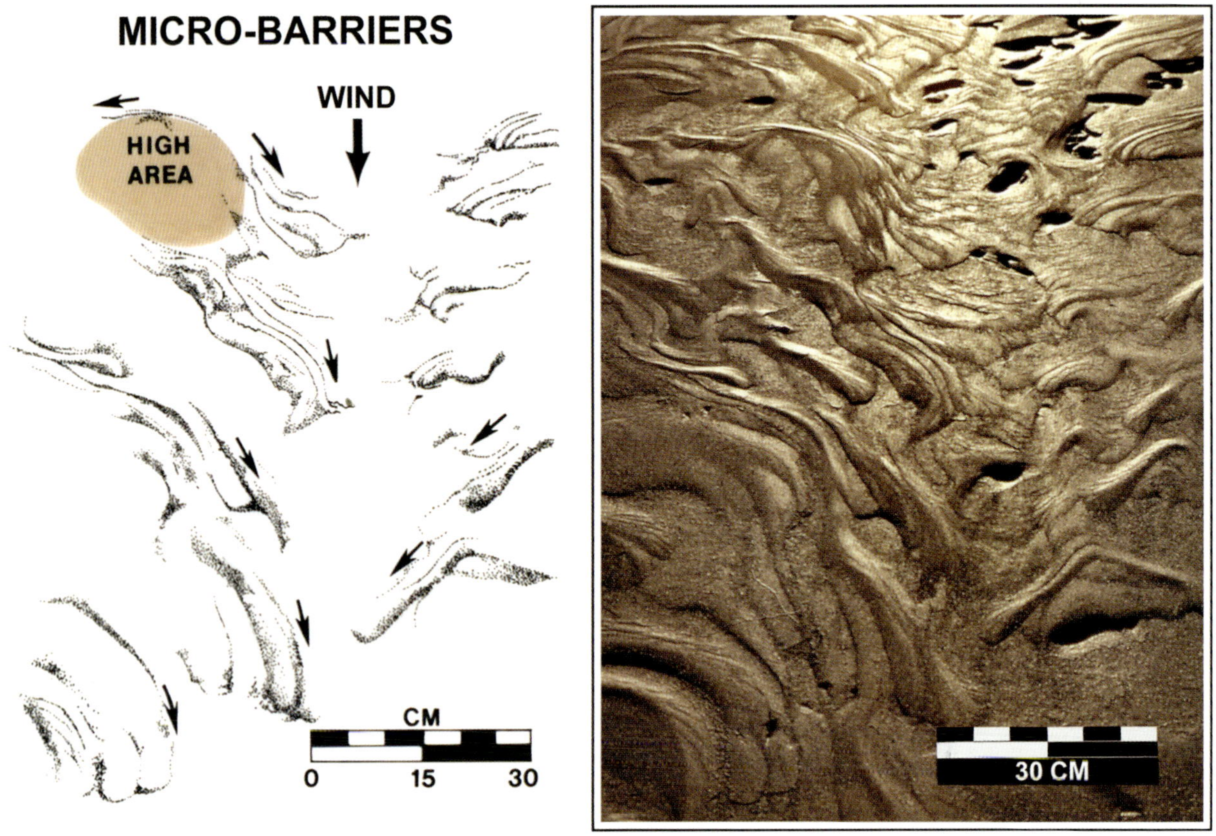

FIGURE 44. *Micro-barriers* forming on a tidal flat in Iceland in the spring off 1973. These, features, which have many of the characteristics of ocean-front barrier islands (tidal inlets, recurved spits, and beach ridges) were formed in a short period of time (less than two hours) during a high wind tide as a result of specific wave conditions on the very flat, flooded surface.

bottom became an extension of the Coastal Plain with forests, plants, and animals. As the sea level began to slowly rise, animal communities were able to retreat from the rising sea and forestlands were inundated. About 5,000 years ago, the rate of sea-level rise decreased from three feet per century to a little less than a foot per century. The Holocene barrier islands began to form, at near their present locations, and, in most cases, welded onto the Pleistocene barriers. Only vestiges of the Holocene barriers remain on Cumberland, Jekyll, and St. Catherines Islands, the major portions having been eroded away by the slow but inexorable rise of the sea.

A relatively complete record of these events is present on Sapelo Island, the details of which were given by Hoyt, Weimer, and Henry (1964). Four of the different steps of this evolution, beginning about 125-130,000 years ago, are shown schematically in Figure 45.

Step One The Pamlico Barrier Island System formed during a highstand that was 20-25 feet higher than the present sea level. This system was located 6-10 miles landward of the present shoreline.

Step Two The sea level retreated for some time and then returned to 10-15 feet above the present sea level at around 85,000 years ago, at which time the Princess Anne Barrier Island System was formed. This system was fairly close to the Pamlico System, the two being separated by *a narrow lagoon-salt marsh* (see map of the systems in Figures 9 and 10).

Step Three Another drop in sea level occurred only to return to another highstand at around 40-50,000 years ago. Another barrier island system, called the Silver Bluff, was formed during this highstand. This highstand was only 2-5 feet higher than the present sea level. *It was during this period that the main portion of* (the present) *Sapelo Island was formed.*

Step Four The Holocene barrier islands on the Georgia Coast were formed by mechanisms already discussed. However, on some of the islands, only thin slivers of the Holocene deposits remain. As shown in

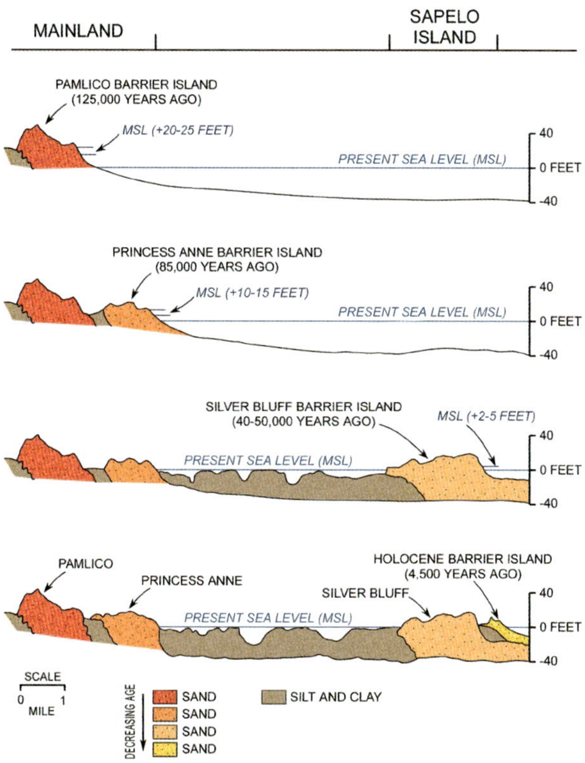

FIGURE 45. Cross sections showing the evolution of the barrier island systems in the vicinity of Sapelo Island, Georgia. Highly modified and simplified after Hoyt, Weimer, and Henry (1964).

Figure 39A, the central islands contain less Holocene sand than those just south of the mouth of the Savannah River. There also appears to be a sight increase in Holocene sand on Cumberland Island (discussed later). The process and the final results for Jekyll Island are shown in the generalized diagram in Figure 46. We hypothesize that the striking difference between the Georgia barrier islands and the ones further north in the Georgia Bight is related to some extent to the downwarping of this shoreline, which allowed the higher water levels to more readily impact the landward shorelines as they slowly sank (at higher relative levels and shorter distances apart). This meant that the horizontal space between the more recent Pleistocene beach ridges, as well as between the Holocene and Silver Bluff beach ridges, is less on the Georgia Coast than in South Carolina and North Carolina. The decrease of the Holocene deposits along the more southern barrier islands is thought by us to be related to a diminishing sand supply in that direction due to increasing distance from the potential sand sources supplied by the Savannah, Ogeechee, and Altamaha Rivers.

Regional Variation

Analysis of the geomorphology and sedimentology of the coast of the Georgia Bight by compartment shows a systematic change of barrier islands around the Bight. Long barrier islands averaging 24 miles in length that are morphologically wave-dominated and stratigraphically transgressive compose greater than 50% of the shoreline on the outer edges of the Bight (Figure 47). Over 50% of the shoreline at the head of the Bight is composed of short (~5-8 miles), mixed-energy, regressive barrier islands that show downdrift offsets and drumstick configurations in plan view. The shorelines of the northern and southern flanks of the Bight, which are positioned across the Cape Fear and Ocala tectonic arches, contain abundant welded barriers and barrier spits (24% of total shoreline of Bight; most common shoreline type) that have narrow to non-existent backbarrier systems. The barrier islands along the Georgia Coast are some of the more complex ones that occur not only within the Georgia Bight, but also around the world, because of the mixing of Pleistocene and Holocene deposits within the same barrier island "host."

TIDAL INLETS

General Introduction

As defined here, most tidal inlets are **major channels that intersect barrier islands,** or separate two barrier islands, usually to depths of tens of feet. Examples of three tidal inlets on the Georgia Coast are given in Figures 36C, 48 and 49A. Tidal inlets are the focus of the most dynamic changes that occur on any component of a barrier island system. Because they connect the open ocean with more sheltered backbarrier settings, including some potential harbor sites, engineers have long been engaged in efforts to control tidal inlets, commonly by constricting their migration with a set of two parallel jetties. In Georgia, two inlets have been so "improved"– at the mouth of the Savannah River on the South Carolina border and at the St. Marys Entrance at the Florida border.

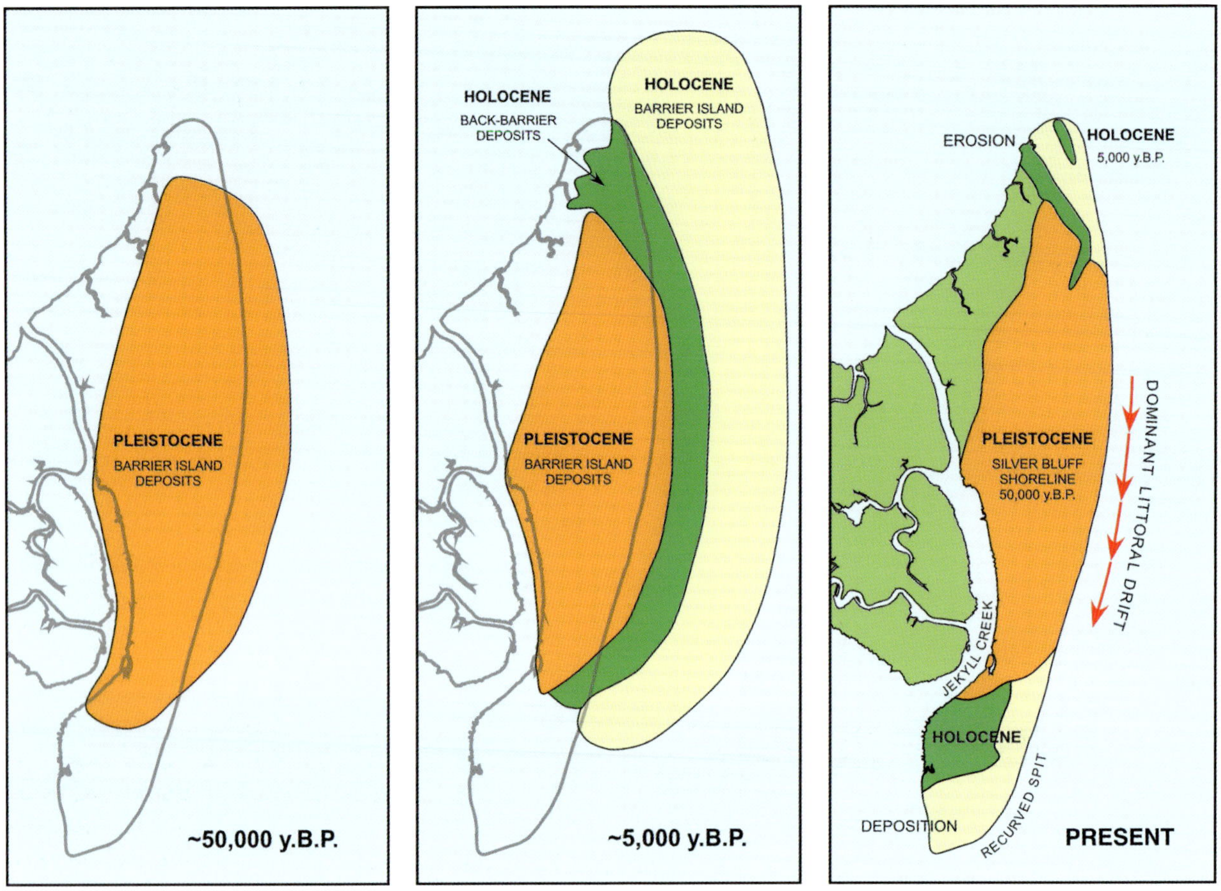

FIGURE 46. The evolution of Jekyll Island. Modified after Henry (2009).

In Georgia, tidal inlets have been formed by one of three processes:

1) Storm-generated scour channels;

2) Closure of estuarine entrances by growth of sand spits by the mechanism illustrated in Figure 43; and

3) Intersection of large tidal channels in salt marshes by the landward migration of transgressive barrier islands.

When the storm surge of a tropical or extratropical cyclone floods a barrier island, water flowing through breaks in the dunes is capable of scouring channels across the island. This may occur both during the elevation of the storm surge and during the storm surge ebb. This is a mechanism by which many new tidal inlets are formed (Hayes, 1965; Shepard and Wanless, 1971); however, most of these channels are closed off by wave-transported and wind-blown sand within a few weeks or months after the storm. A few may survive for the long term. Studies of the Outer Banks of North Carolina by Moslow and Heron (1978) showed an abundance of tidal inlet-fill deposits preserved underneath a long, landward-migrating and wave-dominated barrier island (Core Banks). These deposits are the record of ephemeral inlets generated by scour during major hurricanes.

As some of the barrier islands in the Georgia Bight evolved by the process illustrated in Figure 43, major tidal inlets were formed by the narrowing of estuarine entrances by the growth of sand spits once sea level reached near its present elevation around 4,500 years ago. Over time, the much-narrowed inlet throats were stabilized near the center of the original valleys, achieving depths of as much as 90 feet in places. The throats of the large inlets in South Carolina formed by this mechanism are essentially anchored in bedrock, and our studies of historical trends of inlet changes (Zarillo et al., 1985) show that they do not migrate significantly. Oertel (1975) indicated that this is also true for the big inlets on the Georgia Coast, such as the one shown in Figure 49A (entrance to St. Catherines Sound).

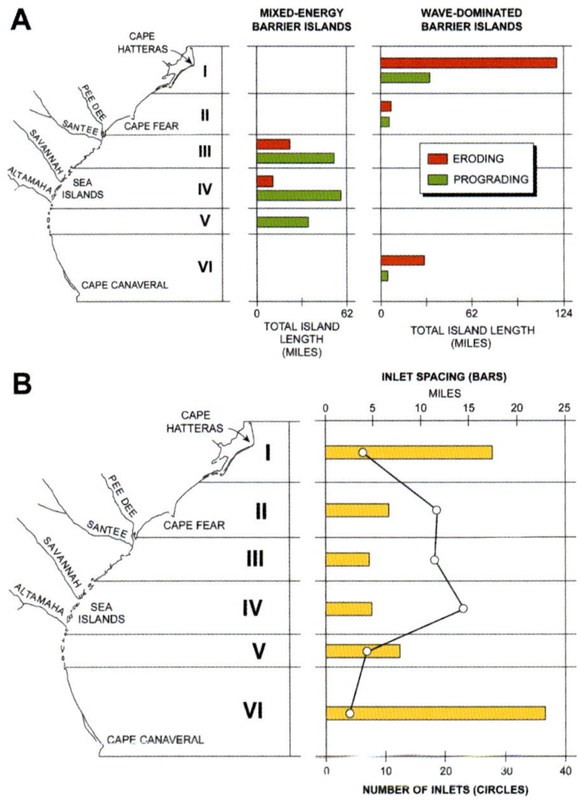

FIGURE 47. Morphological compartments of the Georgia Bight. (A) Distribution of barrier islands. Wave-dominated barrier islands are restricted primarily to the outer margins of the Bight, compartments I, II, and VI, whereas mixed-energy barrier islands occur at the head of the Bight, compartments III, IV, and V. The horizontal axis gives the sum total length (combined) of all barrier islands of that type in the compartment. (B) Number of tidal inlets and inlet spacing.

For the most part, transgressive barrier islands migrate landward over salt marsh systems that always contain a number of tidal channels (Figure 34B). When the landward migrating island passes over a major tidal channel, it is common for a new tidal inlet to develop where the channel was formerly located, the third major mechanism by which tidal inlets are created in the Georgia Bight. As a result, many of the transgressive barrier islands are relatively short (<2-3 miles).

The ultimate morphology, or shape, of a tidal inlet in a barrier island complex is the product of the constant interaction between tidal and wave-generated forces. Longshore currents generated by wave action move sediment along shore in such volumes that would normally fill the inlet in the absence of tidal currents. Sediment carried into the inlet throat (see general model in Figure 50) by the wave-generated currents is constantly swept away by tidal currents that scour the deep throat between the barrier island beaches (that encroach upon it from both sides). Sediment carried away by flood currents deposit a lobe of sediment on the landward side of the inlet called the **flood-tidal delta** and that carried seaward by the ebb currents form a seaward lobe called the **ebb-tidal delta**.

Regional Variation

As shown in Figure 47, inlets on the wave-dominated, microtidal outer flanks of the Georgia Bight are widely spaced (20-22 miles), and those at the head of the Bight are closely spaced (3-6 miles). This trend is in keeping with information presented earlier on the wave and tidal condition in the Bight (Figure 23). The ratio of wave- to tidal-energy flux controls the morphology of the tidal inlets in the Georgia Bight (Hubbard, 1977; Nummedal et al., 1977; Hubbard, Oertel, and Nummedal, 1979). Inlets are commonly differentiated into tide-dominated, wave-dominated, and transitional types, which are illustrated in Figure 51. Wave-dominated inlets, which are characterized by dominant landward transport, typically have large, multilobate flood-tidal deltas and relatively small ebb-tidal deltas. These types of inlets occur at the outer margin of the Bight. Inlets at the head of the Bight tend to have large ebb-tidal deltas and very small or non-existent flood-tidal deltas, as is illustrated for McQueen Inlet on St. Catherines Island in Figure 48. The transition inlets, which usually occur along the two flanks of the Bight, have abundant shoals choking the inlet throat (Figure 51C).

Therefore, the type of tidal inlets represented by the general model in Figure 50, which shows relatively equal volumes of sediment in the two tidal deltas, is rare in the Georgia Bight, and also rare in nature as a rule. Thus, in microtidal, wave-dominated areas like the Texas Coast and the North Carolina Outer Banks, the flood-tidal delta is much larger than the ebb-tidal delta, because of the influence of wave action in both moving sediment inside the inlet and in erosion of the outer lobe (among other factors). On more tidally influenced coasts, such as along Georgia and South Carolina, ebb-tidal deltas tend to be much larger than flood-tidal deltas. In fact, flood-tidal deltas are rare on the Georgia Coast, as is illustrated by the two images of McQueen Inlet on St. Catherines Island given in Figure 48. There is also no

FIGURE 48. Ebb-tidal delta at McQueen Inlet on St. Catherines Island, a classic example of a tide-dominated inlet. (A) Vertical image. Note complete absence of a flood-tidal delta. Digital orthophoto acquired in June 2008, courtesy of USGS. (B) Oblique aerial view looking landward during a low spring tide on 28 April 2010.

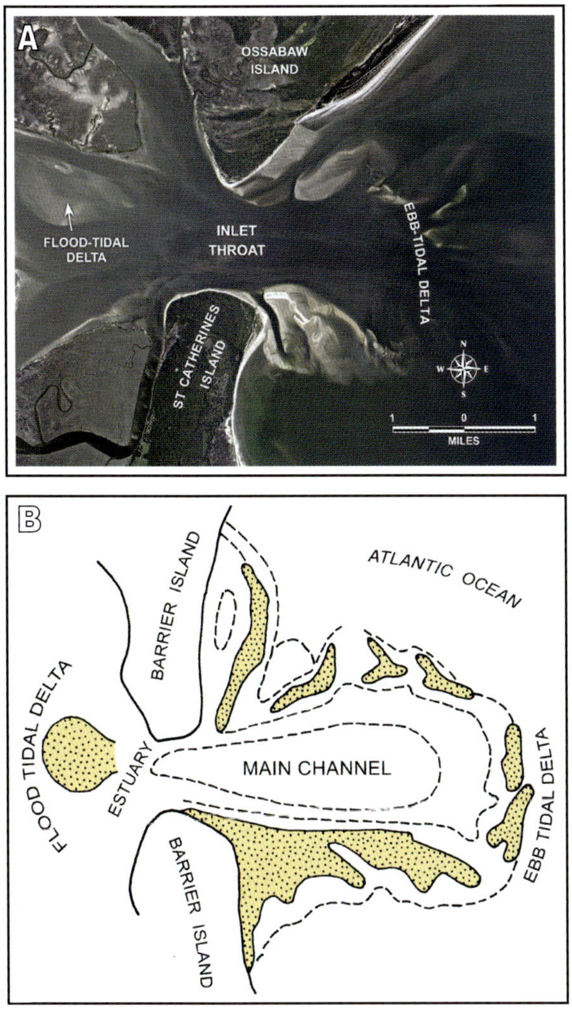

FIGURE 49. Ebb-tidal deltas on the Georgia coast. (A). The tidal inlet at the entrance to St. Catherines Sound. Note the large ebb-tidal delta. Digital orthophoto acquired in 2009, courtesy of USDA-FSA Aerial Photography Field Office. (B) General model of ebb-tidal deltas on the Georgia Coast as illustrated by Clayton et al. (1992). This model is a close match for the inlet shown in A.

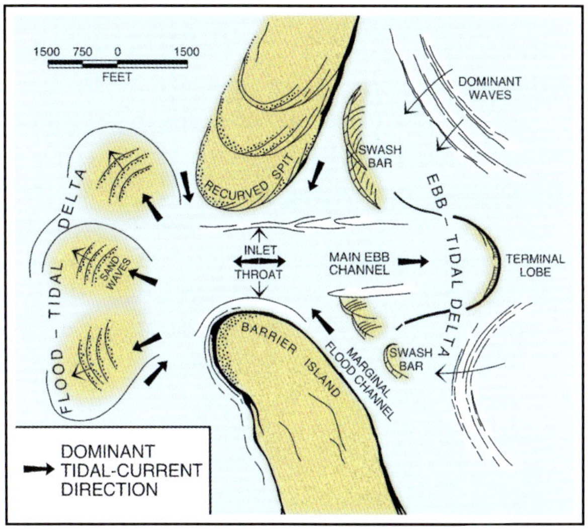

FIGURE 50. General model of the morphology of tidal inlets.

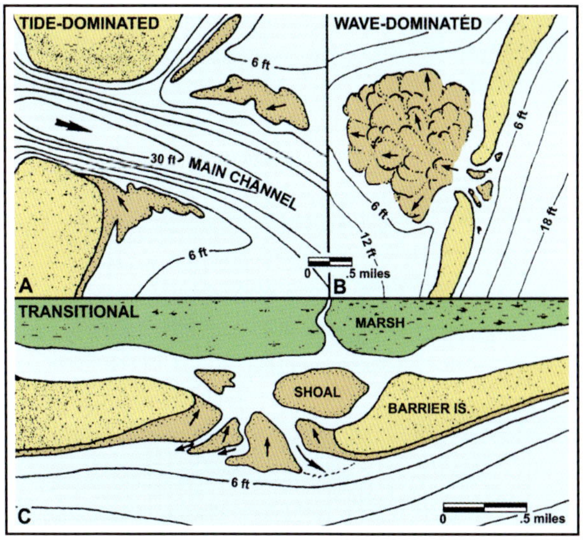

FIGURE 51. Types of tidal inlets in the Georgia Bight (slightly modified after Hubbard et al., 1979).

flood-tidal delta in the small inlet off the south end of Blackbeard Island (Cabretta Inlet; Figures 30 and 36C). There is, however, a small vestige of a flood-tidal delta at the St. Catherines Sound entrance, but it is minor in comparison to the amount of sand in the offshore ebb-tidal delta (Figure 49).

This seemingly strange absence of flood-tidal deltas in this area has been studied in some detail. In the simplest terms, because of the complex nature of the backbarrier regions of the barrier islands in Georgia (extensive marshes and tidal flats cut by numerous tidal channels), water "running uphill" into these areas during the flood portion of the tidal cycle takes longer to fill the backbarrier region with water at high tide than it takes that region to empty when the tide falls. This means that the flood phase is considerably longer than the ebb phase. FitzGerald (1977) found that at Price Inlet, South Carolina, the flood phase averaged 44 minutes longer than the ebb phase. With essentially the same amount of water going in as comes out, stronger currents are generated during the falling tide. These strong ebb currents are able to sweep sand brought into the inlet by the wave-generated and flood-tidal currents out of the inlet and deposit it on the ebb-tidal delta.

Ebb-Tidal Deltas

The large ebb-tidal deltas on the Georgia and South Carolina Coasts contain huge volumes of unconsolidated sand (77% of the total volume on the coast of South Carolina; Hayes, Sexton, and Sipple, 1994; Sexton and Hayes, 1996) and, as noted by Clayton et al. (1992), *Georgia is world famous for having extremely large ebb-tidal deltas, extending up to 4 miles offshore*. The occurrence of these large ebb-tidal deltas was also documented by Oertel and Dunstan (1981).

The general model for ebb-tidal deltas shown in Figure 52 was originally proposed for the New England Coast by Hayes and his students at the University of Massachusetts. However, this model has a broad global application.

The morphological components of ebb-tidal deltas of the type illustrated by the general model include a **main ebb channel** (with slightly stronger ebb currents than flood currents) flanked by linear sand bars on both sides and a **terminal lobe** at the seaward end. The main ebb channel is bordered by a platform of sand (swash platform) dominated by swash bars (intertidal bars built up by wave action). These swash bars are clearly illustrated by the oblique aerial photograph of the ebb-tidal delta of Cabretta Inlet at the south end of Blackbeard Island shown in Figure 36C. The swash platform is separated from adjacent barrier beaches by **marginal flood channels** that are dominated by flood currents.

There have been many significant studies of the tidal inlets in Georgia in the early 1970s by George Oertel, a fellow coastal geomorphologist and friend (e.g., Oertel, 1971; 1972; 1973a; 1973b; 1975a; 1977; 1988; Oertel and Howard, 1972). He described the sediment types found in the massive ebb-tidal deltas located off the huge sounds on the Georgia Coast. Oertel recognized the same general morphology and dynamic processes as those that led to the Hayes model (Figure 52), but he proposed somewhat different nomenclature. The overly huge size of these ebb-tidal deltas off the sounds along the Georgia Coast is clearly shown by the map in Figure 53. Each of the ebb-tidal deltas off the two inlets at the

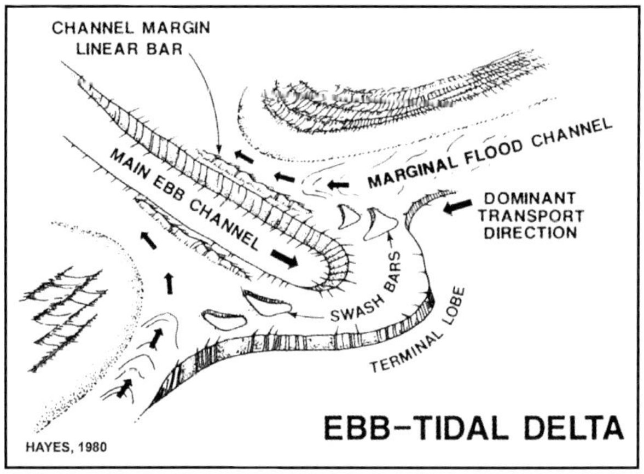

FIGURE 52. Typical morphology of ebb-tidal deltas in mesotidal settings (after Hayes, 1980). Arrows indicate dominant direction of tidal currents. This model was derived for the ebb-tidal deltas of the New England area, but it applies quite well to the inlets of other barrier-island systems around the world.

entrances to St. Catherines and Sapelo Sounds covers an area of around 33 square miles, whereas the one off Doboy Sound covers 17 square miles. The vertical image of St. Catherines Sound in Figure 49A also gives a good sense of the size of these ebb-tidal deltas, which contain an overwhelming amount of the sand in the longshore sediment transport system along the Georgia Coast. Note that there is a small flood-tidal delta associated with St. Catherines Inlet (Figure 49A), which conforms to the general model presented by Clayton et al. (1992) in Figure 49B that is based primarily on the different publications of Oertel.

Speaking of Oertel, some of his comments on these large ebb-tidal deltas on the Georgia Coast, from his 1988 paper, follow:

> *Inlet systems along the Georgia Coast have deep gorges (-20 m to -30 m/* -66 feet to -98 feet*) and large ebb-tidal deltas with linear shoals and ramps that extend 8 to 12 km (*5 to 7.4 miles*) offshore (Nummedal et al., 1977). Oertel (1975) described the deltas of these tide-dominated inlets as Type-D deltas and suggested that their pronounced offshore extension (beyond the upper shoreface) was caused by relatively large inlet discharges with respect to the diminutive longshore-current flows. The generally low longshore sediment flow is controlled primarily by regionally low wave energy (Nummedal et al., 1977 and Walton and Adams, 1976).*

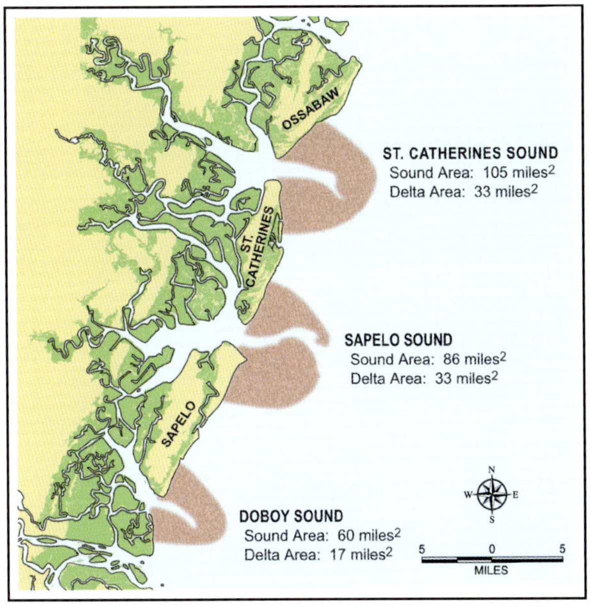

FIGURE 53. Simplified map of the central portion of the Georgia Coast showing the areal distribution of the huge ebb-tidal deltas. The area covered by each of the ebb-tidal deltas is given in square miles. Modified after Nummedal et al. (1977).

Numerous studies have shown that changes in the configuration and sand storage on the ebb-tidal deltas in the Georgia Bight have a profound effect on the erosional/depositional patterns of the adjacent barrier islands (Hayes et al., 1976; Oertel, 1977; FitzGerald and Hayes, 1980; Zarillo et al., 1985). The erosional/depositional patterns on the Georgia barrier islands are no doubt highly affected by the big ebb-tidal deltas (Oertel, 1977).

Inlet Migration

The tidal inlets of the Georgia Bight are highly variable with respect to their patterns of inlet migration, but there are two distinct end members: 1) The already described large inlets that are anchored in bedrock and do not migrate appreciably (about 25% of the total); and 2) Small inlets with throats that are quite shallow, typically less than 20 feet deep (approximately 50% of the total). The remaining 25% are mostly transitional forms (Figure 51C). The inlet throats of the smaller inlets usually cut into sandy beach-ridge sediments and they migrate rapidly. When the overextended main ebb channel of such an inlet "loses its gradient advantage" (that is, the channel is too long and flat for the ebb flow to be maintained) the neck of the recurved spit that was deposited updrift of the migrating inlet is cut through during a storm (usually) and the old inlet is abandoned. Two examples of this type of inlet migration are illustrated in Figure 54.

Captain Sams Inlet, South Carolina, an example of the shallow, sand-floored inlets, typically follows a 40-50 year cycle of migrating downdrift around three miles or so before the associated recurved spit is cut through (Figure 54A). Moving toward the southwest at a rate of over 200 feet per year, this is one the fastest migrating inlets in the Bight, but a few others have remarkably similar morphology, such as the inlet at the mouth of Blackbeard Creek, Georgia, located at the south end of Blackbeard Island (Cabretta Inlet). This particular recurved spit and inlet system is shown on the vertical image in Figure 30. Griffin and Henry (1984) demonstrated that this inlet also migrates south with the neck of the recurved spit being cut through on occasion (Figure 54B).

Numerous studies have shown that sediments deposited by migrating tidal inlets constitute a significant portion of Holocene barrier-island deposits (Hoyt and Henry, 1965; 1967; Kumar and Sanders, 1974; Moslow and Heron, 1978; Tye, 1981; 1984; Hayes et al., 1984; Moslow and Tye, 1985; Sexton and Hayes, 1996). All but the Kumar and Sanders study were carried out in the Georgia Bight. Over 40% of the sand volume of the prograding, mixed-energy barrier islands of central South Carolina is inlet-fill deposits (Sexton and Hayes, 1996). Figure 55 illustrates the geomorphology of an inlet system of this type, as well the character of the deposits that result from inlet migration, such as that depicted in Figure 54. This observation was first made on the Georgia Coast (Hoyt and Henry, 1965). The process is as follows:

1) The ebb-tidal currents create a scoured-out zone downdrift of the end of the recurved spit. In South Carolina and Georgia, this scoured-out zone, or channel, migrates to the southwest (at the rate of >200 feet/year in the case of Captain Sams Inlet).

2) As this scoured zone (hole) migrates to the southwest, it is gradually filled in by sediments associated with: a) The updrift side of the channel; b) A component of the updrift portion of the ebb-tidal delta called the channel margin platform (Figure 55A); c) The beach

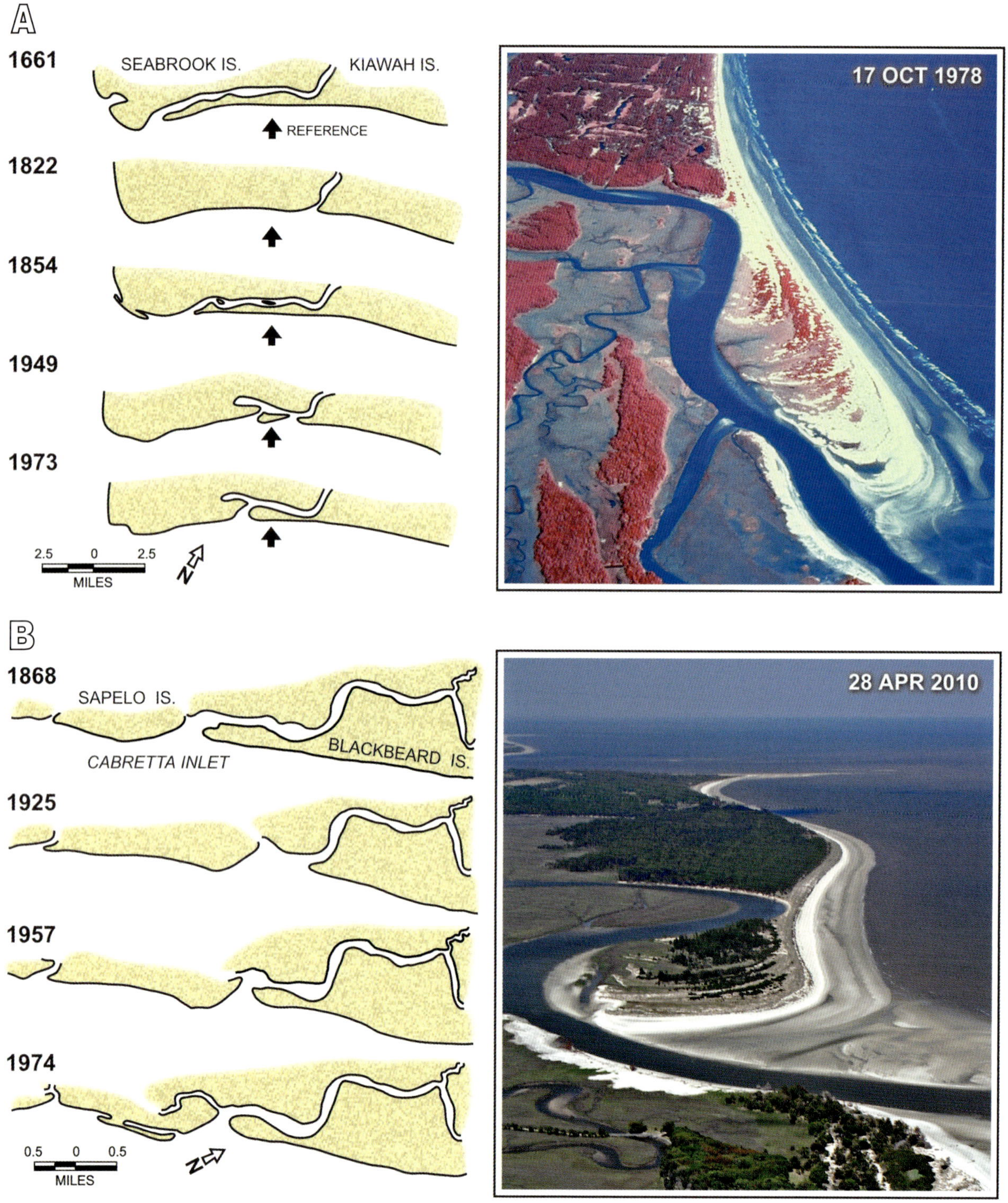

FIGURE 54. Two migrating tidal inlets in the Georgia Bight. (A) Historic changes of Captain Sams Inlet, South Carolina between 1661 and 1973, based on a combination of historical charts and aerial photographs, showing three periods of breaching of the spit: 1822, 1922, and 1949. Breaches occurred at numerous other times, of course, at intervals we estimate to be about every 40-50 years. From Hayes et al. (1976). (B) Historic changes of Cabretta Inlet, Georgia (see also the vertical image in Figure 30). Note that the spit at the south end of Blackbeard Island had been cut through sometime either not too long before or during 1925. From Griffin and Henry (1984).

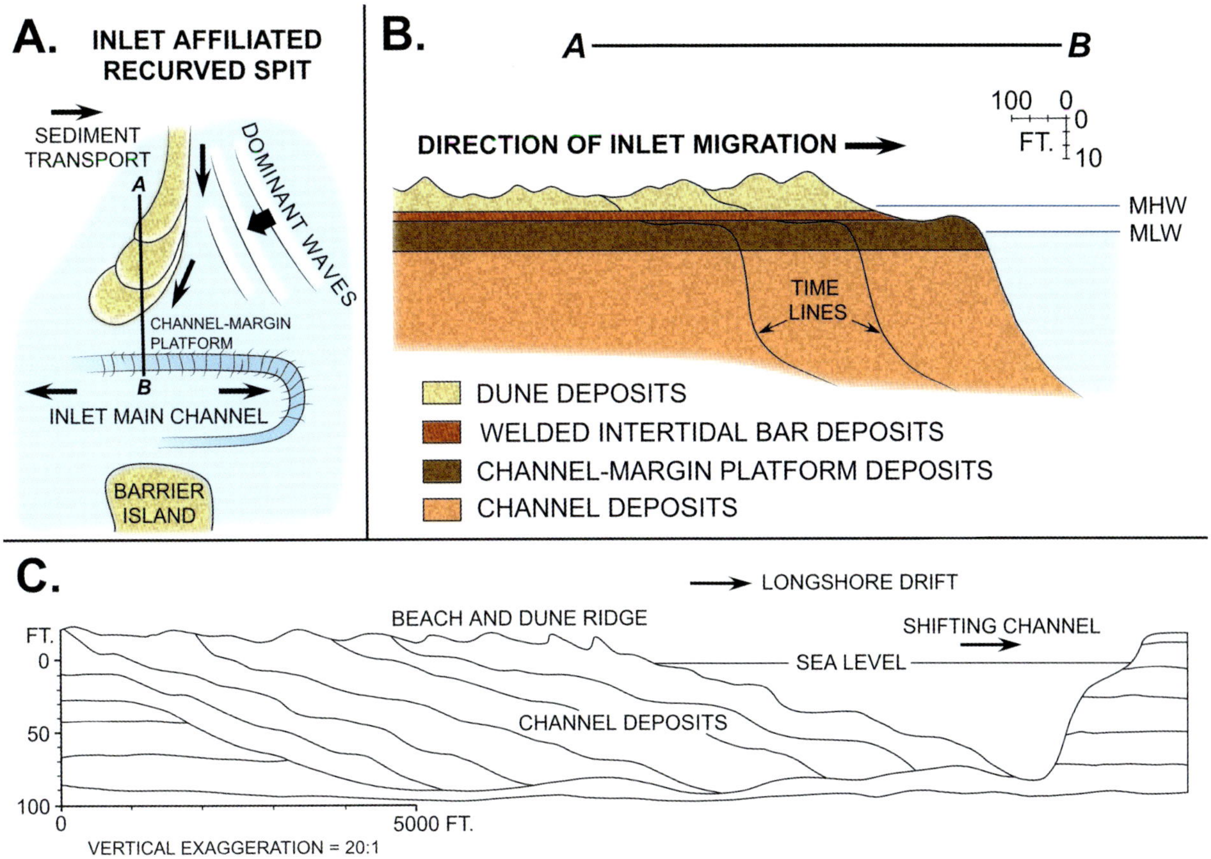

FIGURE 55. Sediment deposits resulting from the migration of a relatively shallow tidal inlet with an associated recurved spit, such as the ones illustrated in Figures 30 and 54. (A) General geomorphology. (B) Resulting sedimentary record. A and B are after Hayes (1980). (C) General cross-section of the sedimentary deposits resulting from inlet migration, based on observations on the Georgia Coast. From Hoyt and Henry (1965).

sediments on the leading edge of the recurved spit; and d) Dune sediments on the recurved spit (Figure 55B).

3) Hoyt and Henry (1965) not only pointed out the significance of such inlet deposits in barrier island rock sequences, but they also showed the diagram in Figure 55C that depicts the record of added sediment layers through time that generally dip in the direction of inlet migration.

Many property losses occur on developed barrier islands as a result of inlet migration or expansion. The effect of inlet migration on downdrift properties is well understood, resulting in remedial measures, such as inlet stabilization with jetties, or inlet relocation, as was done at Captain Sams Inlet, South Carolina in 1983 and 1996 (Kana and McKee, 2003; Hayes and Michel, 2008). Studies on the central South Carolina Coast following hurricane *Hugo* (1989) showed that several inlets expanded to accommodate the 12-20 feet storm surge associated with the hurricane (Figure 37). Price and Capers Inlets became 25 percent wider than they were before the storm; however, within a few months, they were back to their original width. The expansion of the inlets and scour of adjacent subtidal areas caused major property losses at the Isle of Palms, Pawleys Island, and Sullivans Island. A generalized model of inlet expansion, based on detailed study of historical charts, maps and aerial photographs, allowed us to establish a meaningful construction setback line for Kiawah Island, South Carolina before the major phases of development began in 1975 (Hayes, 1977). Kiawah Island was essentially undamaged by beach erosion during hurricane *Hugo*, partly because the developers of the island had adhered to the original setback line.

ESTUARIES

Definition

According to the original definition, which was basically a description of Chesapeake Bay (Pritchard, 1967), an **estuary** has three defining qualities:

1) A flooded river valley that was formed during the lowstand of sea level that was at a maximum during the last major glaciation (20,000 years ago) and ended about 12-15,000 years ago;
2) A water body with a substantial freshwater influx; and
3) A water body subject to tidal fluctuations.

The second and third criteria are self explanatory, but the first requires some understanding of sea-level fluctuations during the Pleistocene epoch. As noted earlier, when sea level was low during the major glaciations, streams flowing across the coastal plain of the eastern United States carved valleys up to a hundred feet deep across the coastal plain and the continental shelf (see Figure 7). In many cases, this type of erosion was confined to the same valleys during each of the four major drops in sea level. When sea level started to rise about 12-18,000 years ago, those valleys were flooded as the shoreline advanced across the then-exposed continental shelf. When sea level essentially stopped this sudden rise (said to be either two or three feet/century by different researchers) and began to rise at a considerably slower rate (not more than one foot/century) around 4,500 years ago, all but a few of the valleys were flooded with sea water for some distance up the valleys. Where the valleys were eventually filled with sediment and a bulge of sediment protruded out into the ocean away from the general coastal trend, the original "estuaries" evolved into what we now call deltas. On the east coast of the United States, this has occurred at only three places–the combined Santee and Pee Dee Rivers in South Carolina, the Savannah River at the South Carolina/Georgia border, and the Altamaha River in Georgia, all three of which are located within the Georgia Bight. Therefore, only the major piedmont rivers (Figure 7) have formed modern deltas in the Georgia Bight, because of their larger sediment loads.

Characteristics

A key factor in defining the characteristics of an estuary is the way in which the salt and fresh water mix. As shown in Figure 56, a highly stratified estuary with a well-defined **salt wedge** occurs where a significant freshwater stream enters a flooded valley with a relatively small tidal range, with the more dense salt water hugging the bottom of the channel during a rising tide (a highly stratified estuary; see upper diagram in Figure 56). Mixing of the salt and fresh water is enhanced by an increase in tidal energy; therefore, with a larger tidal range, the two water masses become partially mixed (as indicated by the arrows in the middle diagram in Figure 56). With an even larger tidal range, the water within the estuary may become vertically homogeneous from top to bottom, completely fresh top to bottom at the head of the estuary and pure salt water from top to bottom at the entrance (lower diagram in Figure 56; a fully mixed estuary). As noted by George Oertel (pers. comm.), most of the estuaries in Georgia are fully mixed because of the significant tides. Two of the estuaries are somewhat stratified, the entrances to the Savannah and Altamaha Rivers.

In this discussion, we will focus on the partially mixed type because they show the widest range of conditions. Figure 57A shows a plan view model of a partially mixed estuary, and Figure 57B shows a cross-section illustrating the circulation within these partially mixed estuaries. We consider the most landward boundary of an estuary to be the place where the tide "stops going up and down," the so-called "head of tides." In the upper portions of this type of estuary, the primary hydrodynamic process is the river flow, whereas at the entrance, waves and tides dominate. As far as sediment sources are concerned, the sediments in the middle and upper portion of the estuary are provided by the incoming river, but in some estuaries, a considerable amount of sediment comes into the entrance of the estuary from offshore under the influence of tidal and wave-generated currents.

Considering the different components of the system as illustrated by the plan view map in Figure 57A, the upper estuary is a zone of relatively diminished tidal flow, and the river flow is still strong in a seaward direction. In the middle of the estuary, tidal flow is increased, the flood tide reverses the river currents when the tide rises, and a complex two-way, two-layer circulation occurs, creating a zone of mixing of the two water masses. In that zone of mixing, demonstrated in the cross-section in Figure 57B, an important process known as **flocculation** of the clay particles in suspension occurs. Flocculation takes place because the abundant clay particles in suspension

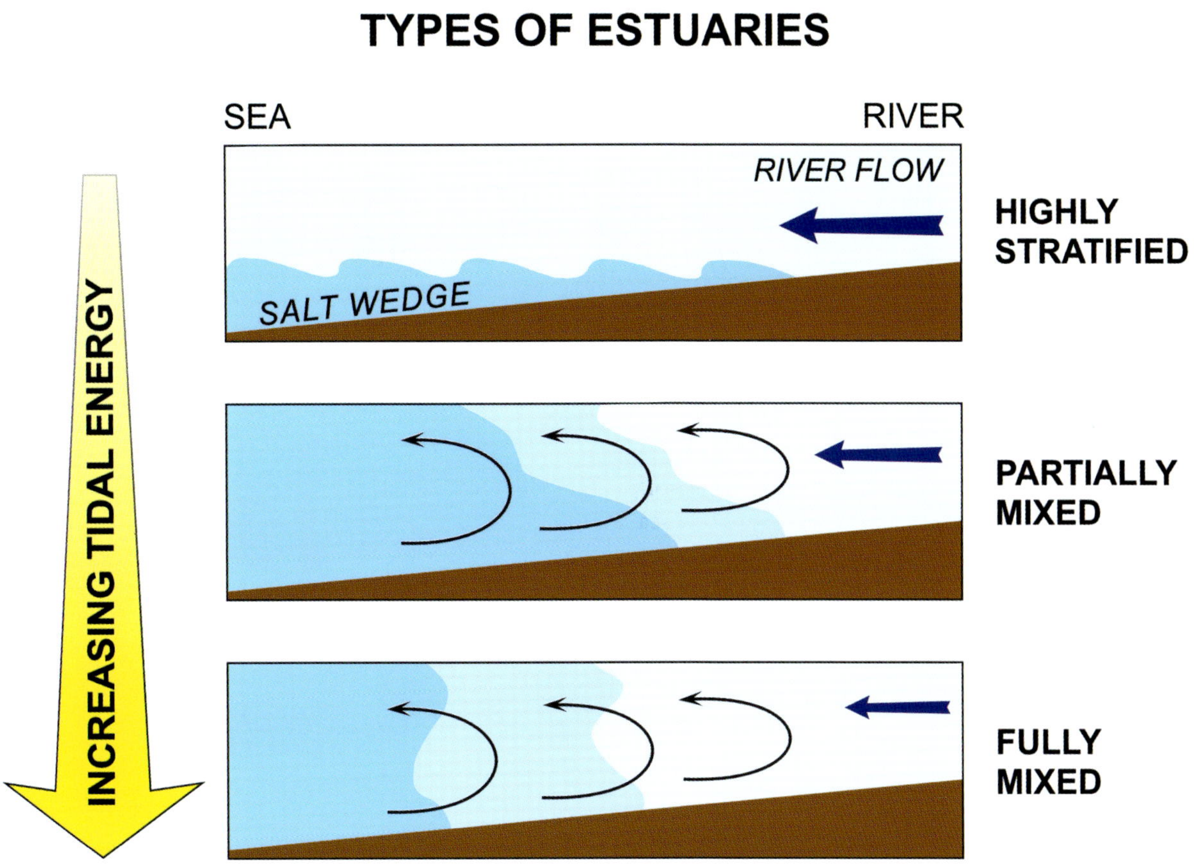

FIGURE 56. Effect of tides on the mixing of salt and fresh water in estuaries (modified after Biggs, 1978).

brought in by the river flow slow down their seaward movement at the point of mixing as a result of the reversing of tidal flow directions twice a day. This allows an increase in the number of clay particles in the water column to the extent that they begin to collide with one another. These colliding particles tend to adhere together under the influence of the increased concentration of cations, such as Na+, Ca++ and Mg++, derived from the seawater. As a result, the clay particles, which may have diameters as little as one micron, cluster into groups (**flocs**) that may have diameters measured in hundreds of microns. Consequently, many of the flocs may sink to the bottom at slack tide (Figure 57B). Once on the bottom, some of the flocs are resuspended by ebb- and flood-tidal currents. The net result is the formation of a zone in the middle reaches of the estuary that contains abundant fine-grained sediment in suspension known as the **zone of the turbidity maximum.** This zone of turbidity can migrate up and down the estuary over time, moving downstream during periods of flood on the river and upstream during low-water periods. A side effect of this process is the creation of a "mud zone" on the tidal flats and in the tidal channels in the middle of the estuary.

At the estuary entrance, conditions take on an entirely different aspect in that:

- Tidal currents are more dominant than river currents with common differentiation between ebb-dominated and flood-dominated channels;
- The waters are more mixed, with diminishing estuarine stratification;
- Wave activity becomes a more important dynamic process; and
- Moving of sediments along the bottom by wave- and tide-generated currents causes the tidal flats and channel bottoms to become much more sandy than those in the central portions of the estuary in the "mud zone."

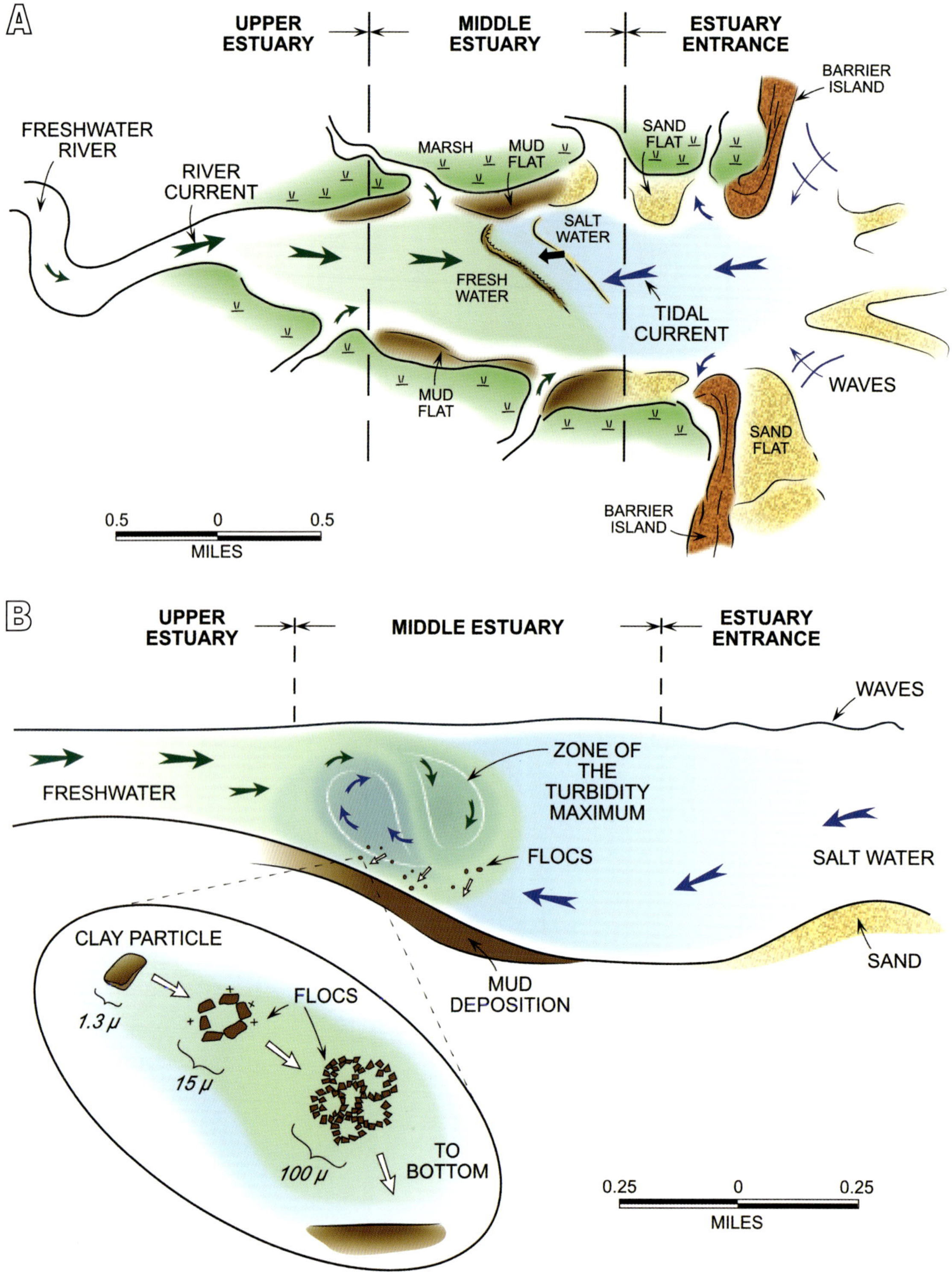

FIGURE 57. Estuaries. (A) Plan view of an estuary in the Georgia Bight that has a significant fresh-water stream entering it. (B) Cross-section of such an estuary showing predominant processes and circulation patterns. Especially noteworthy is the formation of a *zone of the turbidity maximum* where the clay particles brought in by the river undergo a process called *flocculation.*

Georgia's Coastal Waterbodies

With some of the largest tides in the southeastern U.S., and the flattest coastal plain, the coast of Georgia is deeply inundated twice daily by the tides. The large embayments along the coast, many of which are called "sounds," are typically bordered by extensive marshes. Howard and Frey (1975), who questioned whether the term estuary could be universally applied to these embayments on the Georgia Coast, classified them into two main categories: **salt marsh** and **riverine**. The salt marsh systems are dominated by marine waters, and riverine systems are fed by prominent piedmont or coastal plain rivers (discussed in more detail below).

The following succinct description of Georgia's major coastal waterbodies was given by Howard and Frey (1975):

> *Estuaries along this coast exhibit considerable variation in geomorphic, hydrographic, sedimentologic, and biologic characteristics... In fact, semantically, some of the sounds included in our study arguably should not be called "estuaries," because they are not "measurably diluted with fresh water derived from land drainage" (Dyer, 1973). Of the 10 major waterbodies along the Georgia Coast, only 5 have true rivers at their heads: (1) Savannah River-Entrance; (2) Ogeechee River-Ossabaw Sound; (3) Altamaha River-Entrance; (4) Satilla River-St. Andrews Sound; and (5) St. Marys River-Entrance. The other waterbodies are essentially contained within an 8- to 11-km-wide (*5- to 7-mile-wide*) zone of backbarrier salt marsh* (see Figure 58) *and receive fresh water only from local small drainage basins and (or) from distributaries of the rivers listed above. These "estuaries" include: (1) Wassaw Sound; (2) St. Catherines Sound; (3) Sapelo Sound; (4) Doboy Sound; and (5) St. Simons Sound. Such sounds have been called "salt marsh estuaries" (Redfield, 1967).*

For some reason, they did not mention Cumberland Sound, which we assume would also be classified as a salt marsh "estuary," based on our studies of the estuaries

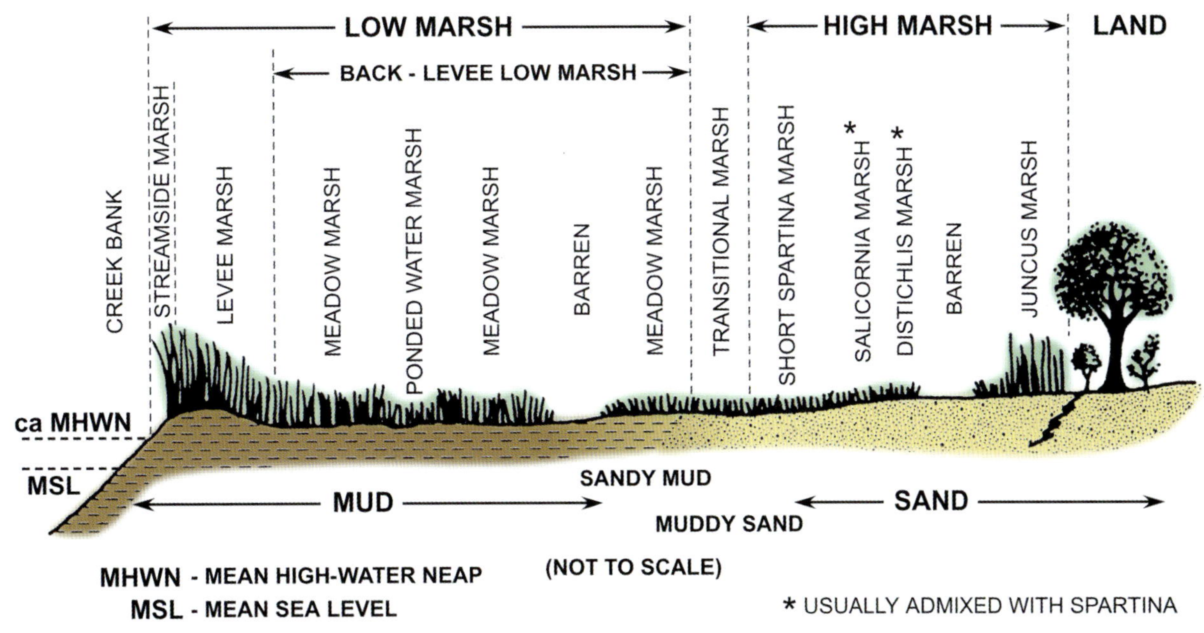

FIGURE 58. Diagrammatic cross section of the salt marsh at Sapelo Island. Modified slightly after Edwards and Frey (1977). The tallest grasses occur in two places, tall smooth cordgrass (*Spartina alterniflora*) along the edge of the tidal channels, where the soil is saline but highly flushed by the tides, and tall black needlerush (*Juncus roemerianus*) at the landward edge of the high marsh, where the soils are less saline. These two plant species are by far the most abundant plant types occurring in the salt and brackish marshes of Georgia. The sediment distribution shown in this diagram is true where the high marsh adjoins a sandy remnant of a barrier island. However, in places where the topographic high is an interfluve between tidal channels, the upper part of the section might be floored with muddy sediments. Elevation and soil salinity are the key controls of the plant populations, not sediment type or water salinity.

on the New England Coast where Redfield (1967) did his field work. Estuary is one of those terms that has gone through a *potpourri* of strangely mangled definitions.

Freshwater River Systems

Considering the freshwater rivers of the southeastern region of the United States as a whole, they can be classified into three fundamental types on the basis of regional physiography and water source/chemistry: 1) Those streams that originate in and derive most of their waters from the Blue Ridge Mountain and Piedmont Physiographic Provinces (here termed **piedmont rivers**; also known as "red-water rivers"); 2) Those streams that have most or all of their drainage basins located within the Coastal Plain Physiographic Province (here termed **coastal plain rivers;** also known as "black-water rivers"); and 3) The **spring-fed rivers** of the Florida Peninsula that clearly require no further analysis in this book. In this discussion, we are excluding those rivers in the coastal zone that carry mostly saltwater (e.g., Little Satilla River).

Two piedmont rivers, the Altamaha and Savannah, deliver sediments to the Georgia Coast, as do three significant coastal plain rivers, the Ogeechee, Satilla, and St. Marys Rivers. The average flow rates of these five rivers are given in Table 3, which shows that the Altamaha, with an average of over 13,000 ft^3/sec is the largest, with the Savannah, at over 11,000 ft^3/sec, being close behind, although the presence of dams on the Savannah may impact its flow rates (record for the table started in 1930). The two largest coastal plain rivers, the Ogeechee and Satilla, have considerably lower rates than the two piedmont rivers at around 2,200 ft^3/sec.

Sediments carried by the piedmont rivers are more immature mineralogically than those of coastal plain rivers, which have passed through more than one cycle of erosion and deposition and longer episodes of chemical weathering. Data by Pilkey and Giles (1965) indicate that coastal plain rivers do not contribute much sand to the modern Georgia beaches.

[NOTE. The term **maturity** as it applies to sediments refers to how long the individual sediment particles (sand grains usually) have been subjected to abrasive processes, such as bumping together while being transported by the wind in sand dunes or by waves on beaches. Most of the sand grains on the Georgia Coast were originally derived from rocks in the Blue Ridge Mountain Province. In the original rocks, such as granites, a variety of minerals were weathered out of the rocks to eventually end up as sand grains. The most common minerals in granite, for example, are quartz and feldspar. Quartz is harder than feldspar, therefore it is apt to survive a lot more abrasion than feldspar. If you pick up a handful of sand on a Georgia beach, the majority of the individual grains will be composed of quartz. This relationship is discussed in more detail later.]

DELTAS

As noted in the previous discussion, an irregular bulge of the shoreline at the mouth of a sediment-laden stream is normally referred to as a **delta**. In their classic summary paper on deltas, Coleman and Wright (1975) discussed over 50 parameters that have an impact on river delta morphology. Factors such as characteristics of the drainage basin, river slope, and coastal climate were acknowledged. Most present-day workers, however, simplify matters by focusing on three basic factors–sediment supply, wave energy, and tidal current energy–in their attempts to classify deltas. Galloway (1975) placed this concept on a ternary diagram, with sediment supply, wave-energy flux (i.e., essentially how big the waves are), and tidal-energy flux (usually determined by tidal range) comprising the three end members. We have superimposed tidal range on Galloway's diagram in Figure 59.

TABLE 3. Average flow volumes of the main freshwater rivers of the Georgia Coast in ft^3/sec. (data provided by USGS).

River	Years of Record	Station Location	Year / Highest Record	Average ft^3/sec
Savannah	1930 – 2008	*near* Clyo	1998 / 18,940	11,454
Ogeechee	1938 – 2008	*near* Eden	1964 / 5,369	2,246
Altamaha	1932 – 2008	*near* Doctortown	1964 / 27,420	13,323
Satilla	1931 – 2008	*near* Atkinson	1964 / 5,920	2,201
St. Marys	1927 – 2008	*near* Macclenny, FL	1947 / 1,838	642

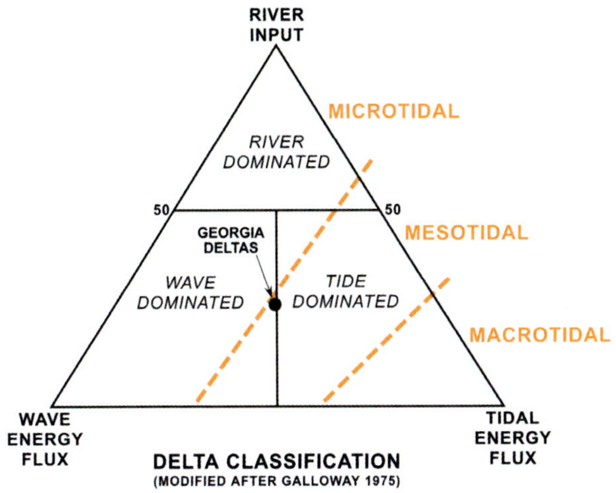

FIGURE 59. Classification of river deltas on a ternary diagram, with three factors controlling the delta morphology: 1) River input (amount of sediment delivered to the coast by the river); 2) Wave energy flux (how big the waves are); and 3) Tdal energy flux (controlled primarily by the tidal range). We have superimposed tidal range on this diagram, which was originally devised by Galloway (1975). The fields within which the three principle types of deltas–river-dominated, wave-dominated, and tide-dominated–would plot are shown. These three delta types are illustrated in Figure 60. Note the approximate location of the two Georgia deltas (Savannah and Altamaha) on the diagram. We refer to these as mixed-energy deltas.

Just considering the end member types, at the top of the triangle in Figure 59 are **river-dominated deltas**, in which a strong river supply of sediment overwhelms the dynamic processes at the shore and a lobe of sediment, such as the birdfoot of the Mississippi River Delta, protrudes out into the sea (see Figure 60A). On wave-dominated coasts, where the rivers have enough sediment to build a bulge out into the ocean, the waves rework the sand into a smooth outer margin composed of prograding beach ridges, such as is illustrated in Figure 60B, a plan view sketch and oblique aerial view of a **wave-dominated delta.** On tide-dominated coasts, the **tide-dominated deltas** consist of a series of funnel-shaped water bodies (estuaries?) at multiple river mouths with a number of shore-perpendicular, tidal sand ridges that extend offshore of the river mouths (see Figure 60C).

There are only two significant river delta complexes located along the Georgia Coast, the Savannah River Delta and the Altamaha River Deltas, which will be discussed in some detail in Section III. We classify both of these deltas as **mixed-energy deltas**, plotting them halfway between the tidal- and wave-energy poles on a ternary diagram in Figure 59. We do so because marine processes mold the outer margins of the deltas, with both waves and tides playing an important role, and the sediment loads of the two river systems are modest in size compared to other rivers around the globe. The locations and general shapes of these two deltas are shown on Figure 39B.

SAND BEACHES

A River of Sand

A **beach** is an accumulation of unconsolidated sediment transported and molded into characteristic forms by *wave-generated* water motion. The landward limit of the beach is the highest level reached by average storm waves, exclusive of catastrophic storm surges, and the seaward limit is the lowest level of the tide. In Georgia, the beach's landward limit typically is a line of wind-blown sand dunes, called the foredune ridge, or in the more developed areas, a man-made structure such as a revetment composed of riprap (chunks of rock of boulder size).

Beaches will form on any coast where clastic particles somewhat resistant to wave abrasion are available and there is a site for sediment accumulation protected from extreme daily erosive wave action, such as wave reflection off rocky cliffs. As noted earlier, on the Georgia Coast, those clastic particles are mostly sand-sized fragments of rocks composed predominantly of quartz. In a few places, the clastic particles are shells and shell fragments (Figure 61A), which may range up to gravel in size. Therefore, quartz-rich grey to tan sand, such as the sample shown in Figure 61B, is the most common sand type on the beaches in Georgia.

Most beaches in the state are a virtual "river of sand" with the sand being in constant motion up and down the beachface and along shore. We call the zone within which the sand moves back and forth the **zone of dynamic change** (ZODC), which is illustrated in Figure 62. Part of the wind-blown dunes that form back of the beach above the high-tide line, called **foredunes**, act as a storage area for sand between storms. Thus, the landward limit of the zone of dynamic change is the point at which the foredunes are eroded during "normal" storms (not

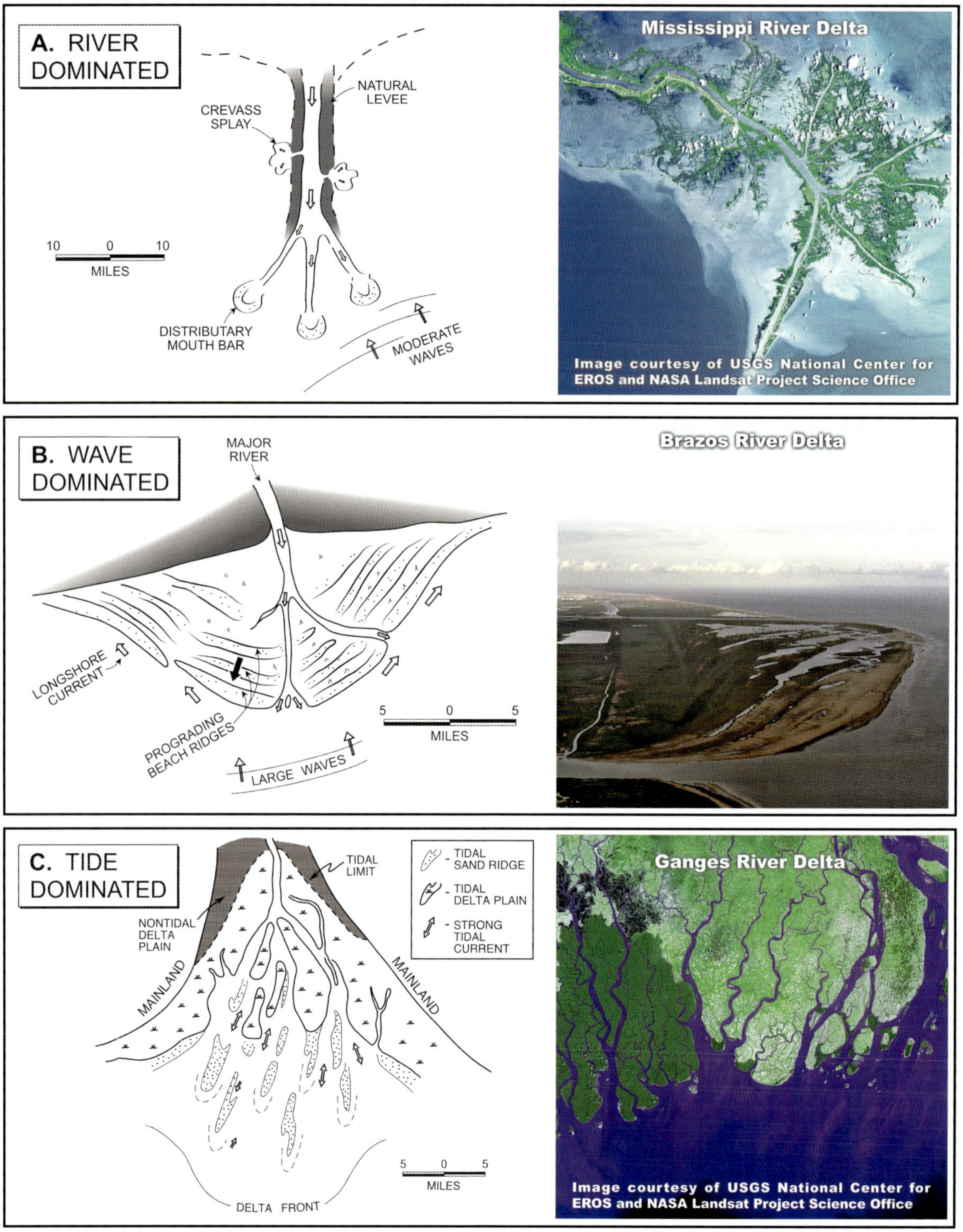

FIGURE 60. River delta models. (A) Schematic sketch in plan view of a river-dominated delta. The main lobe of the delta projects far beyond the shoreline of the adjoining coast. During floods, channels (sometimes referred to as crevasses) may be cut across the natural levees and crevasse splays (minor delta-like lobes) are deposited in adjoining waters. (B) Wave-dominated delta. Note prominence of prograded beach ridges throughout the delta plain. (C) Tide-dominated delta. Note the presence of tidal sand ridges at the offshore entrances to the numerous estuarine complexes along the margin of the delta.

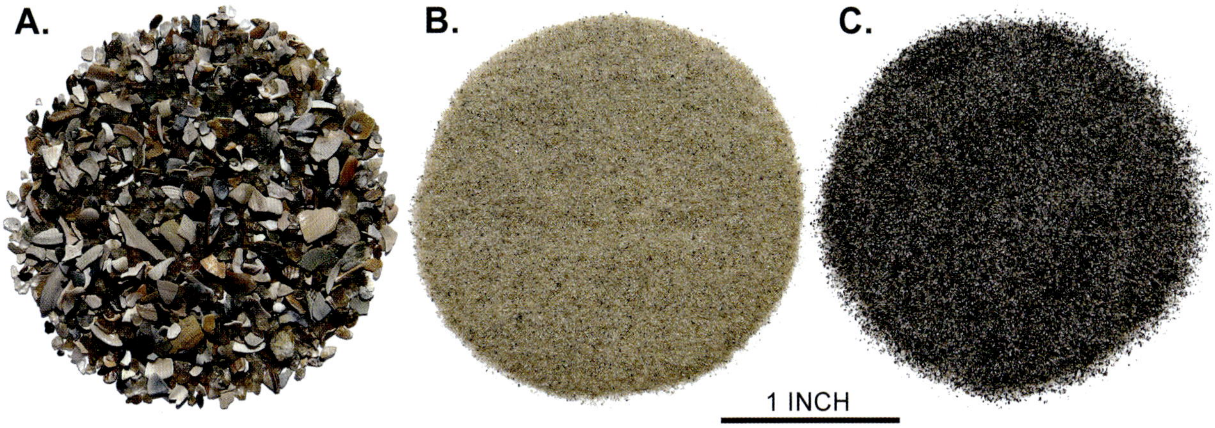

FIGURE 61. Beach sediment samples from Georgia beaches. (A) Shell-rich sample from near the terminal groin on the north end of Tybee Island. (B) Quartz-rich dune sand sample. (C) Heavy mineral layer on the upper beachface. Samples B and C were collected near the north end of the Holocene component of St. Simons Island.

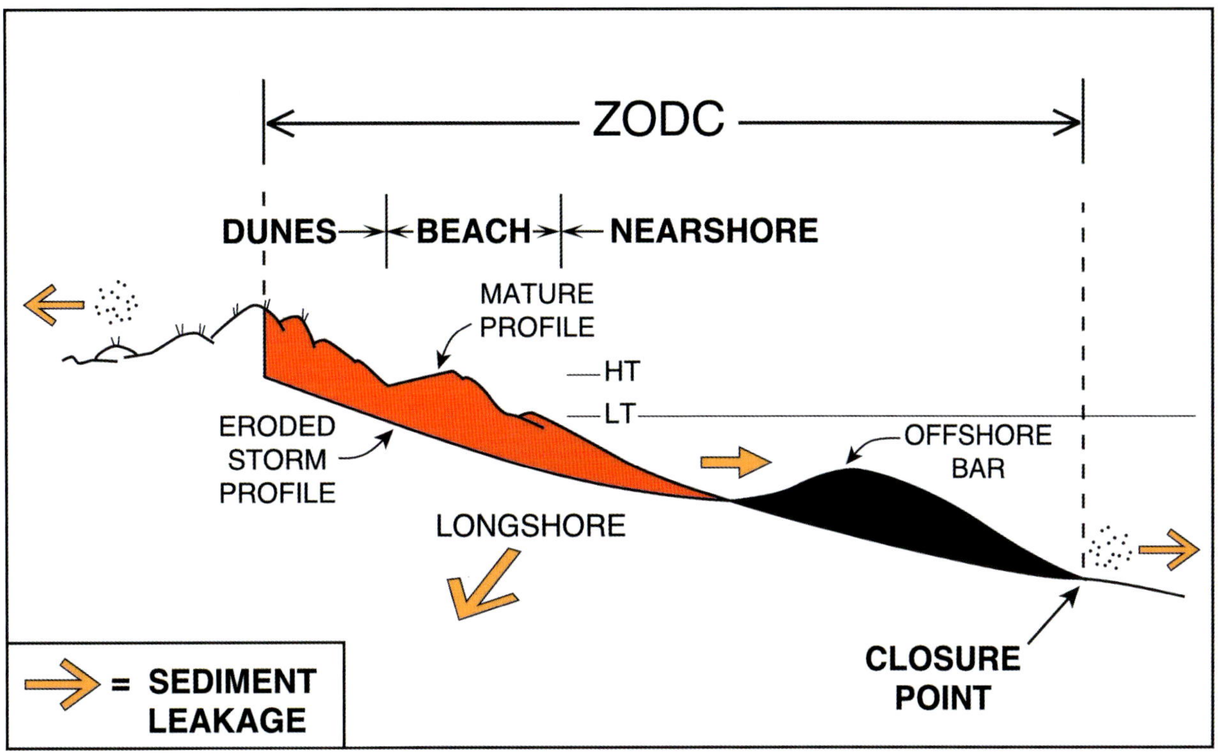

FIGURE 62. Zone of dynamic change (ZODC). The area in red depicts sand that may be eroded from the beach area and deposited offshore during storms. Much of this sand is deposited on the offshore bar (colored black). In between storms, much of the eroded sand is returned to the beach/dune area. The yellow arrows indicate the three mechanisms by which the sand might be lost from this particular ZODC: 1) Carried offshore beyond the closure point during large storms, such as hurricanes; 2) Blown landward out of the foredune area during periods of high wind activity; and 3) Transported alongshore out of the ZODC by beach drift and longshore currents as illustrated in Figure 28.

hurricanes or major northeasters). Accordingly, during these "normal" storms, the sand formerly deposited on the intertidal beach, as well as up to several feet of the foredunes, is usually transported into the subtidal region that, in some areas, may contain an offshore bar (see Figure 62; the red zone represents the portion of the beach/dune sand that potentially could be eroded during a typical, moderate storm). Some of the sand deposited on the offshore bar will return to the beach (and dunes) in between storms. The seaward limit of this zone is what engineers call the **closure point**, the most seaward distance for sand to move in an offshore direction under relatively *normal adverse conditions* (not major hurricanes!!). As indicated in Figure 62, all of the sand grains in motion do not stay within the zone of dynamic change at a specific geographic site on the beach for an infinite amount of time, but most likely will eventually "escape" from the zone by one of three primary types of "leakage:"

1) **Alongshore** out of the zone, under the influence of longshore sediment transport by wave-generated currents, as illustrated in Figure 28;
2) **Offshore** beyond the "point of closure" during major storms, such as hurricanes; and
3) **Onshore** by wind action, beyond recovery by offshore winds or by erosion during storms.

[NOTE: Many coastal geologists and some engineer modelers think the designation of a closure point, a concept commonly used by coastal engineers designing beach nourishment projects, has somewhat limited value, citing the obvious far offshore transport of sand during major storms (Hayes, 1965) as a shortcoming. We present it here because of its common usage, a term you may encounter from time to time, especially if you are somehow involved with or affected by beach nourishment projects.]

Another way sand grains may "escape" from the "river of sand" for a long period of time is for the shoreline to build out seaward to the extent that more lines of foredunes develop until they are located far enough landward to be out of the reach of normal storm waves. This is, in fact, the case in a few places along the Georgia Coast, such as on Wassaw Island, a prograding Holocene barrier island (Figure 38).

Beach Profile

When coastal geologists or engineers study beaches to determine such issues as their potential for erosion, they usually measure a topographic profile across the beach surface using a variety of surveying techniques. This is usually repeated over months or years in order to determine the cycle of erosion and/or deposition on that particular beach. Such studies have been carried out along many of the coastlines of the world, including the coast of Georgia (e.g., Pilkey and Richter, 1965). Some modern studies use a remote sensing system called LIDAR (Light Detection and Ranging) to produce topographic images of the beach and nearshore zone. This system is more expensive than simple horizontal-leveling surveys carried out by two or three junior surveyors; therefore, they are usually spaced at least six months apart.

For purposes of discussion, it is necessary that the different morphological components of the intertidal beach be defined. In the early days of research on the California Coast, the profile shown in Figure 63A was proposed as typical for that area. Under that scheme, the beach was divided into two segments: 1) Backshore–the landward sloping surface of a broad depositional feature called a **berm**; and 2) Foreshore–the seaward face of the berm also called the **beachface** in that scheme. The subtidal area was divided into the surf zone, breaker zone and offshore. This type of profile most commonly forms on a windward shore subject to ocean swell and a relatively large average breaker height, such as occurs on the California Coast.

Although there have been some efforts to force fit this California nomenclature to the sand beaches of the east coast of the United States, we prefer to use the nomenclature in Figure 63B. The east coast is a leeward shore dominated by storm waves and the average breaker height is considerably less than that experienced in California. Also, the influence of the relatively large tidal ranges on parts of the east coast is an important consideration. The upper profile shown in Figure 63B is the flat and featureless surface left after a storm. The middle profile is the one that most commonly occurs on the Georgia Coast during intervals between storms. It consists of three segments: 1) **Backbeach**, or the landward sloping portion of the depositional berm; 2) **Beachface**, the zone where the waves swash back and forth at high tide (seaward face of depositional berm); and 3) **Low-tide terrace,** which commonly contains one or more **intertidal bars** with associated **intertidal troughs**. Examples of intertidal bars are shown in Figure 64. Oblique aerial views of intertidal bars on Georgia

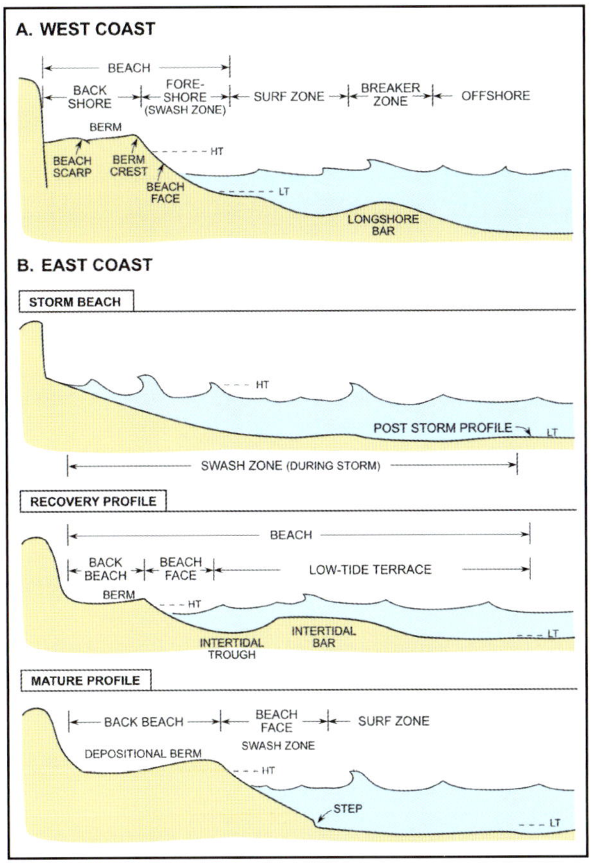

FIGURE 63. Representative beach profiles from the west (A) and east (B) coasts of the United States.

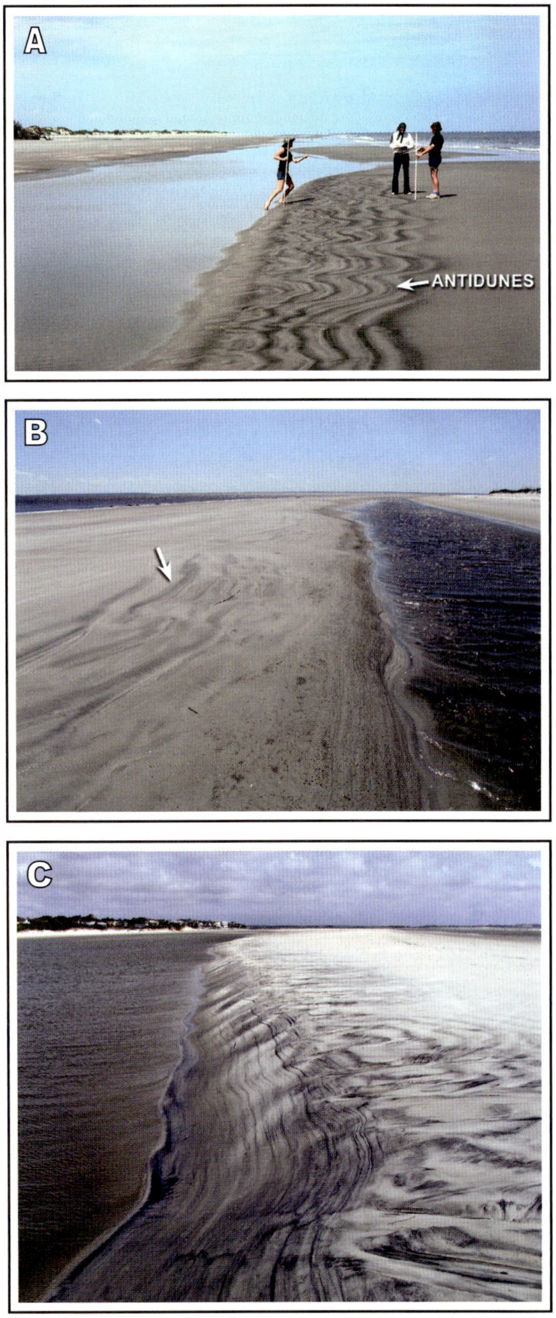

FIGURE 64. Intertidal bars. (A) Low-tide photograph taken in May 1974 showing a large intertidal bar on the north end of Kiawah Island, South Carolina. Antidunes in the foreground were formed when shallow sheets of water pushed forward by breaking waves (at high tide) flowed rapidly down the landward side of the bar. (B) Intertidal bar on the southern end of Jekyll Island, Georgia exposed at low tide on 26 March 2010. Arrow points to preserved antidunes. (C) Intertidal bar in the middle of St. Simons Island, Georgia exposed at low tide on 27 March 2010.

beaches are also shown in Figures 34C and 65. A general sketch showing a typical intertidal bar on the outer beach of a barrier island on the Georgia Coast is given in Figure 66A, and ground views of the beach on Sapelo Island are given in Figures 66B and 66C. The sketch in Figure 66A, and the ground photos in Figures 64C and 66B, show that the landward side of the bar has a steeply dipping surface called a slip face that commonly slopes toward the land at angles of 15-25 degrees. As the bar continues to migrate in a landward direction, a preserved series of steeply dipping beds result, as is shown in Figure 67.

In situations where the depositional cycle is complete, or where the beach borders a tidal inlet, the configuration shown in the bottom diagram in Figure 63B may be present, which is similar to the typical California profile. More detailed definitions of some aspects of beach morphology follow:

1) **Berm**: A wedge-shaped sediment mass built up along the beach by wave action. It typically has a relatively

FIGURE 65. More intertidal bars. (A) Intertidal bar in the middle of St. Simons Island exposed during a low spring tide on 28 April 2010. A ground view of this intertidal bar system is given in Figure 64C. (B) Intertidal bars on the south end of Sapelo Island exposed during a low spring tide on 28 April 2010.

FIGURE 66. Intertidal beach. (A). Typical morphology (at low tide) of a barrier island beach on the Georgia Coast. Note the prominent intertidal bar. Also shown are the locations where antidunes, ghost shrimp (*Callinassia major*), and ghost crabs (*Ocypode quadrata*) typically occur. (B and C) Low-tide view of intertidal beach on Sapelo Island in May 1970. Note large intertidal bar.

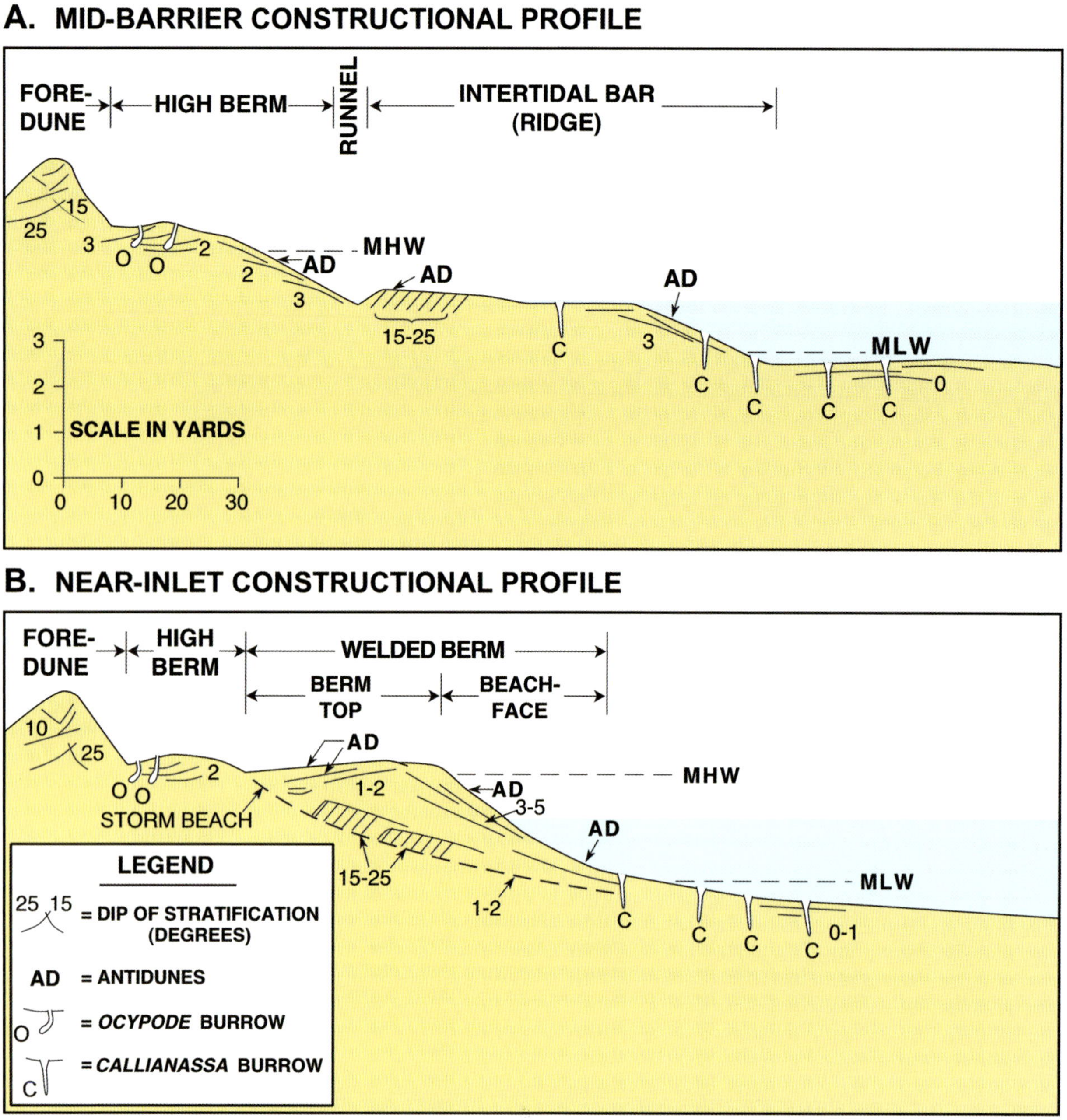

FIGURE 67. Representative morphology and sedimentary features occurring on the beaches of the prograding, mixed-energy barrier islands at the head of the Georgia Bight. Based on inspection of hundreds of beach trenches. (A) Features found on a typical mid-barrier beach in a constructional mode (modified after Hoyt and Weimer, 1963). This is a view at low tide. When the tide rises, the intertidal bar migrates landward, and the steeply dipping surface of the slip face (see example in Figure 64C) also continues to build landward, creating the landward dipping beds (at angles of 15-25 degrees) illustrated in this sketch. Also shown are the three zones where antidunes form: 1) In the swash on the upper beachface during high tide; 2) On the low oceanward side of the intertidal bar at low tide; and 3) On top of the intertidal bar as the tide rises (see example in Figure 64A). Also shown are the zones of occurrence of the two most conspicuous intertidal organisms, *Ocypode* (ghost crabs) just landward of the high-tide line, and *Callianassa* (ghost shrimp) in the lower intertidal zone. (B) Features found on beaches near tidal inlets after the intertidal bar has welded to the beach, and a broad depositional berm has formed. MHW = mean high water; MLW = mean low water.

steep seaward face and a gently sloping landward surface (2-3 degrees). A sharp peak (berm crest) usually separates the two oppositely slopping planar surfaces on the top of the berm (see the example of the welded berm in Figure 67B). There may be two berms present during all tidal phases except unusually high spring tides, a **high berm**, the most landward, oldest berm, and an **active berm** (termed the welded berm in Figure 67B), the most seaward and most recently activated berm.

2) **Beachface**: The zone of wave uprush and backwash during mid to high tide. It usually slopes seaward at an angle of around 5 degrees and is commonly the seaward face of the berm. However, the berm may be washed away during a storm. In that situation, the beachface would be a flattened-out, eroded surface subject to the uprush and backwash of the waves during mid to high tide.

3) **Beach Cusps**: The scalloped seaward margin of a berm consisting of a regularly spaced sequence of indented bays and protruding horns. As noted earlier, they are thought to be primarily depositional features (i.e., on the seaward flank of a prograding depositional berm) that form as a result of the interaction of shore-normal standing waves (edge waves) with shore-parallel, reflective waves, the same general process that produces rip currents (but at a smaller scale) (Guza and Inman, 1975). See two examples and the general model in Figure 29.

4) **Low-Tide Terrace**: A relatively flat surface that is located seaward of the toe of the beachface and slopes offshore at a small angle (2-3 degrees). This feature is illustrated in the recovery profile in Figure 63B. The landward margin is usually near mean sea level and the seaward margin near mean low water. The sand that composes the terrace is commonly water saturated throughout, essentially representing the intersection of the water table with the beach surface. During periods of recovery of the beach following an episode of erosion, **intertidal bars** migrate across the low-tide terrace in a landward direction. An **intertidal trough** is commonly located on the landward side of the elevated intertidal bar (Figures 63B and 66A).

A key aspect of any beach profile is the slope of the beachface. Beach research as early as during World War II focused on this issue, due to its relevance to beach landings by amphibious watercraft. Those early researchers found that the slope of the beachface in any given area will attain an equilibrium profile (that includes a near constant slope of the beachface) under what they referred to as "steady" conditions, that is, relatively unchanging, day-to-day wave characteristics. During storms, sand beaches are commonly eroded down to a flat surface that essentially parallels the water table in the beach with sand typically being transported offshore (see upper profile in Figure 63B). When sediment returns to the beach after the storm, the depositional profile attained by the beachface is steeper than the beach was immediately following the storm. Komar (1976, p. 303) explained why:

> *"Due to water percolation into the beachface and frictional drag on the swash, the return backwash tends to be weaker than the forward uprush. This moves sediment onshore until a slope is built up in which gravity supports the backwash and offshore sand transport."*

(NOTE: This early post-storm process happens regardless of the grain size of the beach sand. There's that law of gravity again!)

Several field studies and wave tank experiments have shown that coarser-grained beaches, especially those composed of gravel, have steeper beachface slopes than those composed of finer-grained sediments. Coarser-grained beaches allow greater percolation of the water brought to the beachface in the wave uprush than finer-grained beaches, thus reducing the strength of the return of the backwash. Consequently, a steeper beachface is required to maintain an equilibrium profile on coarse-grained beaches (gravity-supported transport). On sandy beaches of comparable grain sizes, those beaches consistently exposed to large waves tend to be flatter than more sheltered beaches exposed to smaller waves, because larger waves produce stronger backwash, which tends to flatten the profile. In this case, with the percolation rate being constant because of similar grain sizes, a smaller relative proportion of the water brought ashore in the larger waves is lost to percolation than for smaller waves (relative to the volume in the uprush), hence the stronger backwash of the larger waves and the resulting flatter beachface slopes. It seems a little counterintuitive that steeper beachface slopes would occur where the waves are smaller, given a similar grain size of the sand, but that is indeed the situation.

On the Georgia Coast, the grain size of the sand on the beaches decreases in size from north to south (Orrin Pilkey, pers. comm.). Also, most of the beaches are composed of fine-grained sand, consequently they tend to be quite wide and flat. An exception occurs where the beach flanks a major tidal inlet channel, where the slope of the beachface is an extension of the slope of the channel margin itself. The steepest beaches in the state occur where the beach is composed of pure, gravel-sized material, mostly oyster shells. These steep shell beaches occur in two types of settings in this area:

1) Along the Intracoastal Waterway, where boat wakes erode oyster mounds in the adjacent marshes and tidal flats with the eroded shells being deposited in beaches along the channel margin; and

2) On the windward sides of some of the more open estuaries, where at high tide the waves erode the oyster mounds in the marshes and tidal flats and a ridge of coarse shells (storm berm) is deposited on the adjacent, eroding marsh surface.

The previous discussion applies to relatively pure shell beaches that have steep beachfaces most of which occur along the Intracoastal Waterway or back in the estuaries. They are usually composed of coarse material such as oyster shells. However, as noted by Dörjes, Frey, and Howard (1986), shell material does occur in the following places on the outer sand beaches of Sapelo and Ossabaw Islands:

1) They called this location (on the outer beaches of the barrier islands) the *landward flank of foreshore runnels*, but we would classify this location to be the **step**, or most seaward part, of the beachface. In places, this can be a layer of shells a few inches thick and a few feet wide. However, this material is not a factor in determining the slope of the beachface itself. This is, in fact, the location from which the shell sample shown in Figure 61A was collected.

2) This location, termed *the lowermost foreshore surface* by the authors in the paper under discussion, is presumably the low-tide line, where wave backwash at low tide would tend to concentrate shells in the same manner as wave backwash at high tide accumulates shells at location 1.

3) The high-tide wrack line and the *upper backshore*, which is in more-or-less the same place near the high-tide line, where mobile shells are pushed (rolled along the surface) by the thin layer of water in the upper levels of the uprush.

In their study, Dörjes, Frey, and Howard identified the shells of 48 different mollusk species on the beaches of the two islands. By far the most common shell was the bivalve mollusk *Mulinia lateralis*, which is small, with the adults averaging about 0.6-0.8 inches in length. It burrows in sediment, part of the so-called infauna, and filters its food (mainly suspended bacteria) out of the water column. The exterior of the shell is cream to yellowish orange, whereas the interior is bright white (Smithsonian Marine Station, 2008). According to Edwards (1997), these so-called dwarf surf clams live in the *clayey mud close to the mouth of the bays* by the thousands. They obviously can survive the wide ranges of temperatures and salinities that occur in the entrances to the major sounds of the area.

Beach Cycle

The concept of the cyclical change of beaches from a flat, erosional profile in winter to a wide, depositional berm in summer had been well ingrained in both the popular and scientific literature for many years following research carried out in the 1940s. The concept originated from detailed studies on the coast of Southern California (e.g., Shepard, 1950a; 1950b; Bascom, 1951; 1954). The idea was that, generally speaking, erosional storm waves are more common in the winter, and flatter, depositional swell waves are more common in the summer on the California Coast–hence the terms *summer* and *winter beaches*. Beach profiles measured at Scripps pier in La Jolla in the 1940s commonly showed wide beaches in the late summer and fall and narrow beaches in the winter and spring.

However, this is an idea that has not withstood the test of time for other parts of the world. In 1969, Hayes and his colleague Cy Galvin observed a striking contrast between beach cycles on the east coast of the United States and those presented in the literature for the west coast. Eroded winter profiles and strongly depositional summer profiles are not a dominant theme on northern Atlantic beaches, particularly in the New England area. The beach cycle in that area is controlled by the passage of individual storms. This also appears to reflect, at least to some degree, less severe wave climate and relatively smaller seasonal changes in wave climate on the northern Atlantic coast. Mean wave heights and periods are 19 inches and 6.9 seconds along the northern Atlantic coast, considerably smaller values than those for the Central California Coast, where the average height of the highest one third of the waves (the so-called significant waves) is

7.6 feet and the average wave period is over 11 seconds (Beyene and Wilson, 2006). These dissimilarities in wave climate result, to some extent, from differences in their locations on western and eastern edges of oceans with predominantly westerly winds. These winds blow predominantly onshore on the California Coast and offshore on the east coast. However, Ed Clifton (pers. comm.) commented that *local winds are less a factor than the fact that the largest storms (typhoons) develop on the western side of the ocean, and that every large weather system in the western Pacific (and the Gulf of Alaska as well) generates long period swell which arrives on the U.S. west coast.* Therefore, the largest waves that occur along the Central California Coast are generated in one of two ways: 1) When major *El Niño* storms cross the coast; or 2) When some of the largest swell from especially huge offshore storms reach the coast (Hayes and Michel, 2010). Both of these events are most common during the winter months.

In the 2002 Coastal Engineering Manual, published by the U.S. Army Corps of Engineers, the following statement appears: *The seasonal onshore-offshore exchange of beach and nearshore sediments ... is now known to apply best to swell-dominated coasts, such as the U.S. west coast, where wave climate changes seasonally.* Therefore, the general consensus today is that the terms "winter beach" and "summer beach" should not be used with reference to erosional and depositional cycles on beaches, although many published discussions of the beaches in California still use these terms, or else refer to "normal seasonal changes," which presumably means more-or-less the same thing.

Based on six years of research between 1964 and 1969, Hayes and his students at the University of Massachusetts found that, on the northern New England Coast, northeasterly storms play the dominant role in the generation of cycles of erosion and deposition of the sand beaches in that area (Hayes and Boothroyd, 1969). Observations at 40 beach profile stations surveyed every two weeks over that six-year period revealed the following stages of low-tide morphology relative to storm occurrences (these profiles are illustrated in Figure 63B):

- **Early post storm** (up to 3 or 4 days after the storm) – Beach is flat to concave upward and the beach surface is generally smooth and composed of medium- to fine-grained sand. Severest storms leave erosional scarps along the dunes.
- **Early accretional, or constructional** (usually 2 days to 6 weeks after storm) – Small berms at the high-tide line and intertidal bars and troughs are quick to form.
- **Late accretional or maturity** (6 weeks or more after storm) – Landward-migrating intertidal bars weld onto the backbeach to form broad, convex berms. On some beaches, it takes a long time for welding to occur and large intertidal bars (up to three feet in height) lie between the backbeach and the low-tide line (see examples from the Georgia and South Carolina Coasts in Figures 64, 65, and 66).

Paul Komar, working on the Oregon coast, suggested the terms **storm profile** (flat beach) and **swell profile** (well-developed berm).

Our work on the South Carolina Coast shows that the South Carolina sand beaches follow a cycle similar to those in northern New England, particularly those close to tidal inlets (Hayes et al., 1975; Stephen et al., 1975). In mid-barrier areas away from inlets, however, the beaches are usually quite flat or contain intertidal bars that migrate landward very slowly. A completed cycle like the one illustrated in the lower diagram in Figure 63B is rare indeed in mid-barrier areas, but they can be seen on occasion in the near vicinity of tidal inlets (see Figure 67B). The pattern of erosion/deposition of the sand beaches in Georgia is similar to that found in South Carolina (Pilkey and Richter, 1965).

Georgia's Beaches

In one of the earlier papers on the subject, Howard and Frey (1980) made the following comments with reference to the beaches in Georgia:

1) Median grain size of beach and dune sediments is finer, and $CaCO_3$ content lower, than anywhere else on the southeastern U.S. coast (Giles and Pilkey, 1965).

2) Because of intertidal to supratidal exposure of beaches and dunes, and consequent rigorous conditions for marine life, the abundance and diversity of organisms generally is lower in beach sediments than in subtidal sediments (Dörjes, 1972).

3) Beachface slopes commonly are less than 2°, and seasonal changes are slight. As we have already noted and illustrated in Figures 64, 65, and 66, intertidal bars are common features on the Georgia barrier islands, a

fact that was pointed out earlier by Wunderlich (1972).

4) We think they implied that the lower intertidal zone has more *Callinassa major* (ghost shrimp) burrows than the upper intertidal zone (see Figure 67).

Howard and Frey (1980) also emphasized that the most dramatic changes in the beach are caused by landward migrations of intertidal bars (Howard et al., 1973). On this same topic, they observed further that high-angle beach stratification is formed on leading edges of migrating intertidal bars (Wunderlich, 1972). This fact, that is illustrated in Figure 67A, was also observed earlier on the Georgia Coast by Hoyt (1967a), and by Hayes and students on Massachusetts beaches (Hayes, 1969).

Physical Sedimentary Features

Because of the relatively large tides and fine-grained sand on the beaches in Georgia, when you visit the beach at low tide you will be able to walk across a wide intertidal zone that contains an abundance of both physical and biological features that will no doubt arouse your curiosity. Although geology has long been considered to be primarily a field science, the greatest breakthrough in the understanding of the smaller-scale physical sedimentary features you will be walking on at the beach was the result of laboratory flume studies carried out by hydraulic engineers (e.g., Simons and Richardson, 1962). In flowing water, sand grains start to move along the bottom at velocities of about 10 inches/second. These workers demonstrated that the sediment bed of a laboratory flume (box-like trough with a sand bottom and a mechanism to generate water flow across the sand bed) could be made to pass through the following sequence of features (called **bedforms)** by simply increasing the velocity of the water flowing across the sand bed or decreasing the depth of the water:

> *flat bed → small ripples → large ripples (megaripples) → washed-out megaripples → flat bed → antidunes*

While doing this work, they came up with an empirical relationship that dictates which of these different bedforms are active at any given time. It is called the **Froude number**, which equals:

$$\textit{Froude number} = \frac{V}{\sqrt{gD}}$$

V = velocity
g = gravitational constant
D = water depth

Therefore, as the water velocity increases, the Froude number also increases, and as depth increases (at same velocity), the Froude number decreases.

The first part of the sequence up to the washed-out megaripples occurs when the value of the Froude number is <1.0. This part of the sequence is called the **lower flow regime**. By increasing the Froude number to values >1.0, by either increasing the velocity of flow or decreasing the water depth (see formula), the sand bed goes into the **upper flow regime**, where either flat beds or antidunes are formed.

The lower flow regime bedforms are asymmetrical, with the sand grains moving in discreet steps by rolling or bouncing up the gently sloping upcurrent side of the bedform and sliding down the steeply dipping surface on the downcurrent side (called the slip face). Thus, these forms move along in the direction the current is flowing, sometimes several feet during a single tidal cycle. On a beach, it is rare to see any bedforms larger than **ripples** (spacing between crests less than 2 feet) and they usually occur in the intertidal troughs on the landward side of the intertidal bars, where the water depths are greater than elsewhere on the beach (see example in Figure 68A). **Megaripples** (spacing between crests 2-18 feet) are seen more commonly in tidal channels, where the water is deeper and the currents are strong (see Figures 68B and 68C). The mode of sediment transport over these types of bedforms is illustrated in Figure 69. Larger bedforms called **sand waves** (spacing between crests greater than 18 feet) do occur on this coastline, but they are mostly found in deep tidal channels. However, as the examples in Figure 70 show, some can be seen in bars adjacent to deep river channels and in the marginal flood channels of the ebb-tidal deltas off the beach.

As you walk across the beach and tidal sand flats at the coast, you will notice some dramatic changes in the nature of the lower flow regime bedforms (at the scale of ripples), even though the changes in the topography are quite subtle. As the diagram in Figure 71 shows, there is a progression of the three-dimensional shapes from

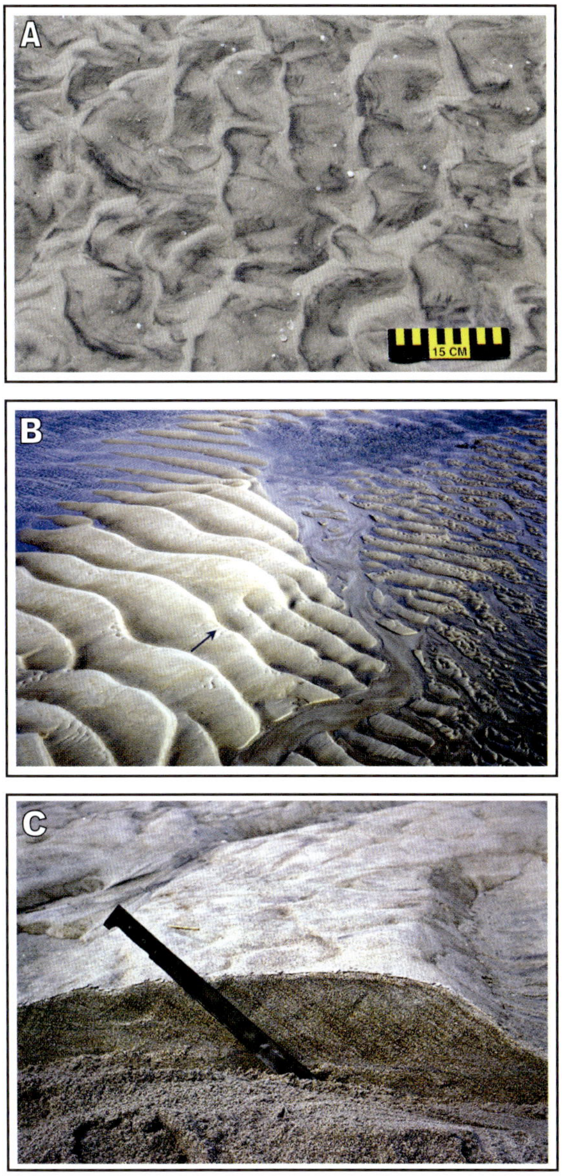

FIGURE 68. Ripples and megaripples. (A) Ripples (spacing between crests <2 feet) exposed at low tide in an intertidal trough near the southern end of Jekyll Island, Georgia (26 March 2010). Scale is ~6 inches. (B) Megaripples (spacing between crests 2-18 ft) exposed at low tide on a sand flat at Hampton Harbor Inlet, New Hampshire. View from bridge. Arrow points to megaripple shown in C. The ebb current that created these megaripples was flowing from the lower left to the upper right. (C) Ground view of the megaripple shown in photograph B. Note the high angle of dip of the beds created by the migration of the bedform by the mechanism illustrated in Figure 69. The machete is about 2.5 feet long. The ebb current that created this bedform was flowing from left to right. Photographs in B and C were taken in November 1965.

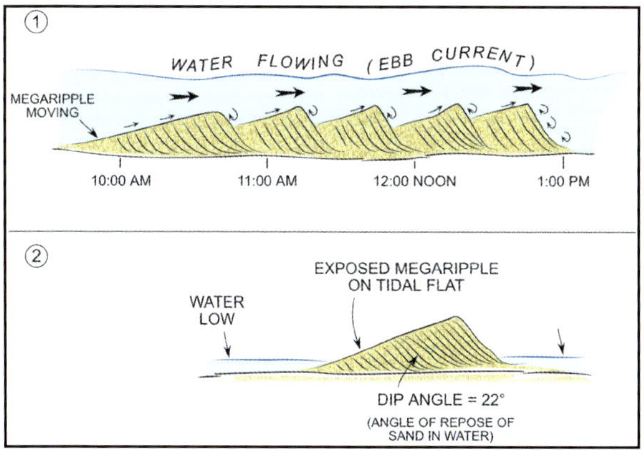

FIGURE 69. Mechanism for the movement of megaripples. Compare the implied slip faces of the bedform as it moves along (in the sketch) with the ones exposed in the excavation in the photograph in Figure 68C. Sand waves show a similar pattern of migration.

straight-crested to undulatory ripples at conditions of relatively low flow strength. As flow strength increases, the ripples take on a cuspate shape. This shape is very common in the intertidal troughs on the beach. As the flow approaches the upper flow regime (Froude number >1.0), the ripples convert to a planed-off rhomboid shape just before the bed goes flat. The photographs in Figure 72 illustrate these bedform types. You will undoubtedly see the pattern illustrated in Figures 71 and 72 repeated in many places as you walk over the intertidal bars or near the edge of tidal inlets during low tide. The pattern will almost always be present where the high-tide waters were flowing down very subtle slopes on the sand bars. Incidentally, the occurrence of rhomboid ripples on the beaches of Georgia was noted early by Hoyt and Henry (1963), and a photograph of some on the top of the intertidal bar on the south end of Jekyll Island (in March 2010) is shown in Figure 72D.

The most common bedform type under upper flow regime conditions is the **flat bed**, over which the flow of the sand grains is more-or-less continuous, streaming in long streaks across the sediment surface. This is the most common type of sand surface occurring on sand beaches, where water depths are typically small and wave-generated water flow is generally rapid. Thus the Froude number is typically >1.0 (see formula again). Under conditions where the velocity is more rapid than normal, for example, when the water flows down a subtle slope (e.g., over the crest of an intertidal bar as in Figure 64 or down a more steeply sloping high beachface),

FIGURE 70. Sand waves (spacing between crests >18 feet) on the Georgia Coast. Both photographs were taken during a low spring tide on 28 April 2010. (A) Beautifully displayed, ebb-oriented sand waves on a sand bar in the Altamaha River near the route US 17 bridge. Downstream is to the right, the direction in which the sand waves were moving during the higher tides (arrow). (B) Flood-oriented sand waves in the southern marginal flood channel of the ebb-tidal delta at Hampton River Inlet on the northern end of Sea Island. Arrow A indicates the direction of movement of the sand waves at higher water levels. This photograph was taken at low tide following the time peak ebb currents were occurring in the main ebb channel of the inlet (arrow B), but these sand waves are still flood-oriented due to their position in the marginal flood channel (i.e., ebb currents are relatively weak in this channel; refer to the model in Figure 52 for orientation).

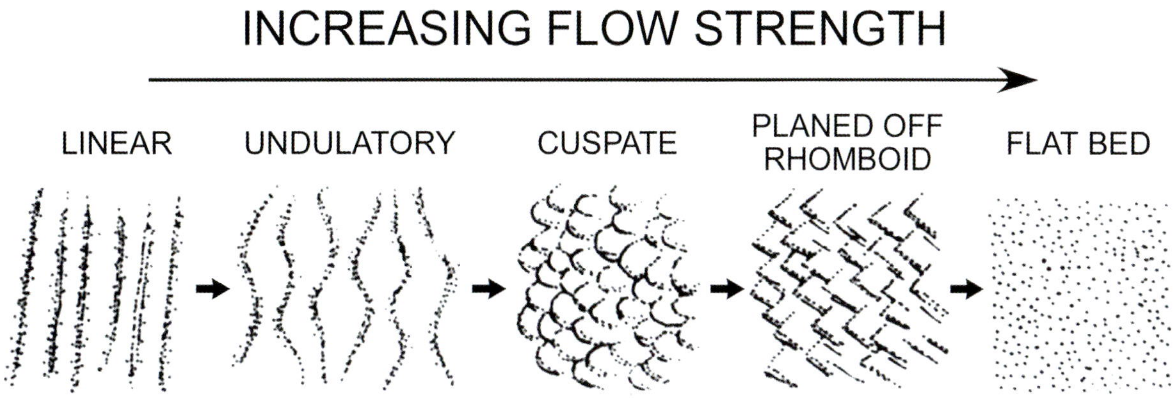

FIGURE 71. Changes in linearity of crests and 3-dimensional shape of ripples under conditions of increasing flow strength. Based on data of Allen (1968), Boothroyd (1969), and numerous field observations by our group on the intertidal bars and sand flats on the coasts New England and the Georgia Bight. In all cases, the water was flowing from left to right.

FIGURE 72. Photographs of the ripple types illustrated in Figure 71. (A) Linear ripples. Scale is one foot. The ebb current that created these ripples was flowing from left to right. (B) Ripples with undulatory crests, formed by an ebb current flowing from left to right. Scale is 15 centimeters (~6 inches). (C) Cuspate ripples, formed by an ebb current flowing from the upper right to the lower left. (D) Rhomboid ripples on top of an intertidal bar on the south end of Jekyll Island (26 March 2010). This ripple form is the last to develop before the bed passes to the flat bed stage in the transitional zone between lower and upper flow regime.

a feature called an **antidune** is formed. Under this condition, the sand surface develops a sinusoidal shape and a relatively thin layer of water flows in phase across the top of these features, which are linear and parallel with a typical spacing of about 15-20 inches.

As shown in Figures 73 and 74, the water flowing down the slope on the downcurrent side of the sinusoidal (trochoidal) antidune picks up sand grains from that relatively steep slope only to drop them out again as the water flows up the slope on the upcurrent side of the next antidune in line. A slice through a preserved antidune, shown in Figure 74, illustrates the deposited sand on the steep upcurrent side of the antidune. At first glance, as you watch antidunes in motion, you experience an optical illusion that the sand grains are moving upstream against the current. Actually, you are viewing the individual trochoidal forms of the antidunes that are moving against the current, because of erosion on the downcurrent side and deposition on the upcurrent side. Although they move in a series of stops and starts, the sand grains always move in the direction the water is flowing.

If you look carefully, you will almost always see antidunes forming on the Georgia beaches in one (or all) of three places:

1) Where water draining from intertidal trough outlets cut through and across the intertidal bar located just seaward of the trough (Figure 66A);

2) Anywhere along the upper or lower part of the beach where the backwash flow gains velocity as it moves back down the beach slope. A photograph of this process is shown in Figure 75A. An example of antidunes preserved on the upper beachface at Kiawah Island, South Carolina is shown in Figure 75B; and

3) On the landward side of intertidal bars where, at high tide, the water flows in a landward direction down the slope on top of the landward side of the bar toward the intertidal trough. During a field visit to the Georgia Coast in March 2010, we observed similar antidunes preserved on an intertidal bar on the south end of Jekyll Island (Figure 64B), and on East Beach (Figure 75C).

It has been our experience that students walking in the swash zone get pretty excited when they feel the antidunes move by past their feet. Sand moving uphill!! Not exactly, but close.

The shallow, rapid flow of water over the antidunes tends to sort the sediment by composition, moving

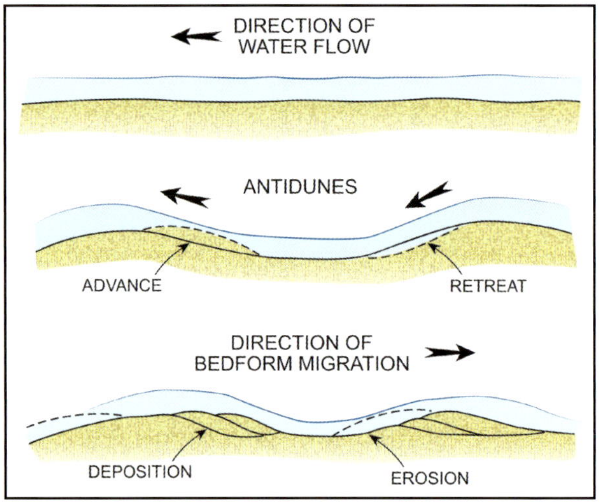

FIGURE 73. Antidune formation and growth. As the shallow sheet of water flows across the linear form of the antidune, water flowing down the downcurrent side of the antidune tends to erode sand from that slope. As it flows up the slope of the next antidune downcurrent, sand is deposited on that slope, causing the feature to migrate upcurrent. This bedform is called an antidune, because it moves in the opposite direction to the current. Lower flow regime ripples, on the other hand, move in the same direction as the current (compare this figure with the diagram in Figure 69). However, the sand grains themselves move downstream in both situations.

FIGURE 74. Trench through antidunes on the beach on Seabrook Island, South Carolina (photograph taken at low tide in June 1975). Water was flowing from right to left when the antidunes were forming. Compare the internal sand layers in the preserved antidune with the sketch in Figure 73. Scale is one foot.

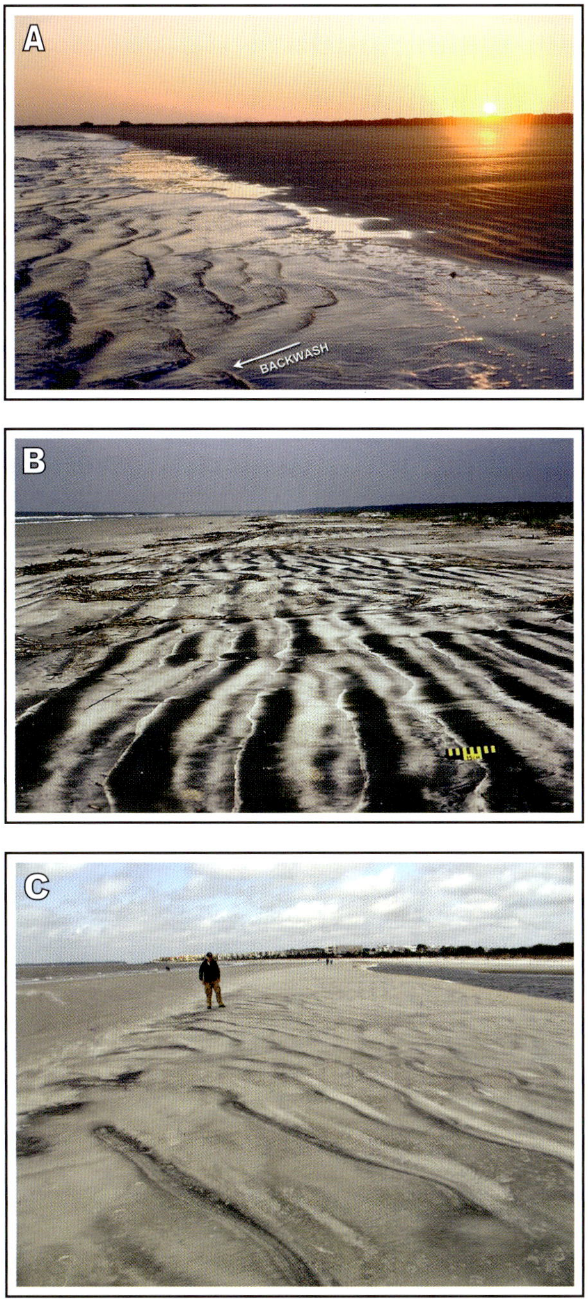

FIGURE 75. More antidunes. (A) In the lower left of the image, antidunes are forming in the backwash of a wave. Further up the beach, preserved antidunes are in evidence. On the beach at Seabrook Island, South Carolina in June 1975. (B) Antidunes preserved on the upper beachface on the northern end of Kiawah Island, South Carolina. Scale is 15 cm (~6 inches). (C) Antidunes (in foreground) preserved on top of a large intertidal bar on St. Simons Island, Georgia. Photograph taken during a low spring tide on 27 March 2010.

out the larger and lighter quartz and feldspar grains and leaving behind more dense and smaller grains, such as hornblende [$Ca_2(Mg,Fe,Al)_5(Al,Si)_8O_{22}(OH)_2$], magnetite ($Fe_2O_4$), and ilmenite ($Fe_2TiO_3$). These more dense minerals, called **heavy minerals**, are darker in color than the lighter minerals, hence they leave linear traces that mark the position of the antidunes in which they were deposited (see Figure 75B). Another place where you will see relatively thick deposits of heavy minerals is at the base of an eroding dune scarp at the back of the beach, where the wind has carried the lighter grains away and left the heavier ones behind. Occasionally, the high part of the beachface is covered with a layer of them, usually after a storm, as shown in Figures 76A and 76C (also see the microscope photograph of heavy minerals in Figure 76B). An example of heavy minerals collected from the high beachface near the north end of the Holocene component of St. Simons Island is pictured in Figure 61C.

In the usual mix of heavy and light minerals on a beach, the heavy minerals will be smaller in diameter than the lighter minerals, because of their **hydraulic equivalence** (e.g., if dropped into a cylinder of water, they would all reach the bottom of the cylinder at nearly the same time despite their difference in size). That is, as they fall through the water in a breaking wave, the smaller, heavier minerals fall at the same rate as the larger light minerals, because the buoyancy of the relatively dense water the grains are falling through creates more resistance on the larger grains than the smaller grains. Once the sand grains are laying on a flat beach, the larger, lighter mineral grains, such as quartz and feldspar, will be transported away more readily by the shallow sheet flow of the water in the wave swash than the smaller heavy mineral grains, such as magnetite and hornblende, because they protrude further up into the water than the smaller heavy minerals (a process called **pivotability**). Over time, this action may leave behind a surface layer of the heavy minerals, because they do not stick so far up into the water column as the large lighter grains, which are pivoted out of position and carried away. The process of the formation of heavy minerals at the toe of a dune scarp or on the surface of the wind-blown sand dunes is different because, in those situations, the influence of gravity takes over again, with the heavier particles settling out first.

Ilmenite, one of the more common minerals in these black, heavy mineral layers on the beaches, is rich in

FIGURE 76. Heavy minerals. (A) Beach trench in heavy mineral zone at north end of Kiawah Island, South Carolina (in May 1978). (B) A 3.5 mm field of view of a sand sample from a heavy mineral layer on a South Carolina beach. The black minerals are dominated by magnetite (Fe_2O_4) and the clear grains are mostly quartz (SiO_2). Courtesy of Ray Torres. (C) Heavy minerals at the high-tide line on St. Simons Island, Georgia (on 27 March 2010). The heavy mineral sample shown in Figure 61C was collected at this site.

titanium, and similar deposits are also abundant in some of the Pleistocene beach ridges landward of the coast, particularly in Georgia and Florida. Pleistocene heavy-mineral deposits in Georgia and northeastern Florida have been mined for more than 90 years. According to Pirkle, Pirkle, and Rich (2013), there are two heavy mineral mines and nine known deposits of economic interest in Georgia. The principal minerals produced are titanium minerals (used to make TiO_2 pigments) and zircon (used as a foundry sand and in the manufacture of refractories, ceramics, and chemicals). There are many other known deposits that have been "lost" to mining because they occur in residential areas, national parks or preserves, or have other environmental concerns. For example, there are large deposits very close to the Okefenokee Swamp National Wildlife Refuge where mining could affect this ecologically fragile area.

Conspicuous Biological Components

One of the most notable biologic features on the Georgia beach, besides the feeding birds and the occasional stranded seashell, are burrows by ghost crabs (*Ocypode quadrata*). Near the base of the dunes and the high-tide line, the burrows of the ghost crabs are usually abundant. According to OperBBA (2012), these crabs are called ghosts because *they are translucent, and because of their ability to disappear from sight almost instantly, scuttling at speeds up to 10 miles per hour, while making sharp directional change.* These creatures have two black eyes, with sharp 360° vision. As you walk along the high-tide line, you will see numerous mounds of sand excavated from the ghost crabs' J-shaped burrows, which tunnel down into the ground at a 45 degree angle as much as 3 feet or so. If you are patient, you will probably see a few scurrying about, but they are more commonly nocturnal. See photographs in Figure 77.

The lower half of the intertidal zone usually contains abundant burrows of the ghost shrimp (*Callianassa major*) (see examples of these burrow outlets in Figure 78). The burrow outlets are usually surrounded by numerous chocolate-jimmie-like pellets composed of organic-rich mud. These are the fecal pellets of this rapidly burrowing shrimp (pictured in Figure 79A), which is difficult to catch without the use of a plunger or similar apparatus. The shrimp use these pellets to line their corncob-like burrows in order to strengthen them. The picture in Figure 79C of some of these burrow

FIGURE 77. The ghost crab (*Ocypode quadrata*). (A) A rare view of one on the surface in the daylight. (B) Sand castings outside one of their burrows. Scale is ~6 inches.

tubes exposed in an eroding bluff of a Pleistocene barrier island along the St. Marys River attests to the ruggedness of the burrows. The location of the ghost crab and ghost shrimp burrows on the beach is illustrated in Figures 66A and 67, with the crabs occurring above the high-tide line and the shrimp entirely within the intertidal zone (mostly from mid tide seaward).

The preserved burrows of ghost shrimp are called *Ophiomorphia* where they are found fossilized in ancient sedimentary rocks. For example, they are abundant in the Cretaceous sandstones of western Canada that are as much as 100 million years old. Consequently, there was a considerable effort by researchers on the Georgia Coast in the 1970s to describe the morphology and distribution of the burrows of ghost shrimp so this information could be used in the interpretation of ancient sediments. The diagram in Figure 80, from a paper by Frey, Howard, and Pryor (1978), presents a profile of these burrows across the intertidal and shallow nearshore zone. The fascinating story these researchers told, illustrated in Figure 80, follows:

We have relatively little information on depth of burrowing by subtidal individuals of Callianassa major. .... However, drilling by means of a water jet along seaward profiles (Fig. 11 – partially reproduced here as Figure 80*), combined with observations along a Cretaceous shore profile, show that the vertical burrows extend downward for 3-5 m near the beach and join a maze of interconnecting horizontal burrows. These gallery systems occur at a continuous, nearly horizontal level, being deepest near the beach and becoming nearer to the sediment water interface offshore. .... The jet drilling also demonstrated that the related animal density is substantially greater than one individual per burrow aperture; at one site more than 100 individuals of C. major were washed to the surface through 8 nearby burrow openings.*

So as you walk over the beach surface in the intertidal zone, especially if you are in a zone with a large number of burrow outlets like the ones shown in Figure 78, there may be a group of these shrimp bustling around in the "catacombs" at depth just below your feet.

Typical Geomorphology

The most common beach profile in the mid-barrier region of the barrier islands along the head of the Georgia

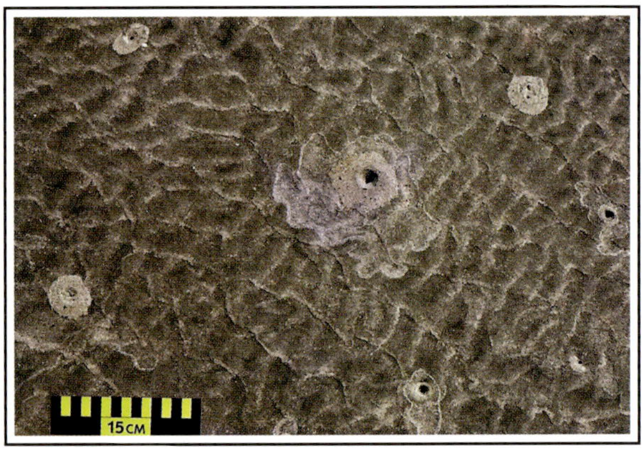

FIGURE 78. Outlets of ghost shrimp (*Callianassa major*) burrows on the lower intertidal beach. Scale is ~6 inches.

Bight contains a very well-developed foredune ridge, which, at times, may have an erosional scarp and be covered with abundant dune grasses, including sea oats (*Uniola paniculata*) on the top of the dunes and sea elder (*Iva imbricatta*), and *Croton sp.* around the base. The beach itself usually contains a modest-sized high berm and a well-developed intertidal bar, as is clearly shown by the sketch in Figure 66A, the diagram in Figure 67A, and the photographs in Figures 64, 65, and 66B.

The diagram in Figure 67A shows the distribution of some of the physical and biological features typical of this type of beach:

- Antidunes on the high-tide and low-tide swash zone areas;
- The low angle of dip of the beds on the swash zones and the high-angle of dip of the intertidal bar's slip face;
- Ghost shrimp burrows near the low-tide line; and
- Ghost crab burrows above the high-tide line.

Figure 67B illustrates the typical late-stage constructional profile (welded berm) commonly found near tidal inlets, as well as the distribution of the physical and biological features found there, including:

- The surface of the beach just after the last storm (dashed line);
- Antidunes on the berm top and beachface;
- The low angle of dip of the beds formed during the later stages of development of the berm and the high-angle of dip of the slip face of the intertidal bar that first welded to the beach;
- Ghost shrimp burrows near the low-tide line; and
- Ghost crab burrows above the high-tide line.

Bubble Sand

In 1964, John Hoyt and Jim Henry published a paper entitled *Development and Geologic Significance of Soft Beach Sand*, which was a discussion of the occurrence of soft sand on the beaches of the Georgia Bight. They attributed this soft beach sand to *a sponge-like, bubble development up to 16 inches in thickness in the upper part of the beach*. This type of soft sand is commonly referred to as **bubble sand** (see photograph in Figure 81A). During our 150-plus training courses held on the South Carolina Coast between 1976 and 2005, our field team dug well over 200 trenches in this type of sand, with two of the most notable areas of occurrence being on the north end of Kiawah Island and in the beaches that border Price Inlet. This aroused our curiosity to the point that Chris Reel was encouraged to conduct an M.S. thesis on the topic at the University of South Carolina Geology Department (under supervision of Hayes), with the aid of the mechanical engineering group at the University (Reel, 1982).

Based on studies of ground water flow in the beach at Price Inlet, plus large numbers of trenches, and cores in the sediment, Reel concluded that bubble sand is best developed on upper intertidal sand flats and beachfaces where air is trapped between the water table in the sand and ocean water coming over the surface as a result of the rising tide. The water table in the beach and tidal flat sediments rises more slowly than the ocean surface does during the rising tide, which creates the trapping mechanism (air between two distinct layers of water-saturated sand). The resulting bubble sand has a sponge-like appearance, with porosities well over 50 percent (Figure 81A). Also, with a little compaction by overlying sediments, some of the bubbles collapse and, in some cases, the escaping air distorts the original sand layers into convoluted, swirling bands, such as the ones shown in Figure 81B.

[NOTE: In this case, **porosity** is defined as the ratio of the open pore spaces between the individual sand grains to the space taken up by the sand grains themselves (when they are stacked together). **Permeability** is a measure of the ease with which a fluid flows through the pore spaces between the sand grains. During an oil spill, for example, oil will penetrate more deeply into a gravel deposit with high permeability than it will into a mud or very fine-grained sand deposit, which typically have much lower

FIGURE 79. The ghost shrimp (*Callianassa major*). (A) Live ghost shrimp with its large "digging claw" exposed. (B) Fecal pellets accumulated around a burrow outlet. Photographs A and B courtesy of Bill Frank, Jacksonville, Florida. (C) Corncob-like burrows exposed by erosion that were fortified with fecal pellets like the ones shown in B. These dense and spectacularly displayed burrow remnants were originally formed in the lower intertidal zone of a Pleistocene barrier island that is now exposed along the banks of the St. Marys River on the Georgia/Florida border.

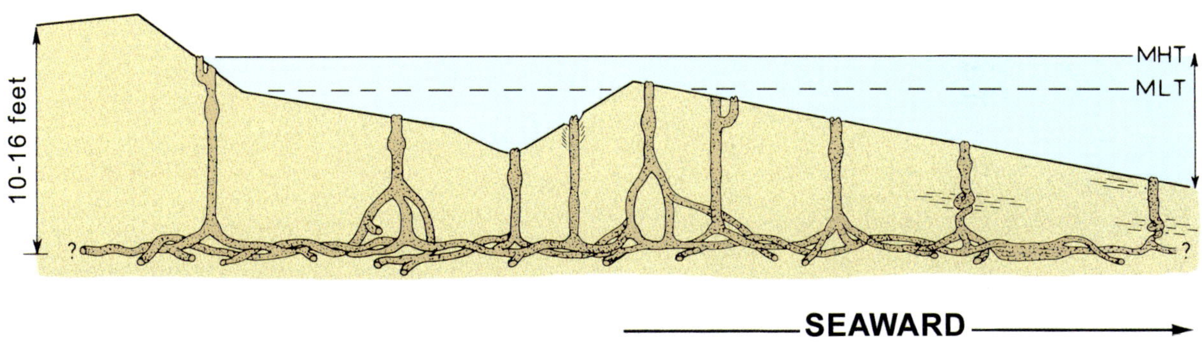

FIGURE 80. A cross-section showing the burrows of ghost shrimp (*Callianassa major*) in the intertidal and shallow offshore zone. Reconstructed on basis of jet drilling on Sapelo Island, Georgia, as well as on beaches in Florida and Mississippi. Observations on outcrops of Cretaceous sedimentary rocks in Alabama were also used. After Frey, Howard, and Pryor (1978).

FIGURE 81. Bubble sand. (A) Shovel dig with bubble sand on the south end of Tybee Island on 28 March 2010. The dark color at the top of the image is the shovel. (B) Distorted layers in a trench in bubble sand on a sand flat on the north end of Kiawah Island, South Carolina in the spring of 1996. Scale is ~6 inches.

permeability than the gravel because of the constriction of flow through the relatively small openings. This last general statement is compromised by the presence of bubble sand (discussed next).]

An amazing application of the discovery of bubble sand came as our group was conducting a survey of the over 500 miles of the Saudi Arabian Coast of the Persian/Arabian Gulf that was oiled during the largest oil spill in history, the *Gulf War* spill of 1991 (40 times larger than the *Exxon Valdez* spill in Prince William Sound, Alaska in 1989, both of which we have researched for years). During surveys conducted in 2002-2003, we found the some of the greatest thicknesses of oil residues in sand flats and beaches occurred in bubble sand (Hayes, Michel, and Montello, 1993). We found that 442 of the 1,336 stations we studied on these types of environments contained oil-residue thicknesses greater than 15 cm (~6 inches; in some places more than twice that deep). Figure 82 shows a 1992 photograph of oiled bubble sand on the Saudi Arabian Coast and an interpretation of the photograph. The occurrence of these somewhat sheltered sandy intertidal environments along the Arabian Gulf Coast provides the proper conditions for the formation of bubble sand, which allowed the oil to penetrate much more deeply than we had found in sand beaches at numerous other spills (where bubble sand was not present). When you visit the Georgia Coast, take a walk back into the more sheltered portions of the sand beaches and sandy tidal flats and notice how soft the sand becomes. Dig up a handful (or shovel full) of the soft sand and you will be able to observe the bubble sand for yourself. The photograph in Figure 81A is of a shovel full of bubble sand dug up on the upper beach at the south end of Tybee Island in March 2010.

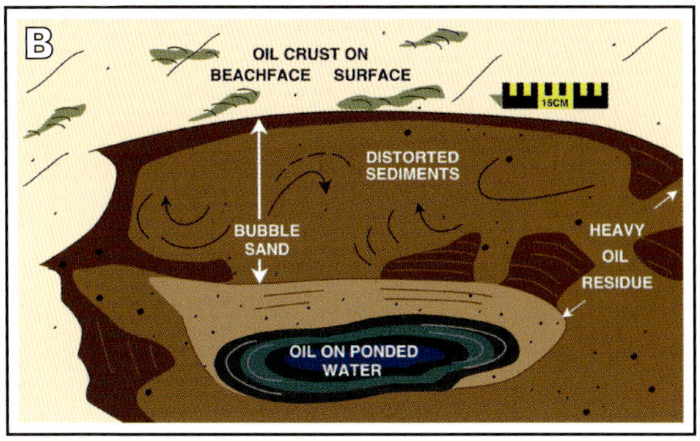

FIGURE 82. Example of bubble sand in sediments on a sheltered beach on the eastern shoreline of Saudi Arabia. This beach was heavily oiled during the 1991 *Gulf War* oil spill. The oil penetrated deeply into the bubble sand because of its high porosity and permeability. (A) Trench in bubble sand dug on 1 April 1992. (B) Interpretation of trench in A, showing the distorted sediments typical of bubble sand (compare with photograph in Figure 81B). Scale is ~6 inches.

COASTAL DUNES

Most of the prograding-type barrier islands in Georgia have well-defined lines of foredunes adjacent to the backbeach. Vegetation plays a very important role in the formation of these dune lines, which, over time, continue to build out seaward such that in places many lines of ancient foredunes perched on top of old beach sediments (called **beach ridges**) cover the surface of those islands that are in the prograding mode (see the example of prograding beach ridges on the north end of Ossabaw Island in Figure 34C).

Figure 83 shows the general model for the evolution of such prograding beach ridges on recurved spits, such as the one shown in Figure 30. A similar, but slower, process occurs in the middle of the prograding barrier islands. The process involves the following steps:

1) An intertidal bar welds to the backbeach and is molded into a depositional berm by wave action, as was described in the previous section on sand beaches. The crest and top of the berm are built above normal high tides.

2) During an ensuing spring tide, a line of vegetative matter called **wrack**, which is usually composed of stems of the low marsh plant smooth cordgrass (*Spartina alterniflora*) accumulates on top of the berm (Figure 84A). The wrack will later serve as a seed bed for beach grasses.

3) Small wind-blown sand dunes, called **wind-shadow**

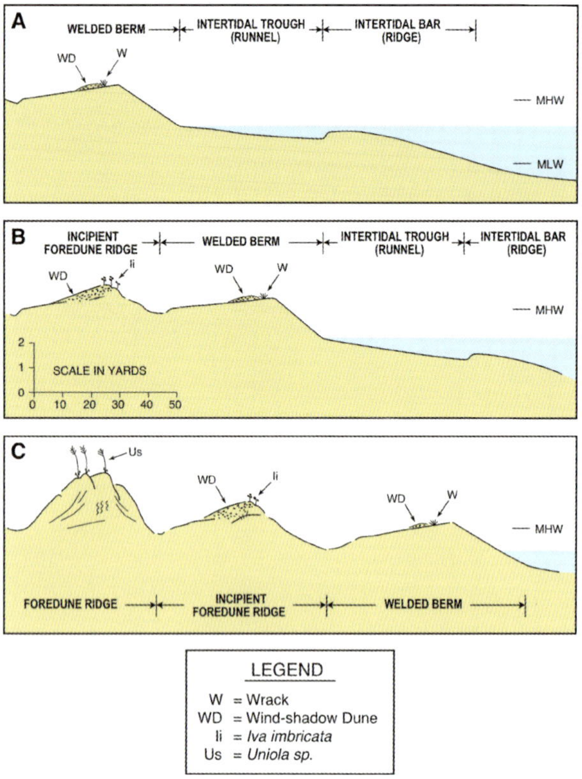

FIGURE 83. Model for evolution of foredune ridges on a prograding, recurved spit on barrier islands (see photograph of one such recurved spit on the Georgia Coast in Figure 30).

dunes, rapidly accumulate in the lee of the wrack as soon as the normal land-directed breezes start to blow (Figure 84B). Shortly thereafter, if it is the growing season, salt-tolerant beach grasses, such as sea elder, grow out of the wrack and become baffles to the wind action as well, creating even larger wind-shadow dunes (Figure 84C).

4) A new intertidal bar welds to the beach and a broad berm forms that provides a source for more wind-blown sand to nourish the original dune line, which will eventually achieve heights up to tens of feet. Sea oats populate the tops of these higher dune ridges (Figures 83C and 84D).

5) The process just described in steps 1-3 takes place on the new berm.

After approximately 15 years, wax myrtle (*Myrica cerifera*) will grow in the swales between the dunes. After approximately 50 years, a climax maritime forest, composed of three dominant tree species–live oak (*Quercus virginiana*), palmetto (*Saleal palmetto*), and loblolly pine (*Pinus taeda*)–will begin to populate the beach ridge (P. Hosier, pers. comm.; Hosier, 1975). These trees and shrubs will be pruned noticeably by salt spray as long as they remain on the front line of the forest. Through time, more beach ridges will accrete, and the original dune ridge will be located well inland of the beach, as is illustrated by the multiple beach ridges that make up the northeast end of Blackbeard Island (Figure 36B), the north end of Ossabaw Island (Figure 34C), and much of Wassaw Island (Figure 38).

FIGURE 84. Formation of the foredune ridge. (A) Wrack and incipient wind-shadow dunes on the top of a "stranded" intertidal bar near First Avenue on St. Simons Island (all photographs taken in March 2010). (B) Well-developed wind-shadow dunes near the same location as that shown in A. (C) Seacoast marsh elder growing along a line in front of a foredune ridge near the convention center on Jekyll Island. Small wind-shadow dunes are forming by the plants. (D) Top of the foredune ridge shown in C. The tall grass is sea oats. This ridge was initiated by the processes illustrated in A and B (and Figure 83), with the growth of vegetation on top of wrack like the kind shown in A and B, being an initial step. Through time, the sand continues to accumulate until a high dune ridge like the one shown in this photograph forms.

6 BEACH EROSION

INTRODUCTION

Erosion is a common phenomenon on many of the beaches of the world. In fact, almost no major coastal area that has been developed by man is free of problems caused by beach erosion. Beach erosion has been of concern for decades: Dean and Walton (1975) reported an average erosion rate of between 1-3 feet per year for the beaches in the United States, and the U.S. Army Corps of Engineers (National Shoreline Survey) in 1971 concluded that nearly 43% of the United States beaches were undergoing "significant" erosion.

Beach erosional and depositional trends on the Georgia Coast were determined early by coastal geologists Oertel and Chamberlain (1975), as they pointed out that a 78-year history of shoreline positions showed that Tybee and Little Tybee islands under went net shoreline retreats of 1.0 and 2.2 feet/year, respectively, during that period, and St. Catherines Island had a net shoreline retreat of 12.8 feet/year. More recent data on these islands and elsewhere are given in Griffin and Henry (1986).

Other islands, such as Wassaw and Ossabaw, showed net advances of the shore. Therefore, the trends of erosion/deposition along the Georgia Coast, which will be discussed later in this Section as well as in more detail in Section III, are relatively complex. Morton and Miller (2005), which will also be discussed in more detail later, reached conclusions quite similar to those expressed by Oertel and Chamberlain (1975).

Clark Alexander has been involved in several studies of beach erosion on the Georgia Coast in the past few years (Langley et al., 2003; Jackson, Alexander, and Bush, 2007; Alexander, 2008; Robinson et al., 2010; Turck and Alexander, 2012; and Jackson, Alexander, and Bush, 2012). Where appropriate, the information in these papers will be discussed in the treatment of the different coastal compartments in Section III. Some of these papers just cited utilize GIS (Geographic Information System) software to advantage. The authors' company, Research Planning, Inc., has been applying these most useful GIS techniques in their coastal mapping projects around the world since 1993.

In a paper published in 1985, Hayes made the following somewhat pompous assertion:

> *"What becomes evident in this discourse is the conclusion that man himself has caused most of his own beach erosion problems. That being the case, perhaps he can do something to solve them. Lessons learned from the mistakes of the past should be applied in the planning phases of every new development in the coastal zone. The costs involved in preventing erosion by wise planning are orders of magnitude less than those required to solve an erosion problem once the structures are in place."*

Now, in reviewing what has been happening in the 25 years since that pronouncement, one would have to conclude that either: 1) Not more than a dozen policy makers read it, or 2) If they did, they had a good laugh, ignored it, and went on about their business.

Getting back to the numbers, a later estimate by Bird (1996) concluded that 70% of the world's beaches are now eroding. Therefore, it should come as no surprise that at least some of the beaches in Georgia are eroding.

Because a significant amount of the Georgia Coast is locked up in wildlife refuges, state parks, and other types of natural areas, any erosion is not generally considered to be a major "problem" except for where turle-nesting beaches are being eroded. Therefore, most of the "problem" areas are in populated regions, where erosion is a result of several factors, to be discussed in Section III.

MAJOR CAUSES OF BEACH EROSION

Introduction

This discussion of the causes of beach erosion will be limited to depositional coasts like those in Georgia. The causes for erosion on leading edge and glaciated coasts are complex and an understanding of them is not relevant for the coastal area under consideration in this book. With that given somewhat limited focus, we can say once more that for any given depositional coastal area not unduly influenced by man, it is the interaction of sediment supply and water level changes that controls beach erosion. In some areas, fluctuations in sediment supply prevail, and in others, rapid changes in water level are most important. The most serious erosion problems occur either where man interferes with the sediment supply or where some phenomenon causes an unusually rapid rise in water level (e.g., abrupt lake level rise in Lake Michigan or around sinking abandoned delta lobes on the Mississippi Delta).

Deficits in Sand Supply

On the Georgia Coast, the sand on the beaches was ultimately derived from erosion of the Appalachian Mountains and brought to the sea by rivers. In many cases, much of the sand stopped somewhere along the way to be incorporated into ancient sedimentary rocks or sediments, some of which were eroded and carried to the sea during later geological episodes. Once in the general vicinity of the present shore, the sand on any particular beach could have been derived from one or a combination of the following sources:

1) **Carried directly to the shore by rivers** during the last few thousand years. With regard to sand beaches in the vicinity of the outlets of the piedmont, or red-water rivers (Savannah and Altamaha), some fraction of this sand came all the way down the rivers from the Piedmont or Mountain regions, but much of it could also have been eroded from rocks or sediments along the river within the Coastal Plain that range in age from Pleistocene all the way back to the Cretaceous (~100 million years ago). All of the sediment delivered to the shore by coastal plain, or black-water rivers, would have been eroded from local sedimentary deposits and not derived directly from the mountains. None of the coastal plain rivers have built deltas on the open ocean shore, thus their contribution to modern beaches at the present time is probably small. Orrin Pilkey (pers. comm.) noted that heavy mineral evidence supports this idea in that the sands of the present beaches have a piedmont river suite of minerals, not a coastal plain river suite.

[NOTE: In the original igneous rocks in the Blue Ridge Mountain Province, such as granites, a variety of minerals were weathered out of the rocks to eventually end up as sand grains. The most common minerals in granite, for example, are two silicate minerals, quartz (SiO_2) and orthoclase feldspar ($KAlSi_3O_8$). The basic building block of the silicate minerals is the **silica tetrahedron** that consists of a single silicon cation surrounded by four oxygen anions. This combination of silicon and oxygen makes up about 75% of the Earth's crust. Silicon cations, which have a charge of +4, are less than 1/3 the size of the oxygen anions, which have a charge of −2. As a result of these size differences, one silicon cation fits "snugly" between 4 oxygen anions, forming the tetrahedron. The tetrahedron is perfectly arranged geometrically, but the resultant ionic charge of −4 is far from balanced. To reduce the negative charge, the tetrahedra must bond with other positive ions, such as Mg^{2+} or link up with other tetrahedra in various ways, such as in sheets (e.g., mica), chains (e.g., pyroxenes), and so on. Because of its particular arrangement of bonding tetrahedra, called framework bonding, quartz is the most stable of all the silicate minerals. In framework bonding, each tetrahedron shares all of its oxygen atoms with adjacent tetrahedra that creates a very strong 3-dimensional framework of Si-O bonds. As a result, the charges are exactly balanced in quartz minerals and no other bonding of the ions is needed (Eggar, 2006). Consequently, quartz is considerably harder than feldspar, being much more resistant to abrasion processes. Also, feldspar is more susceptible to chemical weathering than quartz, because it may react with acids to produce residual clay minerals plus dissolved ions of metals (potassium, sodium, or calcium, depending on the type of feldspar). Being composed of only SiO_2, quartz does not contain metals that might be susceptible to leaving the ionic structure under certain weathering conditions. Therefore, if you pick up a handful of sand on a Georgia beach, the majority of the individual grains will be composed of quartz. However, some the sediments brought to the coast by the piedmont rivers have been subject to only one erosion cycle in the river or on the beach/dune complex of the last highstand of sea level during the present (Holocene) epoch. Therefore, beaches near the mouths of the two piedmont river systems in Georgia will contain some feldspar and perhaps a few grains composed of other minerals less-resistant to abrasion and chemical weathering than quartz, although these rivers also carry an abundance of recycled sand. On the other hand, the sediments delivered to the coast by the coastal plain rivers were subjected to abrasion in streams and on beaches and dunes, as well as to weathering in soil horizons, during the Pliocene, Miocene, or, possibly along even older sequences. Sediments deposited during the Pleistocene epoch, which were also subjected to some episodes of abrasion and chemical weathering, are another important local source along the coast in some situations (Howard and Frey, 1975).]

2) **Eroded from coastal deposits of different kinds to the northeast** of the site in question and moved along the shore as part of the longshore sediment transport system into the present site, as illustrated in Figures 28 and 31. These more northerly sites could have been older Pleistocene coastal sediments, active and abandoned delta lobes, eroding barrier islands, and so on.

3) **Transported from the continental shelf** to the present beach. The sand on the shelf was most often placed there as river deltas, lowstand valley fills, or other coastal deposits at times during the Pleistocene epoch when sea levels were lower than at present. The landward transport of this sediment to some of the active beaches is made obvious by the composition of the sand and by observations on the beach after storms, but the actual process of landward transport is poorly understood. A good example of an offshore sand source is an abandoned delta lobe offshore of Bull Island, South Carolina, which has provided a source for the sand on the three barrier islands immediately to the north of Charleston (Griffin, 1981; Hayes and Sexton, 1989). Also, Schmitter and Freeman-Lynde (1988) showed that sand can be transported from offshore to the beach in South Carolina and Georgia based on hornblende content of the sediment.

4) **A variety of processes at tidal inlets**. Studies by our group of tidal inlets in South Carolina show that erosional/depositional patterns of adjacent beaches are controlled closely by morphological changes of the inlets (e.g., Hayes et al., 1976; FitzGerald and Hayes, 1980). Oertel (1975b; 1977) made similar observations on the Georgia Coast, some of which will be discussed in Section III. One major way for deposition on a beach to occur is for large individual swash bars on the ebb-tidal deltas of tidal inlets to migrate onshore, as illustrated in Figures 36A, 36C, and 85. As the inlet changes, and the bars shift position, the location of the focused wave energy will change. This allows major shifts in deposition on the beaches on either side of the inlet over time (Kana, Hayter, and Work, 1999).

5) **Artificially added to the beach** by engineers in what is termed **beach nourishment**. According to Schmitt and Haines (2003), *a section of Tybee Island GA (18,000 feet) has been nourished multiple times since 1976 at a total cost of over 10 million Federal dollars.* Tybee is the only public beach in Georgia that has been renourished to date. Sea Island, a private beach, has a sand recycling program that will be discussed in more detail later. The sand for these projects can come from a variety of sources, with dredging of offshore sand sources being one of the most common.

To understand which of the four natural sources is most important at a given site, a sediment budget would have to be calculated in which the sediment additions (credits) and losses (debits) are determined and equated to the net gain or loss. A general sediment budget that covers all the beaches of the Georgia Bight has not been done. Clark Alexander (pers. comm.) stated that *there have been budgets done for all major inlets that the USACE maintains, so there must be sediment budgets for them. I know that one exists for the Savannah Harbor.* A partial one involving beach volume changes along the central South Carolina Coast was completed and published in 2001 by Kana and Gaudiano. A somewhat surprising conclusion of that study stated that *while conventional wisdom may suggest that all beaches are doomed to erosion, Charleston's barrier islands include a number that have gained sand for generations. Inlets, some with seemingly limitless sand reserves (compared to the volumes eroding from the beach), have replenished our coast.*

There are a number of natural processes that could produce a deficit of sand at a particular beach, including:

1) **Switching of the mouth of the main river** in a delta system. This is a common process on river-dominated deltas such as the Mississippi Delta, pictured in Figure 60A.

2) **A variety of processes at tidal inlets**. For example, waves refracting around the ebb-tidal deltas create a zone

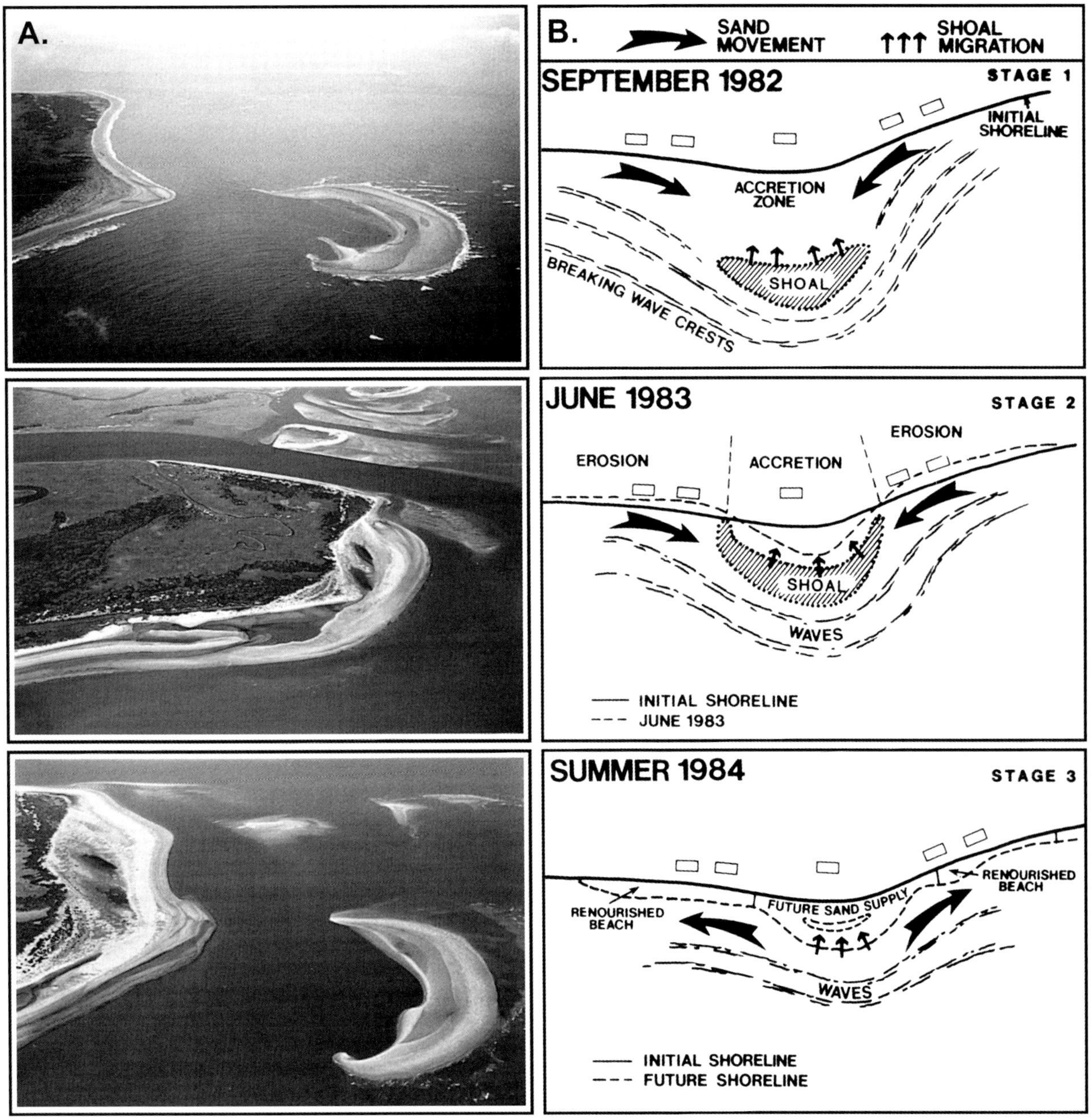

FIGURE 85. Shoreline changes associated with tidal inlets in the Georgia Bight. (A) Large swash bars detaching from the ebb-tidal delta at Stono Inlet, South Carolina and merging with the beach. Upper – Early stage of detachment in 1977. Middle – Attached to shore in 1983. Lower – A new swash bar detached in 1986, which attached to the beach in 1990. From Kana, Hayter, and Work (1999). (B) Illustration of the three stages of bar welding from the same article.

of deposition downdrift of the inlet, thus diminishing the flow of sand further down the beach (Figures 36A, 36C, and 85). Bars affiliated with the ebb-tidal delta also create local zones of erosion by focusing wave energy through wave refraction (see June 1983 diagram in Figure 85B). And, of course, as the inlet migrates, the shoreline downdrift is eroded away and the sand on the beach on the downdrift side of the inlet is carried offshore where it is stranded on the cbb-tidal delta for a period of time, which may be years in duration.

3) A number of natural processes that remove sand from the immediate vicinity of the beach zone, such as:

a) Transport of sand offshore during storms beyond the depth from which the sand is normally returned to the beach during calm periods; b) Wind transport of the sand landward out of the normal zone of the beach cycle; c) Lost down shore as part of the longshore sediment transport system; and d) Washed over the barrier island far into the backbarrier region during major storms.

There are also man-induced changes that could produce a deficit of sand at a particular beach, such as:

1) **Dams on rivers.** Sediment contribution of any river discharging on the coast is of importance to the sediment budget of that coast. If a dam is built on the river to create a reservoir for storage of water, the current velocities in the river are reduced to such an extent that practically all sandy sediment carried by the river settles in the reservoir. The water discharged from the dam contains very little sand; therefore, the sand supply formerly delivered to the coast by that river is cut off. This is the overwhelming cause of beach erosion on shores with narrow or absent coastal plains, such as Japan and California. Even in South Carolina, the construction of a dam on the Santee River in 1941/42, which created the Santee/Cooper lakes (famous for their largemouth and striped bass fishery), caused the beach to erode by hundreds of yards at the river's mouth after the dam was built (Hodge, 1981). There are three major dams on the Savannah River system, the first one near Augusta being built in the early 1950s that will be discussed in more detail in Section III.

2) **Diversion of rivers.** Some engineering projects call for the diversion of rivers away from their original channels. In addition to the construction of the dam across the Santee River in 1941/42, approximately 90% of the river flow was diverted into the Cooper River, which flows into Charleston Harbor. This assured that even most of the fine-grained sediment carried by the river was diverted from the river's mouth. This also contributed to some extent to the erosion of the delta front.

3) **Sand mining.** This is not a practice that is accepted on the beaches in Georgia, but it is a common practice in many parts of the world. The extraction of sand from sand bars in the nearshore zone, such as ebb-tidal deltas, for beach nourishment projects is another practice that could be called sand mining. If such sand is left in its natural configuration, some of it might eventually end up on the shore and its removal could expose the beach to erosion in places. Even mining of sand from shoals located further offshore can possibly influence beach erosion. Dredging of these shoals may possibly have two side effects: a) Alteration of depth contours that changes wave refraction patterns and may cause a focusing of wave energy at locations on the shore different from the original; and b) Elimination of a natural breakwater, which acts to reduce wave energy arriving at the shore. Clark Alexander (pers. comm.) noted that mining ebb-tidal deltas for beach nourishment is no longer allowed in Georgia, but *that is what they used in the first nourishment of Tybee Island in the mid 1970s.*

4) **Tidal inlet stabilization.** As a general rule, the ultimate geomorphic form of tidal inlets is the result of an adjustment to the dynamic action of both tidal currents and waves. Also, the volume and pathway of littoral drift (volume of sand moving along the shore) contributes to the configuration of a tidal inlet. From the studies of O'Brien (1931), it is well known that the stability of an inlet is governed by the relative volume of sediment in the longshore sediment transport system and the tidal prism volume (amount of water that flows in and out of the inlet during a single tidal cycle). Some of the sand moving along the shore bypasses inlets by bar migration under the influence of waves or by tidal current transport of individual sand grains; whereas some of it enters the main channel of the inlet. On coasts with a large volume of sand transport, there is a tendency for the inlet to migrate downdrift in the direction the sand is moving along the beach and nearshore zone due to the infilling of the updrift side of the main inlet channel (inlet throat or main ebb channel; Figure 50) that forces the inlet throat to migrate down the shore. Almost any man-made modification to a tidal inlet, therefore, may have significant influence on the erosional/depositional patterns on adjacent beaches. For example, artificial deepening of an inlet to develop a harbor could change the bar-bypassing nature of the inlet, depriving the downdrift coast of its supply of sand and thus leading to erosion.

5) **Construction of shore perpendicular structures.** **Jetties** are long parallel structures built at a river mouth or a tidal inlet to stabilize the channel, prevent its filling by sand, and protect its entrance from waves. The jetties at the mouth of the Savannah River on the South Carolina/Georgia border are pictured in Figure 86. Jetties are also designed to direct or confine the flow to help the channel's self-scouring capacity. Jetties

FIGURE 86. Jetties at the mouth of the Savannah River on the South Carolina/Georgia border. Photograph taken during a low spring tide on 28 April 2008.

usually extend through the entire nearshore zone to beyond the breaker zone to prevent the deposition of sand from the longshore sediment transport system in the main channel. Therefore, the jetties act as barriers to the longshore movement of sediments, causing a significant volume of the sediment to accumulate on the updrift side of the jetties. At the same time, on the downdrift side, the sand transport processes continue to operate and cause the sand to move away from the jetties, resulting in the erosion of the shore on that side. In some instances, the updrift jetty may have a weir system whereby sand is trapped in a confined area so it can be either hauled offshore or passed across the jetties to the downdrift side of the inlet.

Groins are another structure built perpendicular to the beach, in this case for the purpose of widening the beach by trapping a portion of the sand moving downdrift. They are relatively narrow in width and may vary in length from about 30 to several hundred feet (see examples in Figure 87). Though their function differs from jetties, they are also barriers to longshore sand transport. If they trap a significant portion of the sand in transport, naturally the beach will erode on the downdrift side. Groins may be used in series (groin fields) to protect a large area, but the zone of erosion will shift down the coast through time. Once the sand builds out to near the end of the groin, some or all of the sand in transit will pass on downdrift. In some cases, engineers place nourished sand on the updrift side of the groins to prevent erosion on their downdrift sides. Poorly designed groins may deflect sand past their ends into deeper water, resulting in its loss to the longshore sediment transport system. Kana et al. (2004) discussed the utility and management of groins on the South

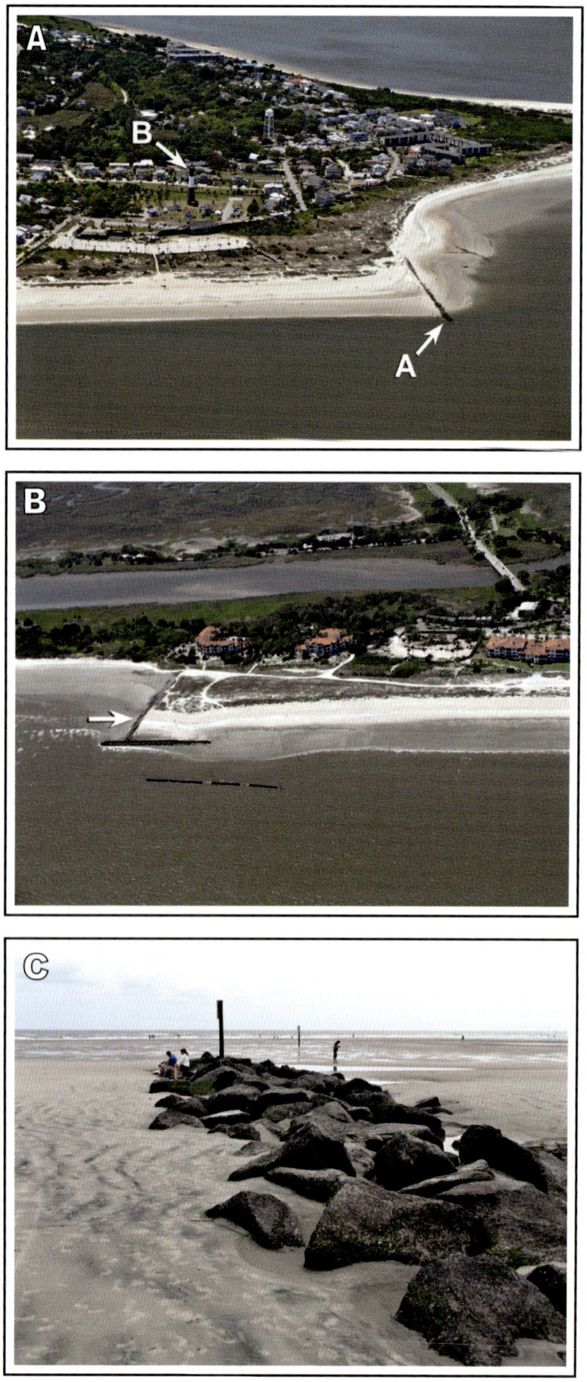

FIGURE 87. Groins on the Georgia Coast. (A) Terminal groin on the north end of Tybee Island (arrow A). Arrow B points to the lighthouse. (B) Terminal groin (arrow) on south end of the developed portion of Sea Island. (C) Groin on south end of Tybee Island on 28 March 2010. The photographs in A and B were taken during a low spring tide on 28 April 2010.

Carolina Coast. They made eight planning guidelines and recommendations, that included: 1) Only use groins on the open coast where the erosion rate exceeds about 6 feet per year; and 2) Construct the groins of a groin field in the downdrift to updrift direction, and nourish the "cells" between the groins "to capacity."

6) **Construction of seawalls and revetments.** Seawalls are vertical, hard structures commonly built to protect man-made structures. Thousands of examples could be cited to demonstrate the effectiveness of such structures in saving property, at least in the short term. However, erosion of the sand beach itself is usually accelerated in front of these structures because of wave reflection from the hard, vertical faces of the seawalls at high tide. According to Silvester (1977), when waves are obliquely reflected from such seawalls, energy is applied *doubly* to the sediment bed and *hence expedites the transmission of material downcoast.* This process is illustrated in Figure 88. Over the past few decades, there has been a hue and cry by a number of concerned scientists against building seawalls along open ocean beaches, to the extent that some states now ban them (e.g., North Carolina). An example of the remnant of a sea wall on the south end of Tybee Island is shown in Figure 89. Revetments usually serve the same purpose as sea walls, but they are typically made of materials such as boulder-sized chunks of rock, called riprap (Figure 90). These features do not reflect waves as severely as sea walls, but beaches seldom build out in front of them, for a variety of reasons, including wave reflection.

Sea-level Rise

When coastal geomorphologists like ourselves use the term sea-level rise, we normally insert the word **relative** as a modifier, because there are three possible factors that create major changes in sea level: 1) Changes in the volume of water in the world ocean; 2) Local changes in the elevation of the shoreline; and 3) Changes in the sizes of the major ocean basins. Changes in the volume of water in the world ocean are mostly related to major changes in the world's climate. For example, during the peak of the last ice age, around 20,000 years ago, the surface of the world ocean was about 400 feet below its present level, because so much of the Earth's water was taken up in the huge ice sheets that covered large areas of some of the continents.

A number of reasons can cause a land mass to rise

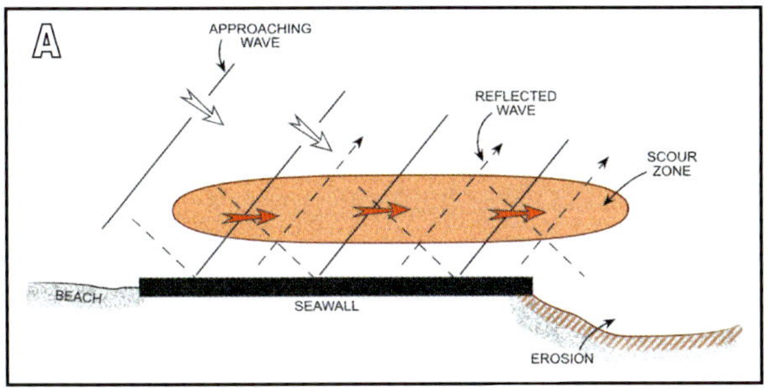

FIGURE 88. Seawalls and beach erosion. (A) Illustration of erosion in front of and on downdrift side of a seawall as a result of waves reflecting from the seawall (modified after Silvester, 1977). The approaching wave crests meet the reflecting wave crests at approximately right angles, generating a **flow parallel with the shore in the downdrift direction**. This current scours a channel a few yards seaward of the seawall. The beach downdrift of the seawall will also erode due to loss of sand in the scour zone, which would normally accrete on the beach in the absence of such a strong current in the nearshore area. (B) Waves reflecting off a seawall at Hampton Beach, New Hampshire. Solid arrow indicates approaching wave and dashed arrow indicates reflected wave moving away from the seawall at a 90-degree angle to the approaching wave. Photograph by A. D. Hartwell taken circa 1969.

or fall, including (to name but a few):

1) Tectonic activity such as mountain building. In areas of Alaska and Baja California, we have observed former beach lines elevated hundreds of feet up the sides of the mountains along the coast.

2) Elevation of the land as a result of the melting of the large ice sheets present during the Pleistocene epoch, because the Earth's surface was depressed under the heavy load of the ice. On Herman Melville Island in northern Canada, we observed numerous lines of beach deposits elevated many tens of feet back of the shore as a result of a process called **glacial rebound** as the land rose slowly in response to the removal of the weight of the ice.

FIGURE 89. Low-tide photograph of riprap revetment in the foreground that adjoins an intact seawall built circa 1939 with the high-tide line well landward of it. South end of Tybee Island on 28 March 2010.

3) Sinking of land where large rivers have dumped huge sediment loads on the continental shelf. This kind of sinking, aided by compaction of the sediments, has caused sea level to rise up to 4 feet per century in areas of Louisiana, where former lobes of the Mississippi River Delta were abandoned as the river shifted positions. Once the delta lobe was abandoned, the river no longer supplied sediment to continue the delta-building process, and as a consequence, the abandoned lobe began to sink.

On the coast of the Georgia Bight, which is located on the relatively stable trailing edge of the North American Plate, sea level is affected primarily by changes in the volume of water in the world ocean. Available data suggest the following historical rates of sea-level rise along the Bight since the end of the last Ice Age (partly from a summary by Pilkey and Pilkey-Jarvis, 2006):

- Between 15,000 and 6,000 years ago – Overall rate of **3 feet per century**, with possible "blips on the curve" of as much as 10 feet per century. Also, there were some fairly well documented stillstands during this rise, at which time river deltas and barrier island systems were deposited, the remnants of which are still present on the continental shelf in places.
- Between 6,000 and 4,500 years ago – Probably around **1.5 feet per century**.

FIGURE 90. Riprap revetments on East Beach St. Simons Island. (A) Near the north end. (B) Near south end looking south from public beach access on Ninth Street. Photographs A and B taken on 27 March 2010. (C) Oblique aerial view of riprap revetment (arrow) near where East Beach and the main body along the middle portion of St. Simons Island join. Photograph was taken during a low spring tide on 28 April 2010. The photograph in B was taken from the area just beyond the lower left end of this image.

- Between 4,500 years ago and the present – Although this time interval is normally referred to as the present "**stillstand**" of sea level, some scientists report that sea level may be continuing to show a general rise, possibly at an average rate of **1 foot per century**. It is during this period of time that all of the modern deltas and Holocene components of the barrier islands on the Georgia Coast have formed. Also, as shown by Figure 8, the curve showing sea-level changes during this period has not been a straight line, but has bumped up and down several times even during this "stillstand" episode, sometimes more than 3 feet. Having worked on the shore of the Georgia Bight for the past 35 years, it is clear to us from simple observation that sea level is rising. How much? We also think probably around one foot per century at the present time (verified by some NOAA data discussed later). Specifically for the Georgia Coast, data for a gauge at Fort Pulaski show a rise from 1933-2012 of just over 3 mm/year (Clark Alexander, pers. comm.), which is, by the way, right at one foot per century.

The theory that a rising sea level is the major cause of beach erosion worldwide prevails in both scientific and popular literature. In fact, a meeting held by a group of coastal geologists on the Georgia Coast in the early 1980s resulted in a publication that expressed this opinion, among others (Concerned Coastal Geologists, 1981). A careful analysis of beach erosion in a number of settings casts considerable doubt on this assumption, particularly with regard to problems of concern in the near future. In fact, we think that, in many cases, man's impact on sediment supplies and unwise construction practices cause the presently ongoing beach erosion problems. However, the opinions of the experts differ markedly on this topic.

Our friend and coastal engineering *guru*, Cyril J. Galvin, challenged the famous report by Concerned Coastal Geologists (1981), stating *sea-level change has negligible effect on shore erosion, compared to fluctuations in longshore sand transport rate.* On a site-specific basis, his assertion is demonstrably true. In Georgia, for example, beaches that are eroding tens of yards per year are adjacent to beaches accreting tens of yards per year. Such differences are usually related to sedimentation processes at major tidal inlets during which a sand deficit is created on one side of the inlet with increased rates of sediment buildup on the other side. Obviously, both the eroding and depositing beaches in these instances were subject to the same conditions of sea-level change. On the other hand, changes in water level clearly impact beach erosion in the Great Lakes, where 11-year cycles of major water-level changes occur (Hands, 1977). By making this statement, we do not mean to minimize the potential of the serious beach erosion that would take place as a result of a relatively sudden, major rise in sea level (discussed in more detail later).

Water-level changes not related to the aforementioned global patterns also affect beach erosion trends. On the coast at the head of the Georgia Bight, with its considerable tidal range, the most severe erosion on the beach occurs at high tide. During spring tides, higher levels of the beach are exposed to wave action than during neap tides, so erosion is at a maximum during spring tides under similar wave regimes. Consequently, coastal storms usually cause maximum rates of beach erosion when they cross the coast during high spring tides.

Water-level changes also show seasonal variations. On the U.S. east coast, sea level is generally lowest in the spring and highest in the fall. This fact is well known by east coast developers who have built too close to the beach, as the high tides in the fall commonly cause them problems. According to Komar (1976), these annual sea-level changes can be attributed primarily to seasonal variations in climate and ocean water properties. However, much is yet to be learned about this process.

Why the issue of **global warming** (referred to more commonly as **climate change** these days) and the potential of this temperature increase to cause sea level to continue to rise in the future is such a controversial political issue is a complete mystery to us, because the evidence is indisputable that: a) Global warming is occurring, based on evidence such as melting glaciers (which we have witnessed first hand in Alaska) and actual temperature measurements; and b) The concentration of CO_2 in the atmosphere is steadily increasing, based on studies such as at the Mauna Loa Observatory in Hawaii over the past 40 years. This issue was discussed in some detail earlier. Please refer again to Figure 15. The relationship of these two processes (CO_2 increase and sea level rise) is treated in detail in Al Gore's book, *An Inconvenient Truth.*

As a matter of general information, a calculation by Bindoff et al. (2007) related the potential rise in sea level to the following factors: 1) Thermal expansion of ocean waters as they increase in temperature (57%); 2) Melting of glaciers and ice caps (28%); 3) Greenland Ice Sheet melting (7.5%); and 4) Antarctica Ice Sheet melting (7.5%).

Pilkey and Pilkey-Jarvis (2006) also discussed this topic in an informative way. They summarized the issue succinctly in the following statement:

> *The evidence for the human connection to global warming is the correspondence of the massive production of excess carbon dioxide produced by burning fossil fuel in the last few decades and the simultaneous atmospheric increase in carbon dioxide. On a purely physical basis, the additional atmospheric CO_2 requires that some greenhouse warming must occur, but how much remains a question.*

So what does the future look like with respect to sea level on the Georgia Coast? Pilkey and Pilkey-Jarvis (2006) have the following assessment that relates to that issue as well:

> *The current most widely accepted prediction of sea-level rise is that its rate will be two to four times the present rate by the year 2100, and at that time the sea level will be two to three feet above its present state. At the same time, atmospheric temperatures will rise four to five degrees Fahrenheit. But these are numbers with a lot of leeway.*

During a recent project involved with the planning for building two new cities along the coast of Abu Dhabi, Hayes recorded the following information regarding potential sea-level rise within the next 100 years (as part of that project):

1) Using satellite altimetry of the ocean surface, which avoids the issue of rising or lowering land masses, a NOAA group observed that sea level rose at the rate of 3.2 mm/year between 1993 and 2008. This is at a rate of about 1 foot/century.

2) The International Panel for Climate Change (IPCC) in a report issued in 2007 gave the following two predictions for sea level in 2100 (as well as several others): (A) +48 cm (1.58 feet) assuming very rapid economic growth, with a global population that peaks in mid century and there is a rapid introduction of new and more efficient technologies, at a balance across all sources; and (B) +50 cm (1.67 feet) assuming conditions similar to B, but the use of fossil fuels continues to be intensive.

3) Stefan Rhamstorf, a German oceanographer and climatologist, predicted a rise of 100 cm (3.3 feet) by 2100, assuming a warming by 2°C (in a report in Science News published on 30 September 2009). He stated further that beyond a 2°C global average temperature rise, the probability of the Greenland ice sheet will disintegrate is 50 percent or more. If that happens, the result will be about 7 meters (23 feet) in sea-level rise in the time frame of about 300-1,000 years.

4) Pfeffer, Harper, and O'Neel (2008) concluded that, on the basis of calculations presented in their paper, an improved estimate of the range of sea-level rise to 2100, including increased ice dynamics (melting?), lies between 0.8 and 2.0 meters (2.6 to 6.6 feet).

Based on all that, it would seem that a rational level to anticipate for the Georgia Coast in the year 2100 would be somewhere around 2.5 to 3.5 feet above the present mean sea level. This, of course, does not take storm surges into account.

Clearly this is a topic of much concern these days. A 2008 report issued by the National Wildlife Federation on global warming in Georgia made the following statement regarding the potential sea level in the year 2100:

> *Scientists project that sea level along Georgia's coast could rise up to 25 inches in some areas, contributing to coastal erosion and loss of wetlands, particularly where sea walls and other coastal development reduce the availability of wetlands to migrate inland.*

That prediction of sea-level rise is slightly less than the one we just gave. The problem, as with a lot of models used to predict change in nature, is the suppositions, or data input, used to create the model. In the earlier models, the data input related to rising sea level was the melting of glaciers, which is well documented as noted above, and the physical expansion of water as it warms. They did not take into account large-scale melting of the big ice fields in Greenland and Antarctica, which is pretty scary when you consider the following statement in a report by an international panel sponsored by the United Nations issued in 2007:

> *In 2002, Antarctica's 1,255 square-mile Larsen B ice shelf broke off and disappeared in just 35 days. And recent NASA data shows that Greenland is losing 53 cubic miles of ice each year – twice the rate it was losing in 1996.*

That type of major melting was addressed as one of the possible doomsday scenarios in Al Gore's book, and was also alluded to in the earlier comment by Rhamstorf.

All of that being said, there is no doubt that an abrupt rise in sea level of two or three feet would wipe out an unimaginable amount of beach homes, hotels, golf courses, and other structures that are now built close to the beach along the Georgia Coast. Question is, what can be done to prevent such a rise? The answer to that question is way beyond the scope of this book.

BEACH EROSION IN GEORGIA

A U.S. Geological Survey open-file report by Morton and Miller (2005) summarized the status of beach erosion in Georgia, making the following assessment:

- Coastal land loss is caused primarily by *erosion of the Atlantic Ocean beaches.*
- Open lagoons are not present landward of the barrier islands and estuaries are small, so *erosion of backbarrier shores and interior marshes is not a significant component of coastal land loss* in Georgia as it is in some other states. Clark Alexander (pers. comm.) disputed this statement, noting that Morton and Miller *only examined the oceanfront.* In a study of the backbarrier shoreline of Cumberland Island, Jackson, Alexander, and Bush (2007) described several areas where archaeological sites are threatened by erosion. Some specific details provided by the Alexander group for different parts of the Georgia Coast will be presented in Section III.
- *Coastal submergence is minor in Georgia* compared to some other states.
- An analysis of long-term historical shoreline change indicates that *65% of the Georgia shore was stable or accreting, with an overall average shoreline change rate of 1.0 +/- 2.7 m/yr* (3.3 +/- 8.9 feet/year).
- The eroding portion of the coast had *an average erosion rate of -1.5 m/yr* (-5 feet/year). *Highest long-term erosion rates (-9.4 +/- 4.0 m/yr;* -31 +/- 13 feet/year*) occurred along Wolf Island.*
- Historical changes in the shore's position were highly variable, with *eroding and accreting segments located adjacent to one another* in places.
- Erosion persists along segments of *Tybee, St. Catherines, Wolf, north St. Simons, north Jekyll, and central Cumberland Islands. Many beach segments*

reported by Griffin and Henry (1984) to be experiencing long-term erosion continued to erode, except for the areas influenced by beach nourishment.

METHODS COMMONLY USED TO PREVENT BEACH EROSION

Engineers have attempted to curtail beach erosion for centuries. Their success ratio has been variable, depending upon the vagaries of the sediment supply, changes in water levels, and storm-wave conditions. For purposes of discussion, we divide the techniques used to prevent beach erosion into two classes: "hard" engineering solutions and "soft" engineering solutions.

In the past, engineers have often dealt with beach erosion by building resistant, permanent features that reflect or dissipate incoming waves. Some of these "hard" solutions include:

Seawalls, revetments, and bulkheads. In places where fixed property, such as highways and large hotels, are threatened by erosion, these three types of features are commonly built. As noted earlier, seawalls, massive concrete structures designed to hold the line against storm-wave erosion (illustrated in Figure 88), are a double-edged sword, so to speak, because, while they tend to succeed in keeping the road or hotel in place, at least until a major hurricane hits, the beach sand itself is usually sacrificed, even under normal wave conditions. A bulkhead is made of pilings, composed of a wide range of materials, driven into the ground. They are usually built in areas of moderate waves. Revetments are constructed by armoring the slope or face of a dune or bluff with one or more layers of rock, concrete, or asphalt. The armor stones of revetments tend to dissipate waves and inhibit reflection better than vertical walls, thus sand removal by wave reflection is not as severe as it is for seawalls. Although seawalls have been built in Georgia since 1942 (Clark Alexander, pers. comm.), riprap revetments appear to be the favorite shore protection structure on the Georgia Coast, with extensive segments being present on Sea Island, St. Simons Island, and Jekyll Islands (see Figures 90 and 91). These types of features, especially seawalls, are severely scrutinized these days by public officials in many states. In fact, in nearby South Carolina, the Beachfront Management Act of 1990 prohibited new seawalls. However, a settlement reached on 9 March 2004 involving some landowners on Daufuskie Island (a few miles north of the Georgia border) appears to have allowed some to remain in place.

FIGURE 91. The north-central beach on Jekyll Island during a low spring tide on 28 April 2010. The riprap revetment (arrows), clearly shown in this photograph, extends along the back of the beach for approximately 3.2 miles along this part of the island.

Groins. These features, which are discussed in some detail in the section on causes of beach erosion (illustrated in Figure 87), are commonly made of rubble stone, but they may consist of wood, sand bags, gabions (rocks or gravel in wire mesh), or other materials. They work best where waves approach the coast obliquely. Whereas they work well if installed properly in the right place, groins have not proved to be an effective solution to beach erosion in many localities, especially where the dominant wave crests approach parallel to the beach. Groins are presently used for shore protection on Tybee and Sea Island, most commonly as terminal groins (see Figures 87A and 87B). The individual terminal groins along the coast, some of which appear to be working quite well, will be discussed in detail in Section III.

Offshore breakwaters. These features are usually composed of riprap or heavy concrete blocks of miscellaneous shapes that are placed offshore, detached from the beach. They reduce wave energy on their landward sides, which causes sand moving along shore in the longshore transport system to accumulate in their lee. Such an obstacle to the movement of the sand along shore will commonly cause erosion on their downdrift side. Thus, many engineers recommend placing nourished sand in the shelter of these structures so that sand can continue moving along the shore. Offshore breakwaters have been used successfully to curtail erosion along a number of coasts. There is only one offshore breakwater in Georgia. It was built just off the southern terminal groin on Sea Island in 1997 as part of a beach nourishment project on that island that is discussed in some detail in Section III (visible in the oblique aerial photographs in Figures 26B and 87B) (George Oertel, pers. comm.).

There are many workers in the area of coastal erosion, particularly coastal geologists, who prefer to use solutions to beach erosion that do not involve hard structures. Two of the lines of reasoning used to support this position are that hard structures, such as seawalls, accelerate sand loss, and once in place, hard structures are difficult to remove, making it virtually impossible to correct a mistake. Some examples of these types of "soft" solutions include:

1) **Setback lines.** The "softest" of the soft solutions is the construction of a setback line behind the shoreline seaward of which building is prohibited. They work best and are easiest to implement in areas prior to development. The criteria used to establish such lines include historical analysis of shoreline trends, preservation of the line of foredunes, defining areas of flooding and storm wave uprush, and so forth. In 1974, Hayes and his associates mapped out setback lines for the developers of Kiawah Island, South Carolina, estimating a 50-year lifespan for the security of structures built behind them. The developers of that island have for the most part adhered to those setback lines and have not lost any significant coastal structures through beach erosion at the time of this writing, 37 years later. George Oertel (pers. comm.) made the following comment relative to setback lines in Georgia: *In Georgia, setback lines are not based on anything this rational. Instead, the GA DNR jurisdiction line is based on the occurrence of the first 20-ft-high native tree.*

2) **Sand bypassing.** As noted earlier, jetties, such as the ones illustrated in Figure 86, constructed to stabilize navigational channels usually block the flow of sand along the shore, with severe erosion commonly developing downdrift of the jetties. This is one of the more common major causes of erosion problems around the world. This type of erosion problem may be at least partly solved by installing mechanical sand bypassing systems, such as land-based dredging plants, pumping systems, and so on, which move the sand from the updrift to the downdrift sides of the jetties. Such systems have been established in many areas, with several examples on the Florida Coast. In Georgia, jetties exist at the mouth of the Savannah River on the South Carolina/Georgia border and at the St. Marys Entrance on the Georgia/Florida border, but neither of them has a sand bypassing system.

3) **Relocating a tidal inlet.** Migrating tidal inlets erode the shore as they move down the coast. Relocating the inlet back up the coast in the direction from which it came would relieve the down shore area of erosion, at least until the inlet migrates back. To our knowledge, one of the major examples, and maybe the only one up to that time, of this was carried out by our group in March 1983 at Captain Sams Inlet on the South Carolina Coast, when the inlet was moved to a new dredged channel approximately 4,000 feet up the coast (to the northeast). This stopped the erosion of the inlet into a development to the south, as well as provided a huge volume of sand to a formerly eroding beach to the south, where the beach built out over a thousand feet when the abandoned ebb-tidal delta of the inlet was driven ashore by wave action. Compared to establishing hard structures or jetties, this was a relatively inexpensive process. The downside is that it has to be repeated about every 14 years or so, but it would take many repeats to reach the costs of other types of protection. Mason Inlet on the North Carolina Coast was relocated in early 2002 (Jenkins, 2010). According to Queram (2012), this inlet, located between Figure Eight Island and Wrightsville Beach, *was relocated about 3,000 feet to the north to rescue Shell Island Resort and other properties in the path of the waterway's decade-long southward march. Ten years after the Mason Inlet move, both supporters of the project and those who were initially wary of the shift say that the move is an unqualified success.* Clark Alexander (pers. comm.) noted that such a project was proposed for the Sea Island spit, but it could not be done because it was thought that piping plover habitat would be threatened by the project.

4) **Beach nourishment.** Beach nourishment with sand is the most commonly used of the "soft" solutions. Numerous projects of this type have been carried out around the world, most of which involve moving millions of cubic yards of sand. The sources of sand for such projects may include: 1) Dredging sand deposits on the continental shelf; 2) Dredging sand bodies associated with tidal inlets; 3) Moving sand by various mechanisms from adjacent beaches with an abundance of sand; and 4) Hauled or pumped in from land-based sources. The major objection to beach nourishment schemes is that they are not permanent. Watching millions of dollars worth of sand wash away within a few years does not appeal to public servants or even casual observers. Therefore, such projects require a careful analysis of monetary costs and benefits balanced against the aesthetic, economic, and recreational value of maintaining a sand beach in place. Some biologists object to this process, because of the potential for it to alter the character of the substrate (sand) on the beach and in the shallow nearshore zone. For example, adding coarser-grained sand and shell to a beach that is naturally finer grained would change the overall nature of the substrate and the population of animals that live in it or utilize it, including the invertebrates, birds, fish, and even sea turtles (Peterson, 2000). Taking this possibility into account, Schmitt and Haines (2003) recommended that the Georgia Department of Natural Resources undertake several measures in reference to potential nourishment projects on Tybee, Sea, St. Simons, and Jekyll Islands. This would include assessing the beach ecology *before, during and at regular intervals after nourishment.* Doing this would allow an analysis of *the short and long term effects on the biota to determine if, and how, the nourishment project is negatively affecting the ecology.* They also pointed out that this knowledge could *be used to guide future nourishment projects.* The specific projects carried out on the Georgia Coast will be covered as those beaches are discussed in detail in Section III. For example, with respect to Tybee Island, at least six major nourishment projects have put over five million cubic yards of sand on the beaches of this small island. The first such project was carried out in the mid 1970s and the last one late in 2008.

The material presented up to this point is a far from complete list of the numerous methods that have been tried to stop beach erosion. But as you may have guessed by now, this issue goes beyond a simple engineering or geological problem, having blossomed into a socio-economic, political, and even cultural phenomenon. The fact that a large percentage of the world's population lives near the coast amplifies interest in the subject.

Over simplifying the story quite a bit, one could say that the scientists and engineers who are supposed to know something about beach erosion fall into two major camps. On the one hand, we have environmentally oriented marine scientists, particularly coastal geologists, who are more inclined to let nature take its course and not crowd the ever narrowing beaches with beach cottages, hotels, etc. They usually think most beach erosion is caused by rising sea level or bad engineering practices and are more or less convinced that the problem will get worse with an increase in global warming. One of the leading spokespersons of this faction is another one of our friends, Dr. Orrin H. Pilkey, emeritus Professor at Duke University, who has been known to use catch phrases such as "NewJerseyization of the coast" and "let the lighthouses fall into the sea." He has also authored and edited many books on living with the coast. He is not even much of a fan of some "soft" solutions, such as beach nourishment, heavily criticizing some of the models engineers use to predict the life of the nourished sand on the shore. Needless to say, Dr. Pilkey is not idolized by many coastal engineers. However, some of his arguments have been heeded by policy makers in different states of the U.S. (e.g., South Carolina and North Carolina's ban of seawalls on the open beach). If sea level does rise abruptly, many of Pilkey's arguments will be awfully difficult to refute.

On the other hand, there are the coastal engineers, who are trying to perform a very difficult job. Our friend, Bill Baird, a world-renowned coastal engineer, has founded a company based in Canada (Baird and Associates) that has the following motto – "Making the world a better place to be!" His company does that by designing ports in developing countries to fuel their economy, preventing coastal villages in Africa from falling into the sea, and so on.

Where do we stand on this issue? To put it simply, our hearts are with the notion that the beaches should be left free of so much development, but our heads love the science of dealing with the problem. There are, of course, some classic cases of mismanagement of the beaches along the coast of the Georgia Bight. We have owned and still own coastal property in South Carolina, but never on the open beach. We were once offered a lot on one of the barrier islands as "payment in kind," but didn't take it because of concerns about erosion during hurricanes. However, when faced with a tricky scientific problem to solve, such as moving a tidal inlet or keeping an African village from falling into the sea, our heads win and we enthusiastically tackle the problem. In fact, we feel pretty good about some of the projects we have done around the world.

As already mentioned, **artificial beach nourishment** is one of those coastal engineering practices burdened with controversy. In our naturalist's guide to the South Carolina Coast (Hayes and Michel, 2008), we presented a previously published debate in a business magazine on projects of this type, specifically those carried out at Myrtle Beach, South Carolina. The two main debaters were Dr. Orrin Pilkey, already introduced, and Dr. Tim Kana, who did his dissertation at the University of South Carolina under Hayes' supervision.

In 1976, Hayes and Kana co-authored a book, *Terrigenous Clastic Depositional Environments,* which presented some of the earlier data on coastal processes and sedimentation. Also, before he formed his present company, Tim was our partner at RPI for several years. Since that time, he has recommended and supervised numerous beach nourishment projects. As professional geologists and members of the same professional societies, as well as through University contacts, we have had many interactions with Orrin Pilkey over the years.

Back to the debate, two excerpts from it that relate specifically to the recent beach nourishment projects at Myrtle Beach, South Carolina follow (this "debate" took place in the fall of 2007):

Pro (Kana) – *Their lots didn't wash away, their seawalls didn't fail, and now they look out over a band of dunes to a white, sandy beach that accommodates thousands of visitors each day.*

Con (Pilkey) - *Beach nourishment encourages high-density development immediately adjacent to the beach. In a time of rising sea level in a location where big storms are a certainty, beachfront construction is at least irresponsible and at worst can only be considered a form of societal madness.*

Kana and Pilkey, who both have an abiding love and respect for the natural environment, obviously approach the subject of beach nourishment in a different way. Kana has a practical approach, which we certainly sympathize with, and he strives to provide practical solutions to problems presented by his clients whenever possible. When RPI followed up on a suggestion our group made to relocate Captain Sams Inlet, South Carolina in 1983 to provide a major, never-before-tried method of beach nourishment, Tim Kana was the on-the-ground supervisor of the project. He also continued to monitor the project in later years and has published several papers on the results (e.g., Kana and McKee, 2003).

Obviously, Pilkey is different, preferring a "let it be" philosophy. He has an agenda, which we also sympathize with, and he sticks to it, much to the chagrin of many coastal engineers who question the authenticity of some of his data and suppositions (as noted earlier). The clearest exposition of his philosophy we can find in his writings appears in the final chapter of his artfully illustrated and informative book, *A Celebration of Barrier Islands,* published in 2003. In this book, he suggested that barrier islands lead a Gaia-like existence. As you probably know, the **Gaia hypothesis** is an ecological theory that proposes that the planet Earth functions as a single living organism (Lovelock, 1972). Beliefs somewhat similar to this go back at least to the great Baruch Spinoza (1677), whose philosophy was cited, but not necessarily believed, by Einstein. Actually, this concept goes back thousands of years (under different names, pantheism being one of them), but you probably get the point.

To quote Pilkey's summary of this matter exactly, he said:

> *Putting barrier islands in a Gaia context does provide a basis for thinking about how to live with them in a way that will keep islands existing and evolving, a way to preserve them for future generations. Perhaps more important, the distinction between life and death, as it applies to a barrier island, should provide guidance to distinguish good and bad development practices.*

Strangely enough, in Hayes' book entitled *Black Tides,* published by The University of Texas Press in 1999, a sentiment similar to Pilkey's was expressed. In that book, which deals with our experiences responding to most of the major oil spills that had occurred since 1974, reference was made in the last chapter to a short article entitled "Ensoulment of Nature" by Gregory A. Cajete. That paper was published in the book *Native Heritage* in 1995. Cajete stated that the importance the American Indians put on connecting with their place of origin (i.e., the environment) is not just a romantic notion out of step with the times, but rather *the quintessential ecological mandate of our times.* He stated further that the *native peoples experienced nature as a part of themselves and themselves as a part of nature.* Under this concept, **ensoulment** means that the human participates with the Earth as if it were a living being. Hayes concluded this subject with the following comment:

> *If we all adopted this frame of mind, I'm sure it would be one more step in the honorable direction toward making the Earth a better place to be, which is something all of us tree-huggers and right-thinking engineers want to do, correct?*

That final chapter in *Black Tides* concluded with a quote from the Bureau of American Ethnology Collection (published in Brown, 1970), which follows:

> *The old men*
> *say*
> *the Earth only*
> *endures.*
> *You spoke*
> *truly.*
> *You are right.*
> *The Earth only endures.*

Of course, the animals and plants that lived on the Earth have not fared so well at times. An example of those hard times being when the Earth was struck by a large asteroid or comet, such as the one proposed to have caused the Cretaceous/Tertiary extinction event about 65 million years ago. But, sure enough, the Earth itself has endured, so far.

[NOTE: One hypothesis for this extinction event was given as follows by apt (2012): *For months, dense clouds of dust blocked the Sun's rays, darkening and chilling Earth to deadly levels for most plants, and in turn, many animals. Then, when the dust finally settled, greenhouse gases created by the impact caused temperatures to skyrocket above pre-impact levels.*

In just a few years, according to this hypothesis, these frigid and sweltering climatic extremes caused the extinction of not just the dinosaurs, but of up to 70 percent of all plants and animals living at that time.

This is one hypothesis, there are several others. However, there is good evidence that a large asteroid did hit the Earth just off the coast of the Yucatan Peninsula around that time.]

7 ESTUARINE AND BACKBARRIER HABITATS

GENERAL MORPHOLOGICAL MODEL

Figure 92A shows a very general morphological model for the environments in the backbarrier and sheltered estuarine regions in Georgia. There are numerous variations and exceptions to this general picture, particularly for those areas located in the zone of the turbidity maximum (see Figure 57). In many estuaries, however, these areas can be subdivided into three distinct environments: 1) *Salt marsh* (from somewhere between mean sea level and neap high tide to the highest spring tide level); 2) *Tidal flat* (spring low tide to somewhere between mean sea level and neap high tide); and 3) *Tidal channel* (low spring tide level to the bottom of the channel). The marsh is commonly divided into low marsh and high marsh (Figures 58 and 92B), and we could similarly divide the tidal flat into low and high segments. Generally speaking, the lower part of the tidal flat is subject to at least a modicum of turbulent water motion generated mostly by tidal currents and occasionally by waves under storm conditions. That part of the flat typically contains some sand exhibiting evidence of sediment transport (e.g., ripples). The high flat contains finer-grained sediment that is riddled with the burrows of multitudes of infaunal organisms, particularly polychaetes. Also, mud snails occur in abundance over the surface of the flats, leaving conspicuous feeding trails. The uppermost part of the flat (still below neap high tide) typically contains numerous oyster mounds (*Crassostrea virginica*). As mud accumulates around the oyster mounds, salt marshes may grow over the mud.

MARSHES

Introduction

The extensive marshes in the estuaries/sounds of the Georgia Coast, illustrated in Figure 58 (Edwards and Frey, 1977), are differentiated on the basis of the salinity of the waters and soils they inhabit–salt, brackish, and freshwater marsh systems. These marshes are well documented as being the primary food sources for the coastal and nearshore ecosystems of the region. Marshes, like bottomland swamps, are flooded with water on a regular basis. The distinction between the two wetland types is that herbaceous plants populate marshes and woody plants populate swamps.

Marshes in estuarine systems occur in areas that were originally tidal flats where sediment accumulated in the early stages of evolution of the system. As the flat was built up to an elevation of around mean sea level to high neap tide, marsh grasses began to take root. Once the grasses grew on the flat, the sedimentation process was accelerated because of the baffling effect of the plants. These present marshes are, in effect, intertidal flats vegetated with **halophytes** (plants that are adapted to grow in salty soil; Basan and Frey, 1977). Marshes can expand very rapidly, up to a yard or so a year, if the slope is flat and sediments are abundant.

The typical estuary in the upper reaches along the Georgia Coast can be divided into upper, middle, and lower (entrance) zones (Figure 57). Each of these zones is characterized by distinct hydrodynamic and sediment characteristics. **Freshwater marshes** are most common in the upper estuary, where there are tides, but very low salinities; **brackish marshes** occur most commonly in the middle estuary, where salinities generally average less than 15 parts per thousand of salt (ppt); and **salt marshes** occur in the lower estuary (entrance), where salinities range from 15 to the low 30s ppt. It should be kept in mind, however, that the estuaries in Georgia show a lot of variability, depending upon how much freshwater is delivered to the estuary by rivers. The more saline-tolerant plants occur much further inland in the estuaries with little freshwater input.

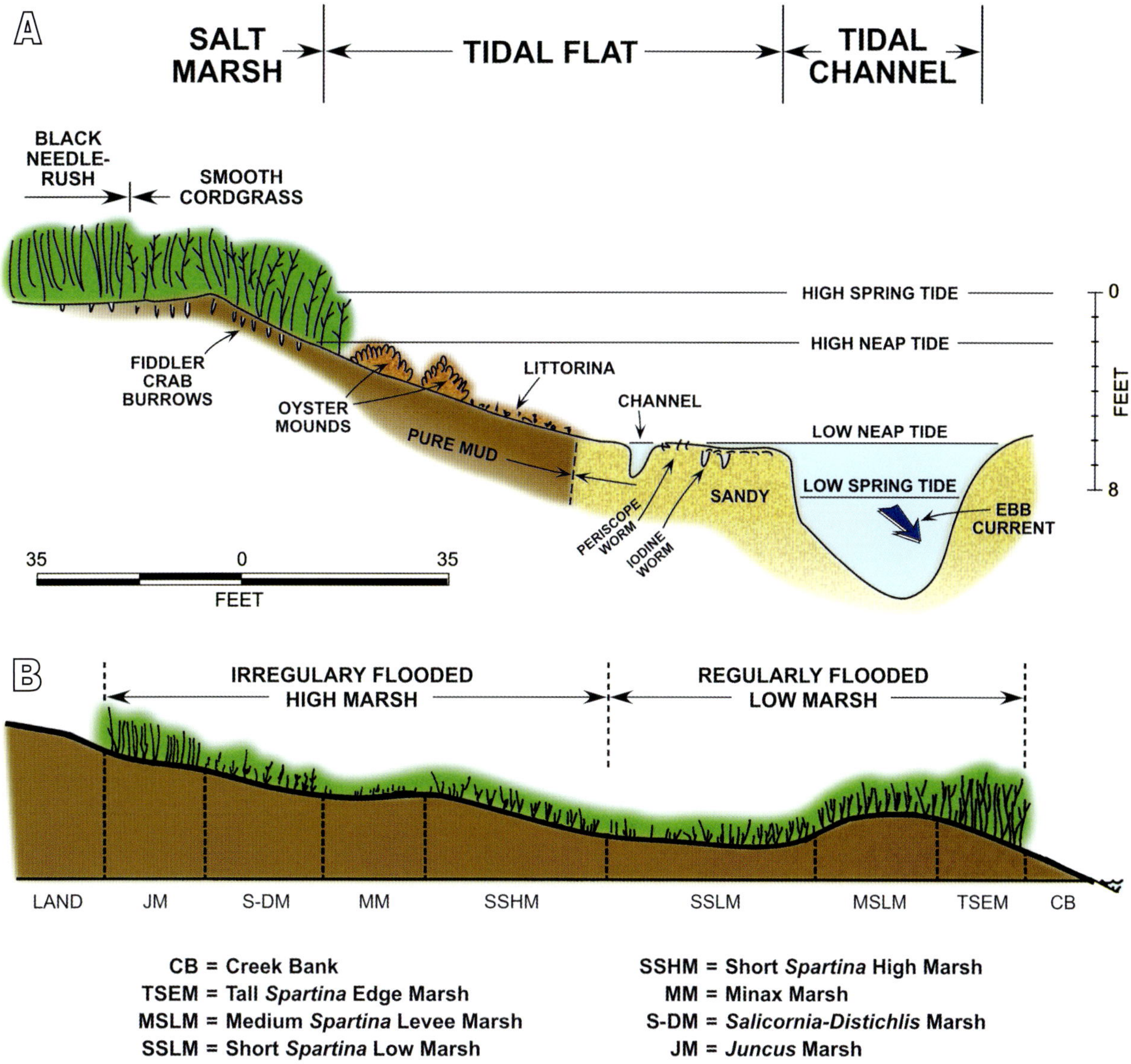

FIGURE 92. Backbarrier habitats. (A) Typical topographic profile of tidal channel and associated habitats of the backbarrier and lower estuary regions along the Georgia Coast. (B) Typical profile of the salt marshes of the Georgia Coast (modified after Teal, 1958). The "Minax marsh" zone is named after the dominant species of fiddler crab found there, *Uca minax*. The dominant plant is extremely short *Spartina alterniflora.* Compare this diagram with the one in Figure 58 for more details.

These lateral salinity changes up and down the estuary have a striking impact on the plant communities. Giant cordgrass (*Spartina cynosuroides*) is conspicuous along the banks of the channels in both the upper and middle estuaries. Black needlerush (*Juncus reomerianus*) is by far the most common plant in the middle estuary, covering many thousands of acres in some places, and smooth cordgrass (*Spartina alterniflora)* typically dominates the lower estuary. See Figure 93A for a delineation of the distribution of the more conspicuous plants throughout the estuaries in Georgia and Figure 93B for sketches of some of the dominant types. This discussion applies only to those estuaries that have a significant amount of fresh water coming into them. Exceptions occur where freshwater input is only minimum or at the mouth of the major rivers.

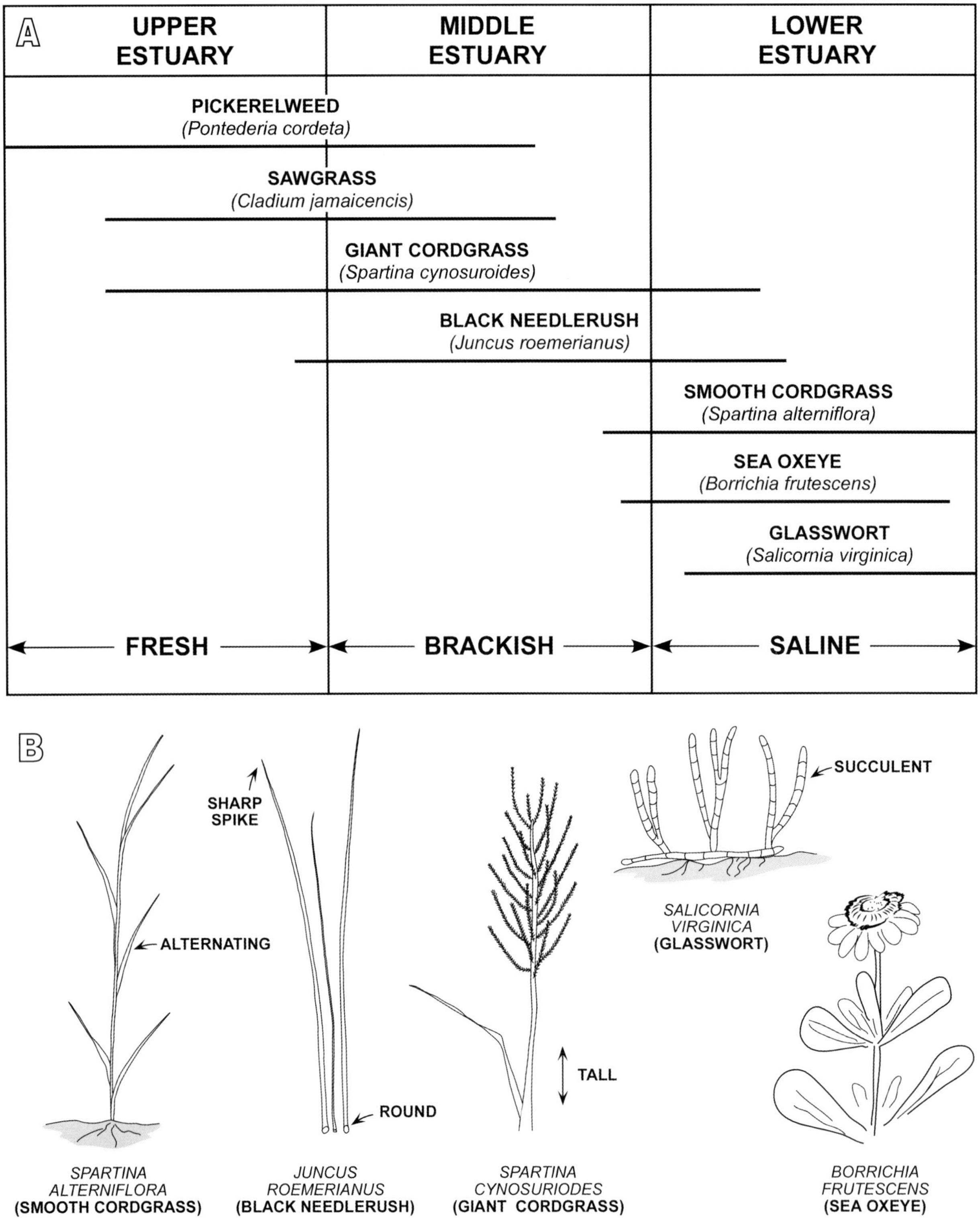

FIGURE 93. Marsh plants. (A) Occurrence of the most conspicuous plants in the marshes of the upper (fresh), middle (brackish), and lower (saline) parts of the estuaries along the Georgia Coast. After Stalter (1974). (B) Sketches of some of the more common marsh plants.

Freshwater marshes

In the upper reaches of some estuaries, as well as along the lower reaches of some of the major rivers, grassy wetlands composed of freshwater vegetation form a border along the major channels. The substrate in these wetlands is seldom exposed, because daily water-level changes are small to nonexistent, but, of course, they can be completely covered during floods along the river. Sawgrass (*Cladium jamaicencis*) is a common plant in these areas. The blades of the grass are almost razor sharp, so beware.

Brackish marshes

The brackish marshes of the larger estuaries along the Georgia Coast, namely those that typically have significant freshwater influx, are dominated by black needlerush. These marshes have filled in broad expanses of the middle estuary regions, crowding in steeply against the banks of most of the rivers and tidal channels. Tidal flats, which were abundant during the early phases of the evolution of the estuary, being in the zone of the turbidity maximum, are notably absent, having been covered by the grasses in most instances. The sediments in these marshes, which are highly mixed (bioturbated) by burrowing crabs, such as fiddler crabs (*Uca pugnax* and *Uca pugilator*), and by the roots of the marsh grasses, are, not surprisingly, composed mostly of silt and clay.

Salt marshes

The lower reaches of the estuaries and landward margins of barrier islands along the Georgia Coast are bordered by salt marshes dominated by smooth cordgrass (Figure 94B). The Georgia Department of Natural Resources estimates that Georgia has 378,000 acres of salt marsh (about 15% of the East Coast's total). Figures 58 and 92B show typical cross-sections of Georgia salt marshes. Examples of Georgia's marshes are illustrated further by the photographs in Figures 94, 95, and 96. Most experts agree that plant distribution in the marshes is controlled by depth and duration of flooding (Barry, 1980); therefore, it is convenient to divide these marshes into regularly flooded **low marsh**, the zone between mid tide and neap high tide, and an irregularly flooded **high marsh**, which occurs roughly between neap and spring high tides (Teal, 1958; Frey and Basan, 1978; 1985; Frey and Howard, 1969).

FIGURE 94. Georgia salt marshes. (A) High marsh in Skidaway Island State Park, Georgia on 28 March 2010. In the foreground, some saltgrass (*Distichlis spicata*) and sea oxeye (*Borrichia frutescens*) are visible. The next zone out contains some glasswort (*Salicornia virginica*). These two zones are labeled as the *Salicornia-Distichlis* marsh in Figure 92B. (B) Closeup of smooth cordgrass at this site, showing new green growth replacing last year's dead vegetation.

Smooth cordgrass is the only plant that normally occurs in the low marsh zone of the more saline areas, which is flooded 2-14 hours/day and has soil salt concentration of 0.5-3.2 parts per thousand (Barry, 1980). The higher portions of these plants can only be completely flooded about 10% of the time or they die (T. Kana, pers. comm.), but the base of the lowest plants, which are located near mean sea level, are flooded much more frequently. Smooth cordgrass is usually quite tall (greater than 6 feet at full growth) in the lower half of the profile (e.g., on creek banks and levees, but becomes

FIGURE 95. Georgia salt marshes, cont. (A) Salt marsh near the airport on St. Simons Island. Arrow points to splay deposits composed of sediment washed onto the marsh through a small channel cut through the natural levee of the main tidal channel. The natural levee is very well-developed at this location. (B) Salt marsh near Darien. Arrow points to a tidal river point bar. Both of these oblique aerial photographs were taken during a low spring tide on 28 April 2010.

FIGURE 96. Georgia salt marshes, cont. Early spring new growth (arrow) of smooth cordgrass along the shoreline of the St. Marys City Park, Georgia on 26 March 2010.

dwarfed (0.3-1.7 feet) in the higher areas (e.g., between drainage creeks; behind levees), as is demonstrated in Figure 92B. The reason for these differences in height is still a matter of debate, but is thought to be related to the degree of flushing of the marsh soils. The vegetation in the lower marsh and creek banks is tall because, apparently, the soils are better flushed by frequent tidal inundations.

In the high marsh, zones of glasswort (*Salicornia virginica*), sea oxeye (*Borrichia frutescens*), and saltgrass (*Distichlis spicata*) grow in soils with salt concentration of 0.3-3.0% (Barry, 1980) that are usually flooded daily but high enough in elevation to where the soils are flushed by rain water. The upper high marsh, which is flooded mostly during spring tides, may be populated by black needlerush in areas of lower salinity, and plants such as saltmeadow cordgrass (*Spartina patens*), marsh elder (*Iva frutescens*), or sea myrtle (*Baccharis* sp.) in more saline areas.

Most marsh sediments are low in oxygen, or reduced (usually dark brown to black in color; Wiedemann, 1972), except in topographically higher areas, such as levees and the high marsh, where water circulation at low tide via burrows and rootlets promotes oxidation (reddish color in some areas). The typically intensive reducing conditions in the sediments help account for the relatively low species diversity in the salt-marsh soils.

Mounds ("reefs") of the eastern oyster play a prominent role in initiating salt-marsh growth in the lower portions of many of the estuaries. On tidal flats exposed to onshore flood-tidal currents, the oyster mounds grow in a row, oriented perpendicular to the current, presumably to take maximum advantage of availability of food particles to be filtered from the flowing water. The mounds are positioned so that their tops are just below neap high tide; therefore, at water levels higher than neap, they act as a baffle to the flood current, producing a current shadow on the landward side. Eventually, a mud flat builds up behind the linear mounds, over which the marsh can grow, and the oysters are eventually buried under the marsh. Oyster mounds also tend to grow across the entrances to smaller tidal creeks draining the salt marsh (Ward and Domeracki, 1978). The dammed creek channel fills up with mud rapidly and the marsh grows over it, totally obscuring the old channels. Exposures of oyster mounds buried in this fashion and later exhumed by migrating tidal creeks are commonplace in the creek banks throughout the region.

TIDAL FLATS

Definition

Tidal flats are nearly flat, unvegetated, intertidal surfaces that are relatively sheltered from direct wave attack. In most places, the sediment grain size decreases landward, away from main tidal channels (Figure 92A), because the sediments are carried to their final resting place by tidal currents that decrease in velocity away from the channels as the water flows up onto the flats. The depositional and erosional rates on tidal flats are much slower than on beaches on the open-ocean front, where the sediment is moved readily by wave-generated and tidal currents.

Exposed Sandy Tidal Flats

Some of the more exposed areas (to wave action) along the Georgia Coast contain broad intertidal flats up to several hundred yards long and wide (see Figures 97 and 98). The fact that these flats are composed of sand indicates that tidal currents and waves are strong enough to mobilize the sediments from time to time, removing the mud. On the lower portions of some of these flats, waves may have formed low ridges that migrate very

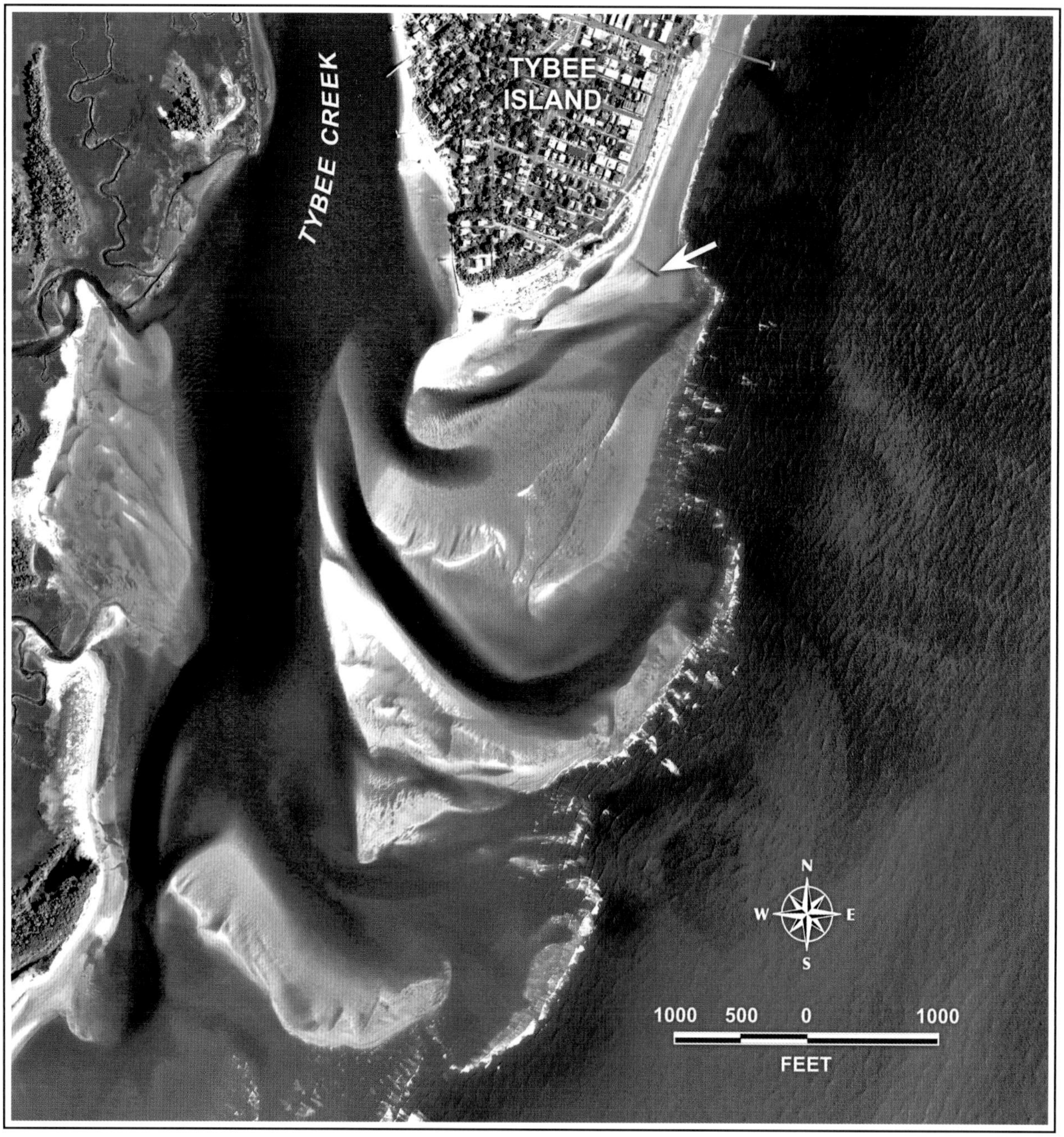

FIGURE 97. Vertical image of the sand flat/inlet entrance shoal complex (ebb-tidal delta) of Tybee Inlet at the south end of Tybee Island (at low tide). Arrow points to groin pictured in Figure 87C. Digital orthophoto acquired in September 2007, courtesy of USGS.

slowly across the flat, and ripples and even larger bedforms may be present in the vicinity of tidal channels that cut across the flat (e.g., the sand waves pictured in Figures 70B and 98A). Despite this kind of sediment motion lower down the intertidal zone, most of the sand and muddy sand on the upper half of these flats is relatively stable; therefore, it usually contains a huge population of infaunal organisms (those that burrow in the sediment). Two of the more conspicuous burrowers (or pit makers) are the Atlantic horseshoe crab (*Limulus polyphemus*; Figure 99) and rays (Figure 100). Others include numerous polychaetes, sparse clams, fiddler

FIGURE 98. Sandy tidal flat, south end of Tybee Island. (A) Oblique aerial view of this exposed sandy tidal flat during a low spring tide on 28 April 2010. Arrow A points to the groin pictured in Figure 87C, arrow B points to the sea wall/revetment shown in Figure 89, and arrow C points to some ebb-oriented sand waves. (B) Ground view at low spring tide of the wide intertidal sand flat at the south end of Tybee Island (on 28 March 2010).

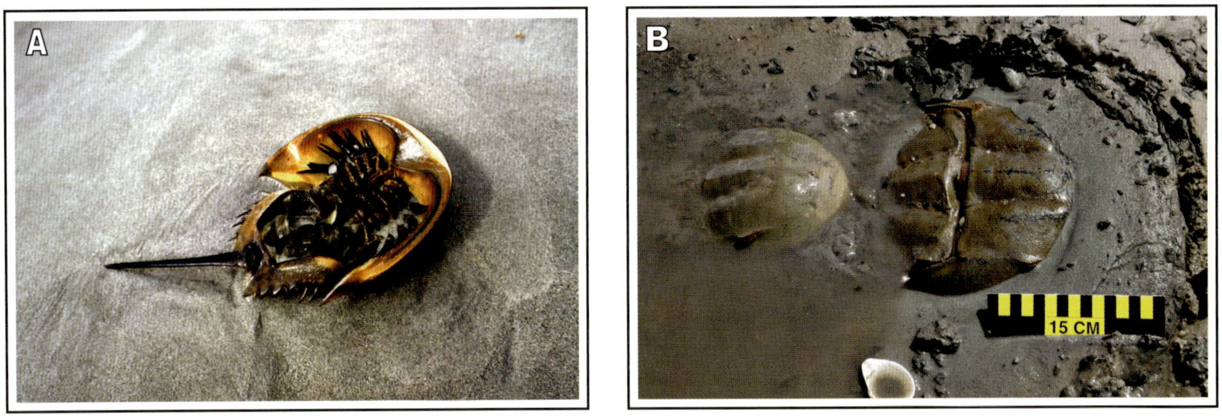

FIGURE 99. Horseshoe crabs (*Limulus polyphemus*) on an intertidal sand flat at the entrance to St. Helena Sound, South Carolina during the spring mating season (April/May). In both cases, these horseshoe crabs were stranded on the flat at low tide. (A) Overturned on its back to show the underparts. This one was carried down to the waters edge and released. (B) Male mounted on top of a larger female, a typical mating arrangement. The female burrowed into the sand a small distance after the tide went out. Scale is ~6 inches. Photographs taken circa 2003.

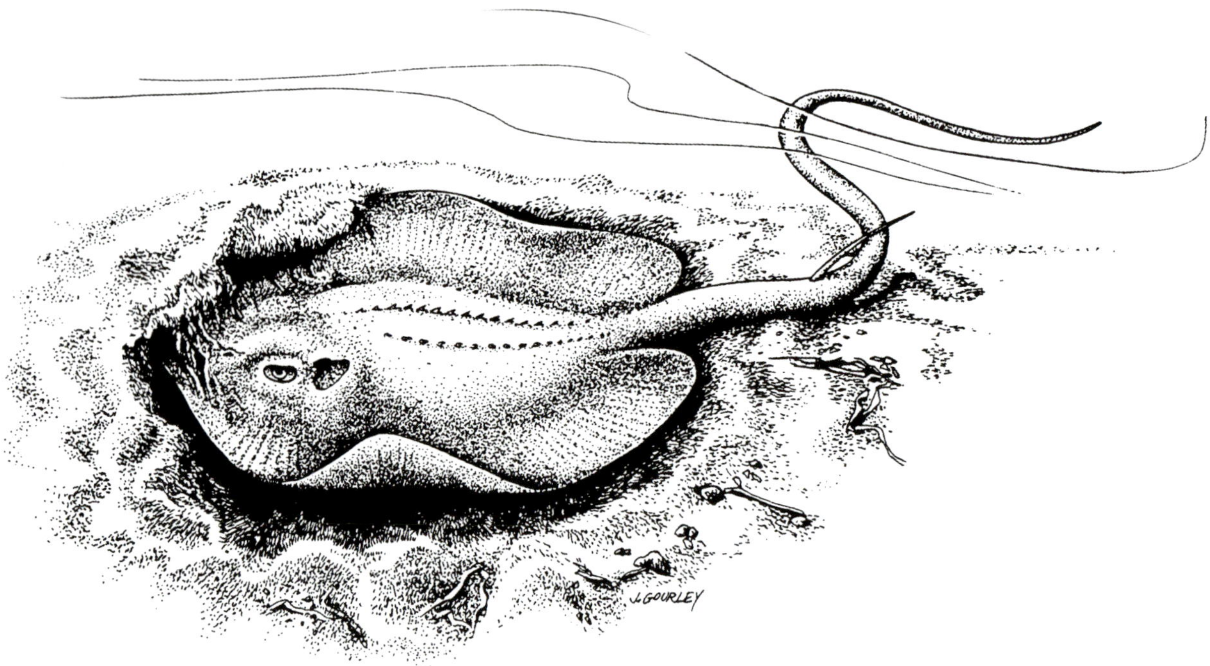

FIGURE 100. Sketch of ray in the process of excavating substrate to feed on organisms in the sediment. Modified slightly after Howard, Mayou, and Heard (1977).

crabs, and the U-shaped burrows of the "iodine" worm, *Balanoglossus.*

A huge population of **horseshoe crabs** congregates in the spring on the large sand flat at the entrance to St. Helena Sound, South Carolina for their annual reproduction ritual. This is no doubt true for other sand flats at the head of the Georgia Bight, but we have observed them numerous times in the spring on this particular sandy tidal flat, because we always visited this flat at low spring tide during our field seminars. These fascinating creatures, which are pictured in Figure 99, are sometimes referred to as "living fossils," because forms almost identical to this species were present during the Triassic period 230 million years ago. They feed mostly at night, burrowing straight down into the sediment in search for worms and mollusks, including razor clams. They grow by molting and emerge 25% larger each molt.

Spawning occurs each spring during high tides of the new and full moon. At that time, males cluster along the water's edge as females arrive during the rising tide. Using glovelike claws, the male, which is smaller than the female, hangs on to the female's shell while she pulls him up the tidal flat toward the high-tide line (see Figure 99B). Once near the high-tide line, the female digs a hole every few feet to deposit eggs (up to 20,000). As the male is pulled over the nest, he fertilizes the eggs.

Not only are these creatures fascinating in their life style, but they also serve a number of important functions, including:

- Their eggs are an important food source for shorebirds, fish, and sea turtles. During the spring migration, several species of shorebirds, numbering in the thousands, time their arrival on the tidal flats of Delaware Bay to match the time when the horseshoe crab eggs are available. The Delaware Bay example is famous in the birding literature (e.g., Lin, 2012).
- The blue blood of horseshoe crabs contains *Limulus ameboycte lysate*, which detects harmful bacteria in drug products and medical devices.
- They play a major role in other aspects of medical research (cancer and vision studies).
- Indians of Roanoke Island, North Carolina tipped their fishing spears with the tails of horseshoe crabs.

According to Howard, Mayou, and Heard (1977), three dasyatid rays, the southern stingray (*Dasyalis Americana*), the bluntnose stingray (*D. sayi*), and the Atlantic stingray (*D. Sabina*), *are the authors of most, if not all, of the ray hole excavations* they observed on the tidal flats in Georgia (Figure 100).

Sheltered Mud Flats

Tidal mud flats occur under the following conditions in the upper shore zones of the Georgia Coast:

- **Filling of abandoned channels**. Because of the fairly intense meandering mode of many of the tidal creeks and rivers on the Georgia Coast, they commonly abandon meander loops that become filled in with mud flats, which eventually are grown over by marshes. The process of channel abandonment, which usually takes a few decades in the smaller creeks, is constantly being repeated throughout the estuaries and backbarrier regions all along the coast, which allows for the accumulation of mud flats. An abandoned tidal channel (oxbow lake) near the north end of St. Simons Island that is presently filling in with mud is shown in Figures 101A and 101B. A mud flat on the side of the channel of the Champney River, a distributary of the Altamaha River in the upper delta plain of the Altamaha River Delta, is shown in Figure 101C.
- **Channel margins, channel confluences, and tidal point bars**. Smaller tidal mud flats occur along channel margins, where a decrease in slope occurs, or at channel confluences, as a result of deposition down current away from the confluence, which is typically a zone of bottom scour. Shallow portions of tidal point bars also contain minor mud flats in some localities.

The problem of why large quantities of mud accumulate on the upper reaches of tidal flats is an interesting one. Several Dutch workers have commented on this issue (Van Stratten, 1950; Postma, 1967), concluding that the mud is deposited during the last stages of the flood flow, during slack water, or at ebb just before water is drawn away, and it requires a stronger current to pick the mud back up than the one that deposited it. The finest particles in the mud are composed of clay minerals, which have a thin, sheet-like shape (like a thin coin). This shape slows down the process of settling of the particles from suspension, which allows them to be transported landward by the waning flood currents beyond the point where more spherical particles of the same density would settle out. Also, once the thin, disc-shaped particles have settled on the bottom, their thin edges provide little resistance to the outflowing water; therefore, they are not easily picked back up into suspension once they have settled on the bottom. This process of relatively permanent deposition is aided further by the fact that the floccules of organic matter and suspended fine-grained sediment illustrated in Figure 57B, as well as slime-secreting diatoms, also readily settle out in the quiet water. Once this mud is deposited, it tends to remain in place as well because: 1) Diatoms move up through the mud and deposit slime; 2) The mud dries out; and 3) Burrowing organisms tend to stabilize the mud. In some cases, mud is trapped between plants on the flat.

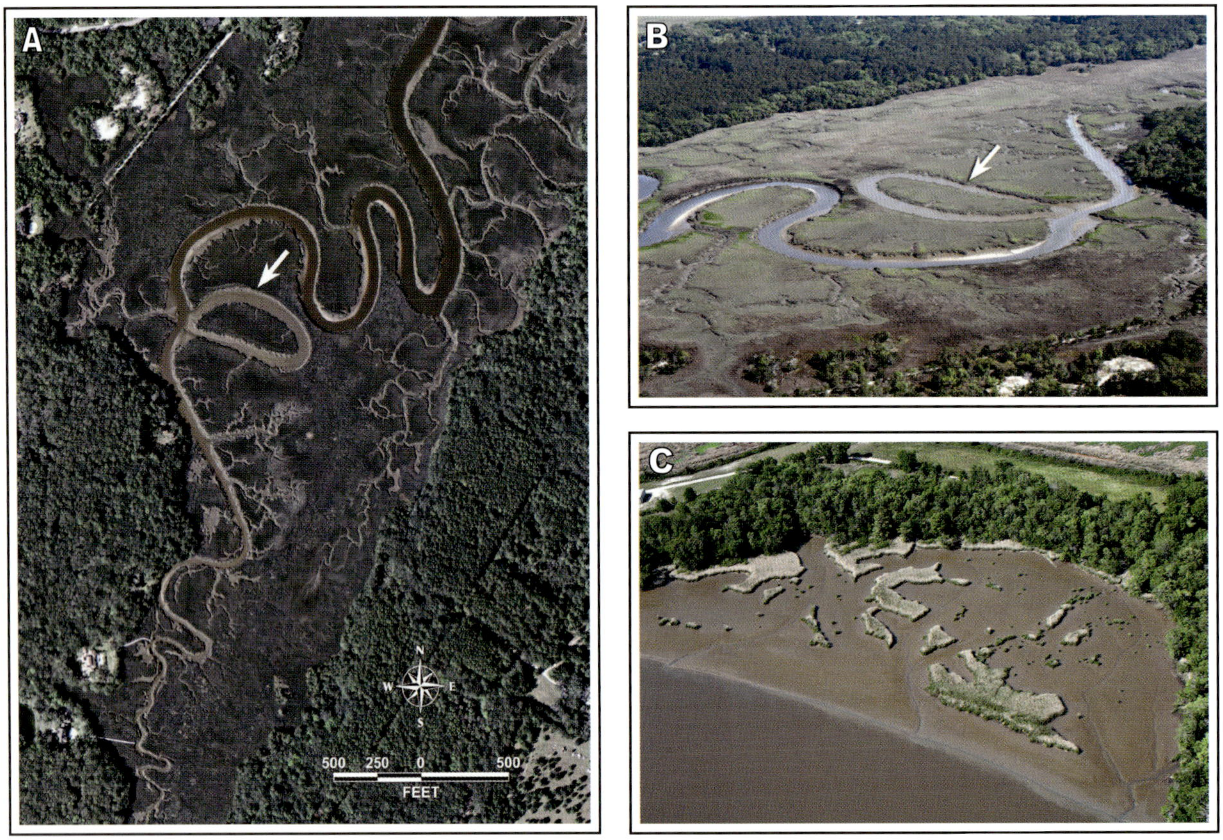

FIGURE 101. Sheltered mud flats. (A) Vertical image of a meandering tidal channel near Butler Point on the north end of St. Simons Island. Arrow points to abandoned oxbow lake that is now almost completely filled with mud. Digital orthophoto acquired in March 2006, courtesy of USGS. (B) Oblique aerial view of the same system shown in A during a low spring tide. Arrow points to mud-filled abandoned tidal channel. (C) Mud flat exposed during a low spring tide on the side of the channel of the Champney River, a distributary channel of the Altamaha River in the upper delta plain of the Altamaha River Delta. Photographs in both B and C were taken on 28 April 2010.

Another factor of prime importance to mud deposition is the super abundance of **suspension-feeding organisms**, such as oysters and clams. During the filtering process, these organisms compress the finely divided clay particles and bind them together in their intestines as fecal pellets, which are excreted onto the flats. Another part of the suspended matter is coagulated in their gills and pushed back into the water as pseudo-feces. These feces and pseudo-feces are easily deposited, even in comparatively turbulent water. This process is clearly made evident around the oyster mounds throughout the estuaries and backbarrier regions in Georgia, that are typically surrounded by accumulations of mud.

Thus, several processes enhance the formation of mud deposits on the tidal flats assuming the individual silt and clay particles are delivered to the shore in some abundance by freshwater streams, which is indeed the case on the Georgia Coast.

Dominant Biogenic Features

The tracks, trails, and types of burrow structures of the animals that live on tidal flats and in their sediments have fascinated geologists for decades, because such evidence is useful for interpreting the depositional environments in ancient sediments (commonly used by geologists interpreting rock layers in their search for economic deposits, such as oil and gas). Knowledge on this topic has been greatly advanced through studies by associates of research institutes on the Georgia Coast at Sapelo and Skidaway Islands (e.g., Frey and Howard, 1969; Howard and Dorjes, 1972). Their work provides a useful guide for the features you may see as you walk the intertidal regions of the Georgia Coast (WITH SHOVEL IN HAND!!).

Some of the more conspicuous biogenic features (commonly referred to as **biogenic structures**) to be seen in the sediments of the backbarrier and estuarine areas of the Georgia Coast include:

Environment	Biogenic Structures
1) High mud flats and sheltered portions of tidal point bars	Miscellaneous Polychaeta tubes (e.g., *Onuphis cremita*); trails of snails (e.g., *Littorina irrorata*)
2) Higher portions of point bars and sandy tidal flats	Burrows of fiddler crabs and sandy tidal flats (e.g., *Uca pugilator*)
3) Low energy, water-saturated portions of sand flats and point bars	U-shaped burrows of the iodine worm (*Balanoglossus*) and vertical tubes of the polychaete *Diopatra cuprea*
4) Exposed sand flats	Burrow pits of the horseshoe crab (*Limulus polyphemys*) and rays, as well as ghost shrimp (*Callianassa major*) burrow networks

Figure 102 shows sketches of some of the more common burrow types. Except in lower reaches of some sand flats and tidal point bars, where highly mobile, large-scale bedforms occur, the tidal flat sediments of the backbarrier and estuarine regions of Georgia are highly burrowed throughout.

TIDAL CHANNELS

Before delving too far into this subject, a brief review of channel systems in general is in order. On a global scale, there are four basic styles of channel morphology that occur in natural river and tidal systems, that are illustrated in Figure 103. **Straight channels** are rare in general, so be suspicious if you see one in the coastal region of Georgia, because it has most likely been dug for navigation or other purposes, such as mosquito control or connections to old rice ponds. As a rule, **braided channels** occur under conditions of: 1) Relatively steep channel slope; 2) High bedload content (sedimentary material on channel bottom; e.g., gravel and coarse sand, that moves along the bottom by rolling and bouncing); and 3) Flashy discharge (high flow during flooding periods and low flow during the rest of the time). None of the conditions that result in braided streams occur in the extreme within the confines of the present lower coastal plain of Georgia, so don't expect to see any braided streams as you sojourn there. However, had you been able to visit this area 16,000 years ago, all of the major stream systems would most likely have been braided, because of the steeper slopes (sea level about 350-400 feet below present) and availability of coarse-grained sediment because of those steeper slopes. **Meandering channels** tend to have: 1) Flatter slopes than braided channels; 2) A high ratio of suspended to bedload sediments; and 3) A more steady discharge. Meandering channels, such as those illustrated in Figures 95A, 101A, 101B, and 104, are by far the most common channel type in the estuaries and backbarrier regions of Georgia. **Anastomosing channels** typically occur on extremely flat slopes in waters with a very high ratio of suspended to bedload sediments. A few anastomosing channels do occur on the Georgia Coast, but their origin is masked and made complex by systems with two opposing current directions due to the tides.

Considering only the tidal channels in the major estuarine systems, they are mostly located in the valleys carved during the last lowstand of sea level (called **lowstand valleys**). As noted earlier, sea level started to rise about 12-16,000 years ago and reached near its present level around 4,500 years ago. You have read these numbers before, but we have learned the hard way that these are numbers that bear repeating. With the relative stabilization of sea level, the old valley was filled up with sediments to a near-level plain that is now mostly covered with marsh grasses, or tidal flats in some situations, through which the channels migrate.

The flat topography of the filled valley surface and associated gentle slope of the river channel, as well as the presence of the suspended muds of the zone of the turbidity maximum, dictate that tidal channels in the most landward reaches of all the estuaries and the two major river deltas on the coast are **meandering**. A good example of this is illustrated in Figure 104A.

As it turns out, meandering is a complex process, but, as noted above, it is clearly associated with flat stream

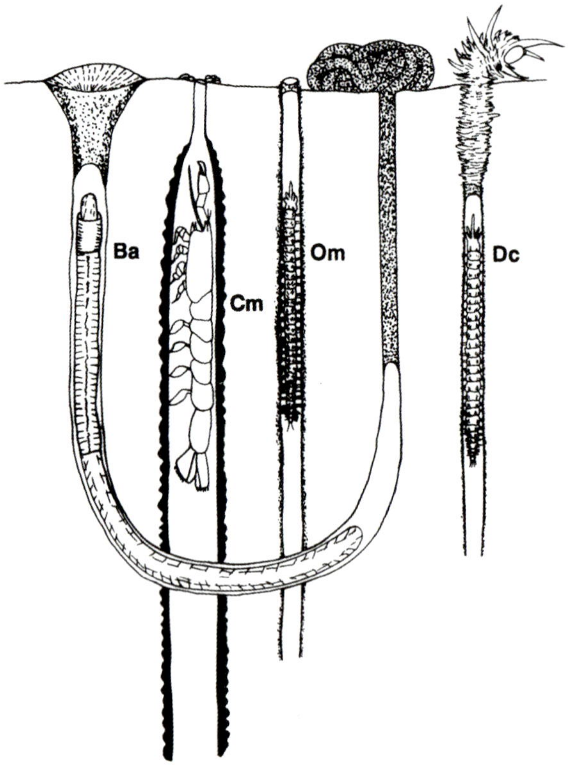

FIGURE 102. Burrowing characteristics of four of the more common organisms on the tidal flats of the Georgia Coast.

Ba – The iodine worm (*Balanoglossus*), which has an open U-shaped burrow. Note the excreted sand pile, a conspicuous feature on the sandy tidal flats where this organism lives.

Dc – The periscope worm (*Diapatra cuprea*), which lives in a vertical chitinous tube with a periscope-like entrance.

Om – The soda straw worm (*Onuphis microcephala*), which lives in a vertical mucoid tube covered with sand grains.

Cm – The ghost shrimp (*Callianassa major*), which lives in a complex vertical to horizontal catacomb-like network of burrows lined with mud-rich fecal pellets.

(Modified after Howard and Dorjes, 1972.)

slopes. Because of the combined effect of constriction of flow by the channel sides and bottom and the Coriolis effect, the pattern of flow of water in most channels is in the form of a **helix**. This results in alternating erosion and deposition of the channel sides in a downstream direction, which ultimately produces a meandering stream. Guess who first made this observation? Albert Einstein! (Einstein, 1954)

A sketch map of the morphology of a typical meandering tidal channel in the upper reaches of the estuaries along the Georgia Coast is given in Figure 105A. As the sketch shows, the outside bend of a meander usually has an associated **cut bank,** where the current undermines the bank and causes it to erode. Typically, the deepest part of the channel is located near the cut bank, where the maximum amount of erosion is taking place. An arcuate-shaped bar, commonly composed of sand, called a **point bar**, occurs on the opposite side of the channel from the cut bank. The main force in creating this channel configuration in the upper estuary is the ebb current, which is stronger than flood currents in that part of the estuary. The reason for this is related to the physics of the tidal wave's progression, which is less efficient (i.e., takes longer) on the flood than on the ebb. Therefore, even with equal volumes of water being

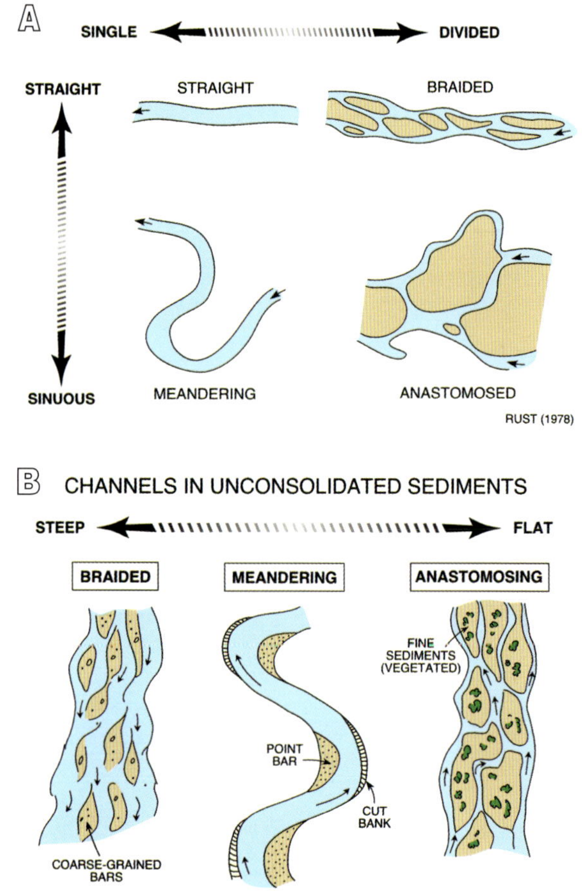

FIGURE 103. Channel types. (A) Classified as to whether they are straight or sinuous and whether they are single channels or a complex divided channel (after Rust, 1978). (B) Contrasting steep and flat channels.

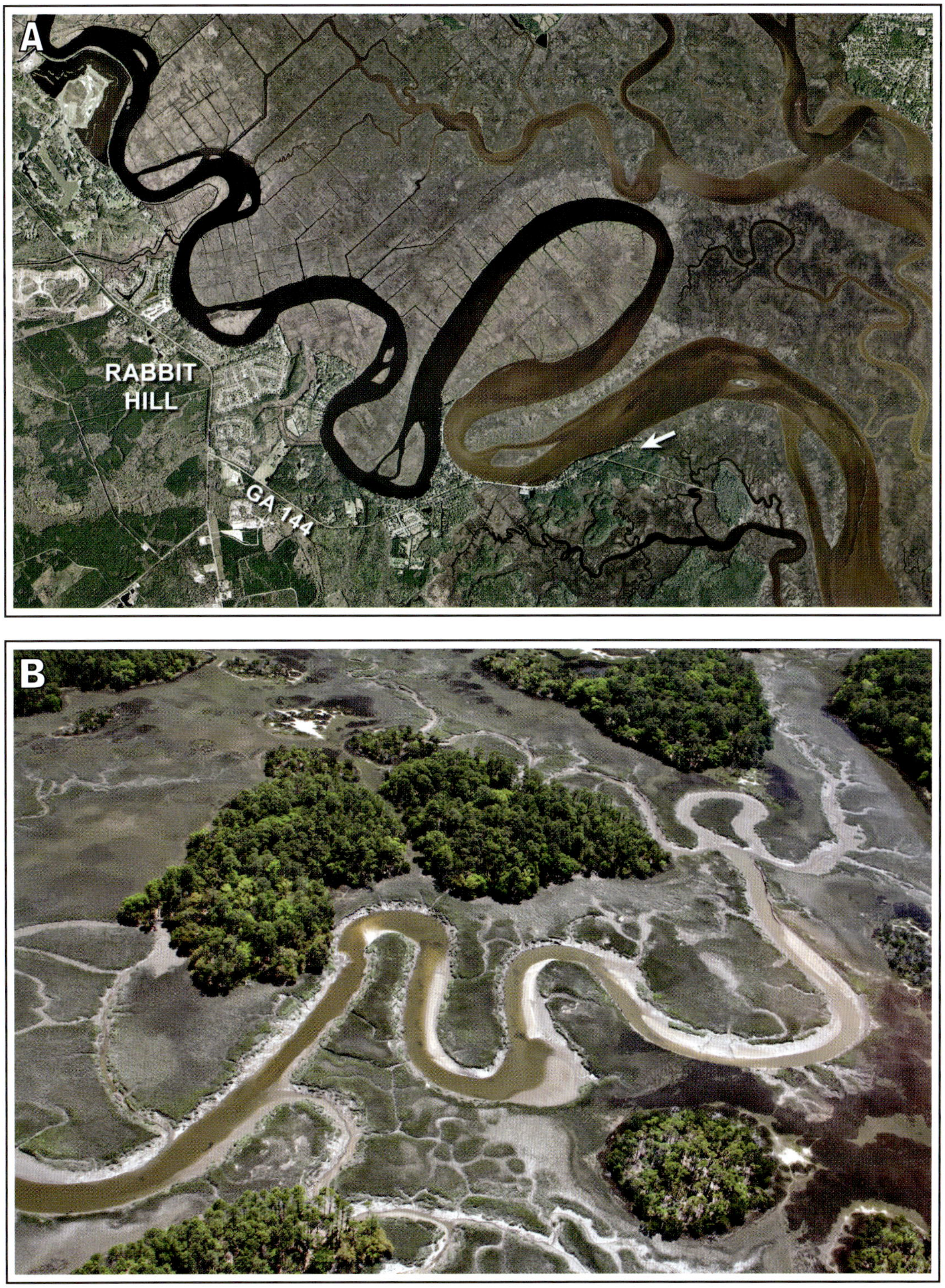

FIGURE 104. Meandering channels. (A) Highly meandering channel of the Ogeechee River, the largest coastal plain (black-water) river in Georgia. This section of the channel is subject to the rise and fall of the tides. Note the rectangular patterns of the old rice fields. The arrow points to Fort McAllister State Historic Park. Digital orthophoto acquired in March 2006, courtesy of USGS. (B) Meandering tidal channel on the lower delta plain of the Savannah River Delta (South Carolina side). Photograph taken during a low spring tide on 28 April 2010.

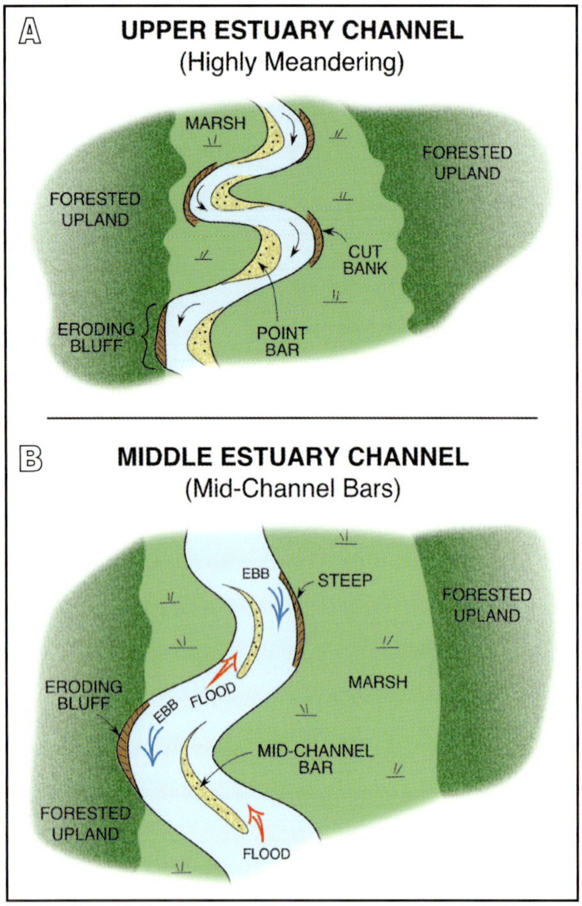

FIGURE 105. General model of the morphology of tidal channels in the estuaries (and sounds) located along the Georgia Coast. (A) Upper estuary. (B) Middle estuary.

exchanged on the flood and the ebb, mean ebb currents need to be higher in order to accommodate the relatively shorter ebb current duration. Also, in most estuaries, and in both of the deltas, the freshwater runoff adds to the velocity of the ebb current. Another factor is the timing of peak flood and ebb currents. Peak flood currents occur late in the flood cycle (between mid and high water) when the water is spread out over a wide area. Peak ebb currents occur near low tide, when the water is more confined to the channels themselves.

Further down, or more seaward, in estuaries (including the sounds in Georgia), the flood current plays a more important role in channel configuration, resulting in the morphology illustrated in Figure 105B. In the middle and lower reaches of the estuaries, a bar tends to form closer to the middle of the main channel and a shallow channel on the side of the bar away from the steeper cut bank at the bend of the channel is dominated by flood currents from mid to high tide. The main channel near the cut bank tends to be host to somewhat stronger ebb currents than flood currents (for same reason as in the upper estuary). As a general rule, the middle estuary channels do not migrate as rapidly as the more highly meandering channels further up the estuary.

Large point bars occur in the middle to lower estuary in some of the larger tidal rivers. Examples of three of these are shown in the vertical image of the Sapelo River (Figure 106). Note that the upstream ends of these point bars are attached to the marsh bank, unlike those in the general model in Figure 105B. The photograph in Figure 107 of one of these **tidal river point bars** in the Julienton River in Georgia shows the distinct dominance of the ebb currents against the cut bank and the dominance of the flood currents in the higher channels on the opposite side of the sand bar. These types of point bars are distinguished from those found further up the major freshwater streams by their more abundant biogenic structures and greater mud content due to their location near or within the zone of the turbidity maximum (Figure 57B).

Near the estuary entrances, the channels are much wider and the ebb and flood currents are nearly equal. The main channels at that location are straighter than elsewhere and linear sand shoals tend to occur near the middle of some of the channels.

The occurrence of complex, tree-like networks that compose tidal channel systems is a fairly common occurrence within the major estuarine/sound systems on the Georgia Coast. Because these drainage networks are contained within individual topographic basins, even though the change in elevation within the basin is relatively small, a major outlet has evolved that drains the basin, ultimately in the ebb direction. Side channels join this main channel like limbs of a tree, similar to those channel networks found in drainage basins for freshwater streams. Three examples of this process from the Georgia Coast are given in Figure 108. This is a complex topic that still is undergoing considerable research, for tidal systems anyway, but these examples are presented here because they are so prevalent and spectacular on the Georgia Coast. This tree-like development of drainage networks was a primary concern of geomorphologists A. E. Strahler and R. E. Horton in the 1940s and 1950s.

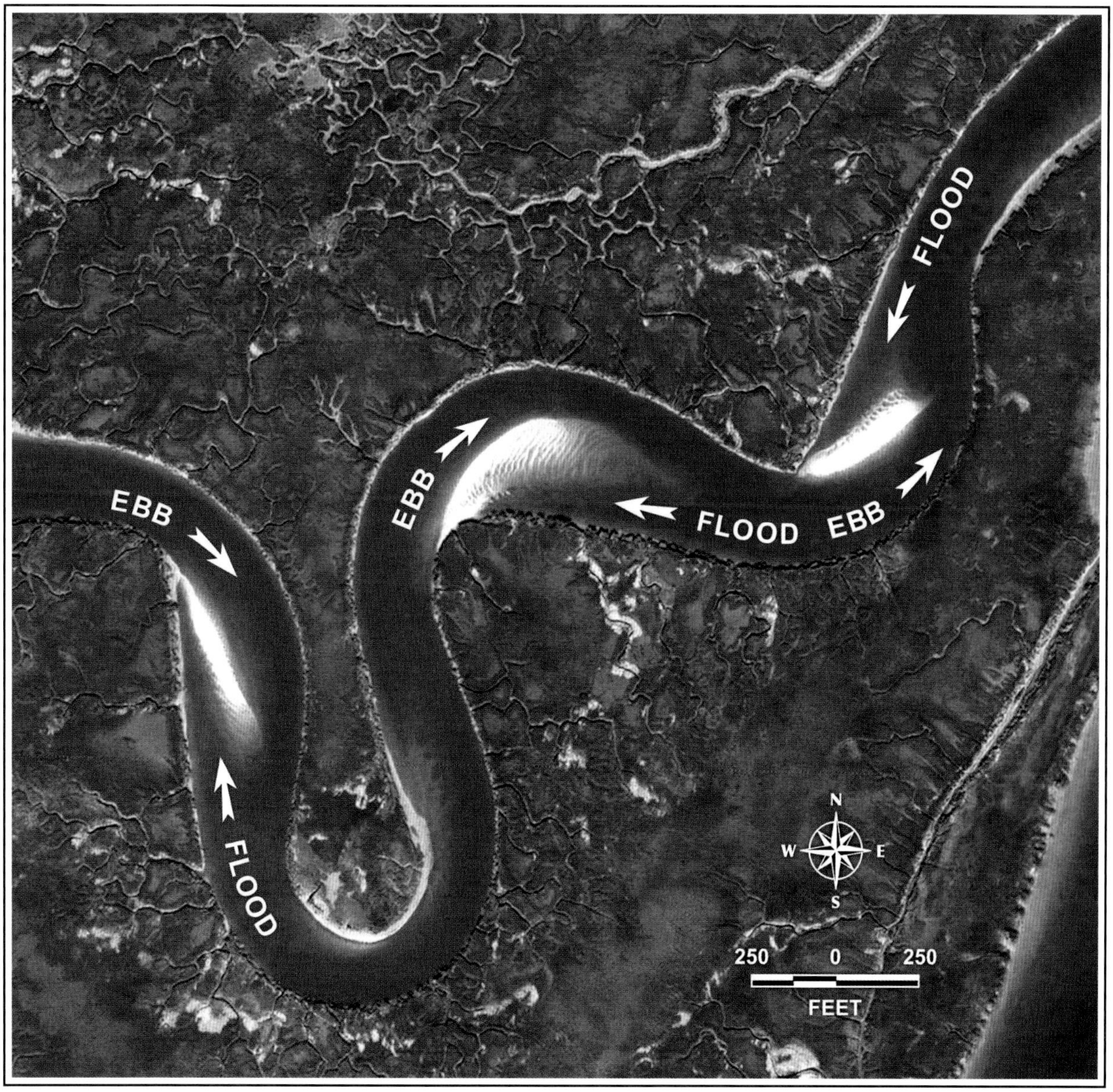

FIGURE 106. Tidal river point bars in the Savannah Cut of the Sapelo River about 1.4 miles southeast of Pine Harbor in McIntosh County (low-tide image). Arrows indicate the dominant current directions, but, of course, both flood and ebb currents do occur on both sides of the point bars. However, the ebb current, which controls the orientation of the bar itself, moves the most sand within each of the systems. Digital orthophoto acquired in December 2007, courtesy of USGS.

Horton produced a number of laws that govern stream lengths, drainage composition, and so on, including a stream order model that ...*convinced researchers that these findings suggested that some underlying factor (or factors) was governing the structure of the various stream attributes in a similar predictable way. Studies of other natural branching networks have revealed patterns similar to the stream order model. For example, the bifurcation ratio of three has also been discovered in the rooting systems of plants, the branching structure of woody plants, and the veination in leaves and the human circulatory system* (Pidwirny, 2006).

This is another topic beyond the scope of this book, particularly with regard to the details of the Horton models. Also, these drainage networks are not something you can easily observe as you drive along the shores of

FIGURE 107. Vertical low-tide image of a large tidal river point bar in the Julienton River, Georgia (near Shellman Bluff, McIntosh County). Arrows indicate dominant current directions. Note the large ebb-oriented sand waves (toward the lower right) on the part of the bar that faces the deepest part of the channel (the cut bank; Figure 105), and the slightly modified flood-oriented sand waves on top of the bar. Digital orthophoto acquired in June 2008, courtesy of USGS.

the sounds and estuaries. However, if they do interest you, they are readily apparent on the Google Earth and Bing images of the Georgia Coast on the Internet.

In the middle and upper reaches of the estuaries, in places, the meandering tidal channels migrate against the valley walls of the original lowstand valleys, which at this time are usually forested uplands. As development pressures from the construction of vacation and retirement homes on the coast have mounted, houses have been built on the sides of the old valleys at elevations normally considered to be safe from all but the most extreme hurricane surges. However, a channel with a cut bank that incises into the forested upland adjacent to the channel (Figure 109) is a cause for concern in some areas. In a number of localities, shore protection structures, such as riprap revetments, have been built to slow the erosion.

As some developers consider this type of erosion problem, two questions related to the understanding of the mechanisms of erosion of the bluffs are: 1) When do the bluffs erode? and 2) What processes may cause them to erode at rates faster than that shown in the past few

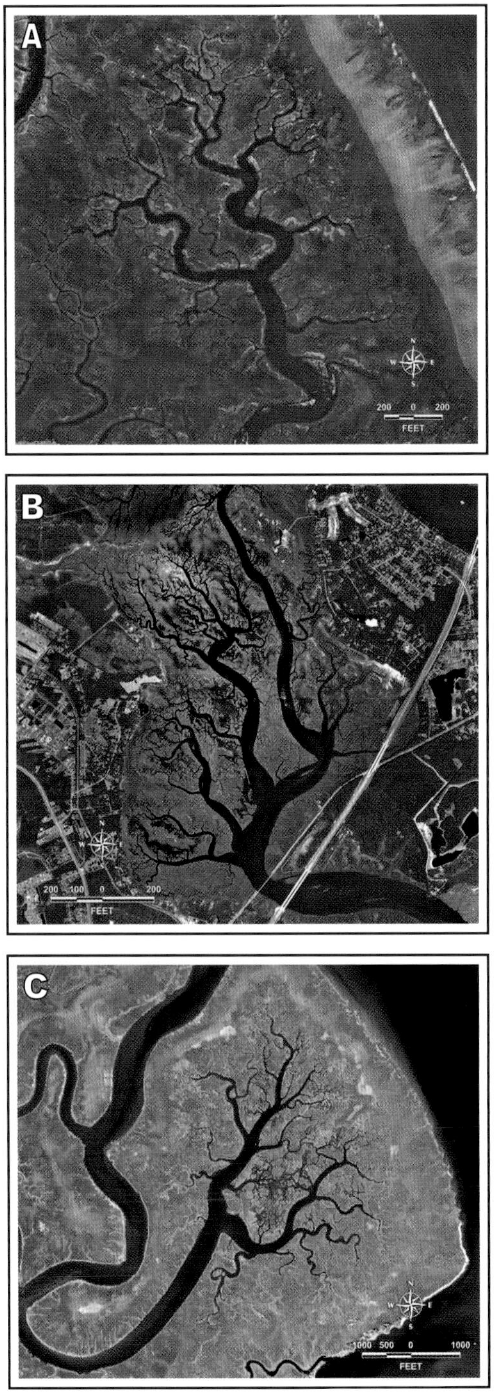

FIGURE 108. Drainage networks in the tidal marshes of Georgia, illustrating their tree-like configuration. (A) Channel network near Brunswick (east of the channel of the Back River). (B) South Brunswick River just north of where the main channel is crossed by the bridges of I-95 and GA 303. (C) About seven miles northeast of Darien on Rock Island (at the border of Doboy Sound). Digital orthophotos acquired in 2008-2009, courtesy of USGS.

FIGURE 109. Bluff being eroded by the Crooked River tidal channel at the Crooked River State Park, Georgia. Photograph taken at low tide on 26 March 2010.

decades? As already noted, the bluff erodes when the meandering channel migrates into it. However, this erosion can only take place when the water is high enough to impact the eroding scarp on the bluff for a significant amount of time. This can happen in the following ways:

1) **During maximum spring tides**, especially the extraordinarily high tides that typically occur during the late fall. It is unlikely that any significant erosion occurs during neap tides.

2) **During storm surges** that may occur during tropical storms or hurricanes. Elevated tide levels also occur during periods of strong onshore winds not associated with tropical depressions, but they would only be a factor during maximum spring tides.

3) **During periods of high fresh-water runoff**. Even in estuaries with a relatively small freshwater stream entering it, periods of very high rainfall may significantly raise the water level in the estuary, creating very strong ebb currents that could erode the bluffs. Once while investigating such an issue, we downloaded the USGS discharge data for the Coosawatchee River, South Carolina (near the Georgia border). These data showed that there were some high flows in 2003, but a much more significant high occurred in 1998 when *El Niño* rains drenched the southeast. Therefore, more erosion of bluffs in the upper reaches of some estuaries can be expected during the *El Niño* years or possibly during high runoff generated by rains associated with tropical cyclones.

4) **During periods of high rainfall in the local area**. The soil along the banks of the channel may become saturated during periods of heavy rainfall causing slope failure (Tim Kana, pers. comm.).

An obvious potential cause for an increasing erosion rate of the bluffs would be an **abrupt rise in sea level**. As noted earlier, sea level is rising at the present time at the rate of around one foot per century. If sea level continues to rise at its present rate, the amount of bluff erosion should not increase significantly as a result of that factor alone (at least not for the next decade or so).

The impact of *El Niños* on both the climate patterns (e.g., rainfall and snows in the winter time) and the resulting physical processes on the coastal regions of Georgia, most particularly as a generator of floods on the piedmont rivers, has not been appreciated enough by local scientists, in our opinion. These events do not occur with strict periodicity, but usually recur every 3-7 years. The characteristics of this phenomenon, a subject of intense research at the present time for obvious reasons, are given below, based principally on two sources: Inman and Jenkins (2005) and Griggs, Patsch, and Savoy (2005).

El Niño-Southern Oscillation (ENSO*)* is the technical name of the *El Niño*. This name is from the Spanish for "the little boy," in reference to the Christ child, because its appearance is usually noticed by the arrival of unusually warm water off the coast of Peru around Christmas time. They are signaled by anomalies in the pressure fields between the tropical eastern Pacific and Malaysia. These anomalies yield an atmospheric signature called the Southern Oscillation Index (SOI) (which reflects the monthly or seasonal fluctuations in the air pressure between Tahiti and Darwin, Australia). If the index is negative (i.e., lower pressure at Tahiti and higher pressure at Darwin), then an *El Niño* or warm ENSO event occurs.

Another associated condition that develops during *El Niño* events is a generally high-pressure anomaly over the northern half of North America and a huge low-pressure anomaly over the eastern North Pacific Ocean that hugs fairly close to the North American shoreline (mostly in the winter time). This forces storm-track enhancement along the southern edge of the low that aims the storm tracks straight at California from which some of them migrate all the way across the continent!

Three factors, the relatively low pressure over the Pacific Ocean, the thermal expansion of warm water, and a weakening of the **trade winds** (during normal conditions, these strong winds generate a very strong westerly flow of the surface waters to the west in the equatorial region) cause an increase in water level on the California coast during *El Niños* (as much as 8-12 inches). Such an increase in water level leads *to significant coastal inundation and sea cliff erosion during the intense storms that are common during El Niño events* (Inman and Jenkins, 2005). During these events, unusually heavy rains typically accompany the increased number of winter storms.

With regard to the Georgia Coast, Hoppe and Smith (2009) reported recently that the general trend for cyclones during the two *El Niño* seasons of 1982-83 and 1997-98 was for the so-called nor'easters (extratropical cyclones) that move across the southern states and up the east coast to be both more numerous and to follow more southerly routes than usual during these *El Niños*. Both of these factors would tend to promote a much greater increase in rainfall, as well as increased storm impacts, along the Georgia Coast during *El Niño* events.

Inman and Jenkins (2005b) noted that *hypotheses explaining the causes of unusual ENSO events are numerous and range from the extrusion of hot lava on the sea floor to variations in intensity of the sun's radiations. Most climate modelers favor a complex set of changes in the interaction of the atmosphere and ocean that cause fluctuations in trade winds, monsoon intensity, and sea surface temperature* (e.g., Pierce, Barnett, and Latif, 2000). Well, enough is now known about the *El Niños* that it is possible for weather forecasters to predict their arrival. However, to complicate things a bit, increases in global temperatures that we are now experiencing may make the ENSO events more intense and less predictable (Federov and Philander, 2000).

8 FRESHWATER RIVER SYSTEMS

GEOMORPHOLOGY AND HYDROGRAPHY

There are four primary piedmont (red-water) rivers and nine fairly large coastal plain (black-water) rivers in Georgia and South Carolina (see Figure 110). There are several fundamental differences between these two types of rivers, in addition to the location of their watersheds (discussed earlier). For example, piedmont rivers (within the Piedmont Physiographic Province) typically have steeper gradients, more flashy discharge, and less abundant wetlands than coastal plain rivers (within the Coastal Plain Physiographic Province). The piedmont rivers also have considerably higher flow rates than the coastal plain rivers, over five times as much in Georgia (see Table 3).

As shown in Figures 1, 7, and 111, the boundary between the Piedmont Physiographic Province and the Coastal Plain Physiographic Province is marked by the presence of what is known as the **fall line**. At that point, the waters of the stream cut down into the older igneous and metamorphic rocks of the piedmont and usually form a zone of rapids. For the most part, the rivers affected by the fall line are piedmont rivers. Seaward of the fall line, the river typically flows over a bed of unconsolidated sediments. During the Pleistocene glaciations, when sea level was hundreds of feet below its present level, the zone where the stream was cutting down into the older bedrock extended much further seaward than it does today. Accordingly, as noted earlier, a valley as much as a hundred feet deep in places was cut across the coastal plain and the present continental shelf as a result of the lowered ocean level. When sea level started to rise again, the carved valley started to aggrade (fill up) with sediments. As sea level rose, elevating the river's **base level**, the rivers in the coastal plain area eventually started migrating back and forth across the old valley, filling it in evenly with sediments. The sediments that have filled in the valley compose what is known as the **flood plain** of the river.

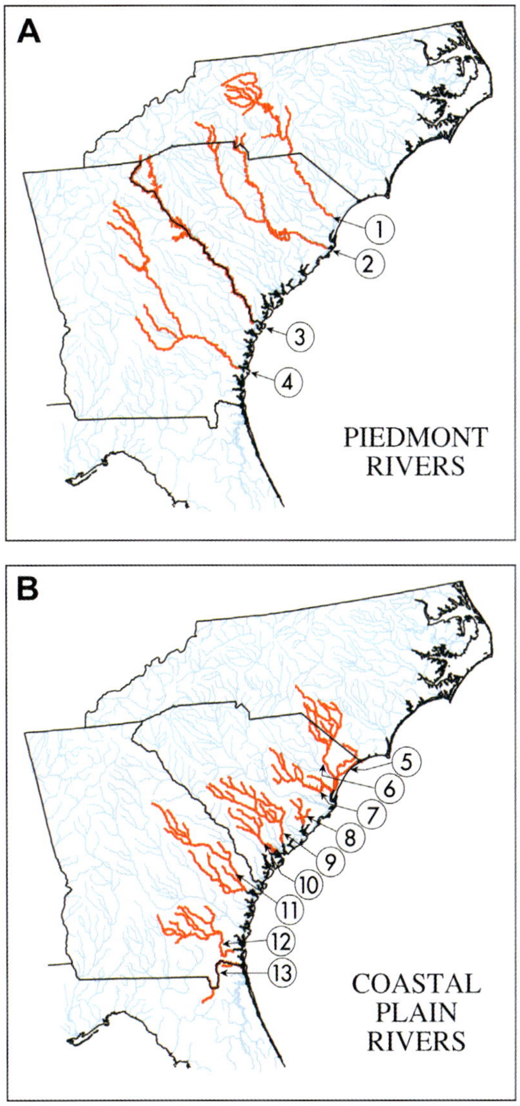

FIGURE 110. Major piedmont and coastal plain river systems occurring within Georgia and South Carolina. (A) Piedmont Rivers – Pee Dee (1), Santee (2), Savannah (3), and Altamaha (4). (B) Coastal Plain Rivers – Waccamaw (5), Little Pee Dee (6), Black (7), Cooper (8), Edisto (9), Combahee (10), Ogeechee (11), Satilla (12), and St. Marys (13).

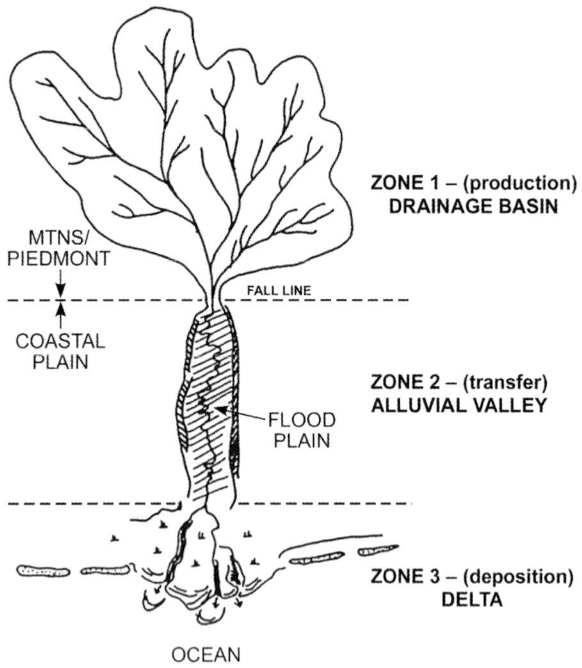

FIGURE 111. The three zones of the direct transfer of sediments from the uplands to the coast. Zone 1 – The watershed or drainage basin located in the Piedmont and Mountain Provinces. Zone 2 – The alluvial valley that cuts across the Coastal Plain Physiographic Province. Zone 3 – The coastal zone where most of the sediments accumulate in the deltas and estuaries. Note that the fall line is the boundary between Zones 1 and 2. The physiographic provinces in Georgia are mapped in Figure 6. Diagram highly modified after Saxena (1976).

Figure 111 illustrates, in a general way, the characteristics of a typical piedmont river as it crosses the coastal plain of Georgia today, particularly with respect to the delivery of sediments from the piedmont and mountain areas to the coast. In the case of the piedmont rivers, much of the sediment is "produced" in the drainage basin (or watershed) of the part of the river located upstream of the fall line. As noted earlier, from the fall line to the shore, the river has by now accreted a flood plain within the lowstand valley that was carved when sea level was considerably lower than it is today. Presently, some of the sediment carried by the river is deposited on the flood plains during major floods. At the same time, a considerable amount of the entrained sediment is carried to the shore by the river. Two of the these rivers in Georgia, the Savannah and Altamaha, have built deltas at the shore.

The geomorphology of the flood plain within these lowstand valleys that extend across the Coastal Plain Physiographic Province is illustrated in detail in Figure 112. Features such as a highly meandering main channel, with its prominent point bars and cutbanks, and abandoned channels (now oxbow lakes) are in abundance. Because of the differences in bedrock erodability and topographic variability, some of the piedmont rivers within the Piedmont Physiographic Province also have minor flood plains, but these narrow flood plains are invariably separated by a reach of rapids located in confined, erosional valleys.

The flood plains of the large piedmont rivers that cross the coastal plain of the southeastern United States are typically quite wide, several miles in some instances. The flood plains of the larger coastal plain rivers may also be more than a mile across. These coastal plain rivers are also located in lowstand valleys (see Figure 7), which are somewhat smaller, and presumably, shallower, than those carved by the piedmont rivers. However, because of their flatter gradients and the limited area of their watersheds (confined to within the coastal plain), none of them has as yet built a significant river delta at the shore.

Hydrographic data obtained from the U.S. Geological Survey are highly detailed and complex, but much of relevance is to be learned from them. The hydrographs for three separate months in 2012/2013 for the Altamaha, a piedmont river, compared to the Ogeechee, a coastal plain river, are shown in Figure 113. Note that the discharge of the Altamaha is considerably flashy, with flood peaks occurring in January and May, but that the Ogeechee discharge was quite steady. During low flow periods in October, both rivers had steady discharge rates.

This contrast in the hydrographs of the two river types is determined by the different rates of infiltration and on-land water storage during and after rainfall. Generally speaking, the soils of the coastal plain, particularly the upper coastal plain, are sandy and quite permeable (i.e., they have a high infiltration capacity), whereas the lateritic clayey soils of the piedmont are not. Another factor is the abundance of swamps and bottomland hardwoods scattered throughout the coastal plain, which act as large "storage tanks" for the runoff from the rains. Extensive swamps are absent from the flood plains and drainage basins of the piedmont. These two factors combined promote rapid runoff in the piedmont and slow runoff in the coastal plain, hence the

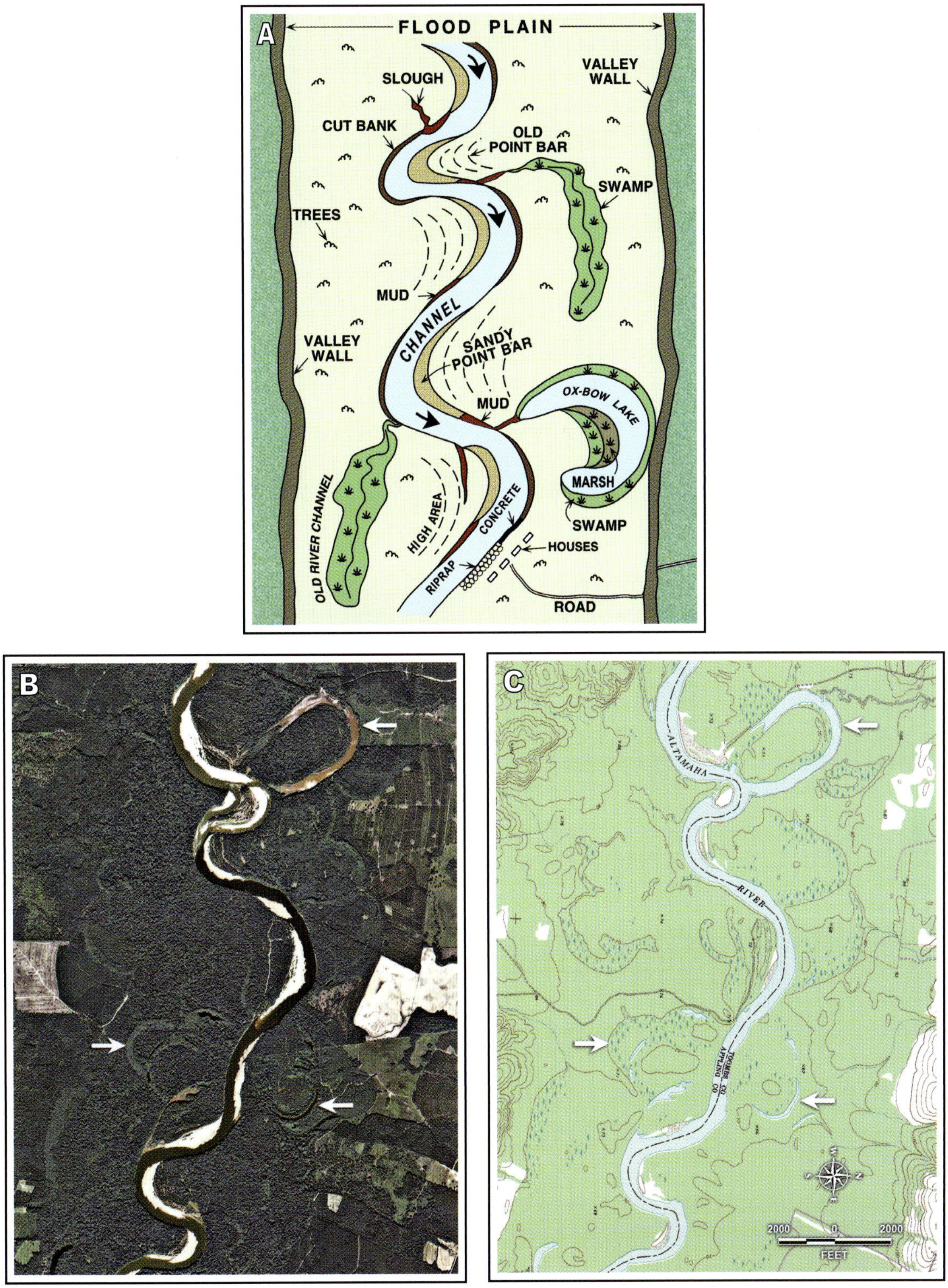

FIGURE 112. The morphology of the flood plain of a typical major piedmont river where it crosses the coastal plain of Georgia. These flood plains are typically several miles across. (A) General Model. (B) Flood plain of the Altamaha River, Georgia near Baxley. Arrows point to abandoned oxbow lakes. Digital orthophoto acquired in 2009, courtesy of USDA-FSA Aerial Photography Field Office. (C) USGS topographic map of the area shown in B.

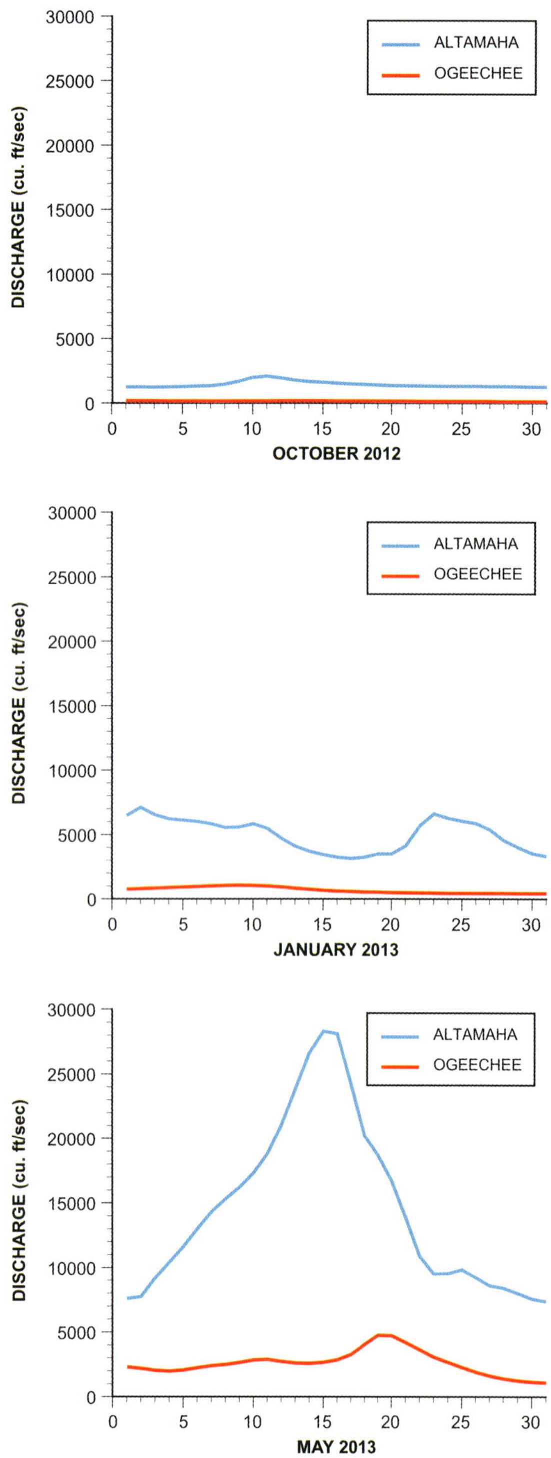

FIGURE 113. Hydrographs for one piedmont river (Altamaha) and one coastal plain river (Ogeechee) in Georgia during three separate months in 2012/2013. Note the flashy discharge of the piedmont river compared with the relative steady flow of the coastal plain river during the same time frame. Data from the U.S. Geological Survey.

striking contrast in the hydrographs presented in Figure 113.

BOTTOMLAND HARDWOODS

Freshwater marshes are rare in the piedmont and coastal plain river flood plains in Georgia, the dominant wetland type being **bottomland hardwood ecosystems**. Clark and Benforado (1981), Mitsch and Gosselink (2000), and Taylor et al. (1990) described the zonation of the bottomland hardwood ecosystems of the southeastern United States relative to the main channel of the associated stream. Six zones were described, ranging from Zone I, the permanently wet stream or river itself, to Zone VI, the transition zone between the floodplain and the uplands that is rarely flooded. Zones II – IV are occupied by bottomland hardwoods.

These zones have very distinct plant assemblages. Zones II and III of the bottomland hardwoods are the most frequently flooded (more than 50% of the years of record and typically >25% duration during the growing season), and Zones IV and V considerably less so. For purposes of discussion in this book, we have combined Zones II and III into one class, referred to as **cypress-tupelo swamps**, because two readily recognizable tree species, bald cypress (*Taxodium distichum*) and water tupelo (*Nyssa aquatica*), are the predominant vegetation of the two classes and their ecological conditions are quite similar. Zones IV and V are also lumped into a single class, **upper bottomland hardwood forests**, that has a much more diverse population of plant species. A generalized flood plain cross-section showing the different zones of the bottomland hardwood ecosystem as we have combined them, as well as descriptions of the wetting conditions and lists of the most common trees in each of the redefined zones, is shown in Figure 114. Photographs representative of the two primary zones are given in Figure 115.

Our observations of piedmont and coastal plain rivers in Georgia revealed striking differences between the flood plains of the two river types. Typically, piedmont rivers in the Piedmont Physiographic Province have high, narrow flood plains, several feet above the mean water level of the stream. No cypress/tupelo swamps are present. Upper bottomland hardwood forests are present in places but are not common. The sediments are usually sandy. Coastal plain rivers, on the other hand, have low, relatively wide flood plains, commonly only inches

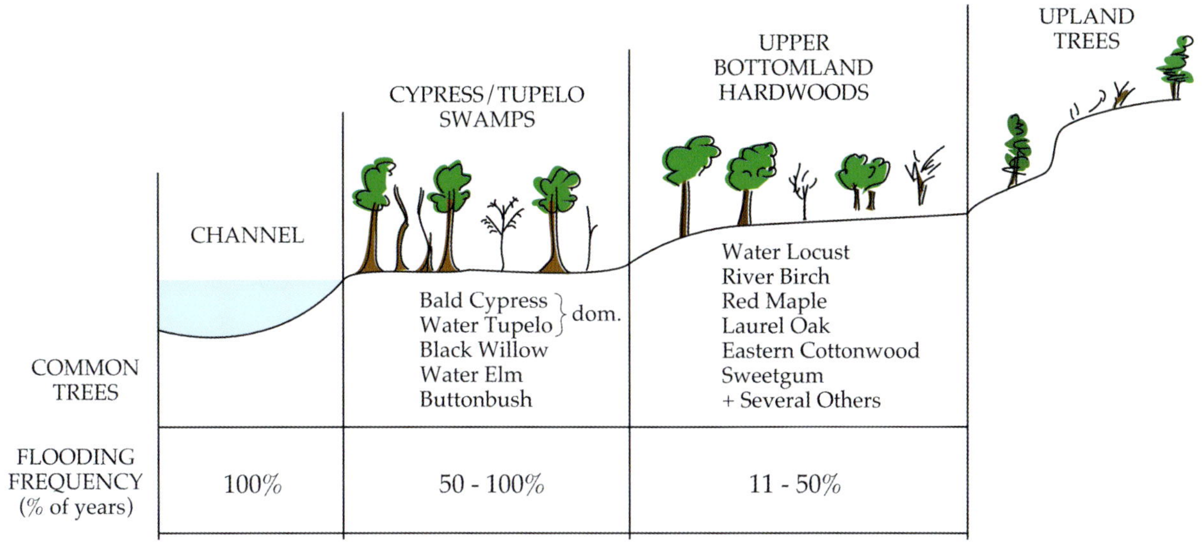

FIGURE 114. Topographic cross-section illustrating the components of typical bottomland hardwood ecosystems in the southeastern United States. Modified after Clark and Benforado (1981), Mitsch and Gosselink (1986), and Taylor et al. (1990).

above the mean water level of the stream, and cypress-tupelo swamps are present in extensive areas across the flood plain.

We believe the high flood plains of the piedmont rivers in the Piedmont Physiographic Province are the result of the flashy discharge of the streams. After heavy rains, the river rises quickly to a high level and sediment is transported up to the surface of the flood plain, building it up because of both the extreme water level and the large transporting capacity the stream has for sediments. As a general rule, the coastal plain rivers have a steady discharge, as illustrated in Figure 113, minus the ultra high water levels, and the flood plains are not built up as high as those on the piedmont rivers upstream of the fall line. With the low banks on the main channel, these rivers frequently "leak" into the adjacent cypress-tupelo swamps.

Where the two piedmont rivers present in Georgia–Savannah and Altamaha Rivers–cross the coastal plain (Figure 110A), they are characterized by: 1) Flood plains

FIGURE 115. Examples of bottomland hardwood ecosystems. (A) Upper bottomland hardwoods. (B) Cypress/tupelo swamp. The closest tree in the center of the photo is a water tupelo (*Nyssa aquatica*) and the large tree in the middle distance is a bald cypress (*Taxodium distichum*). Note the abundant cypress knees around the cypress tree.

several miles wide; 2) Multiple abandoned channels which contain oxbow lakes in places; 3) Extensive cypress-tupelo swamps where the abandoned channels have been partially filled with sediments; 4) Very extensive upper bottomland hardwood forests, some with old growth trees; and 5) Meandering channels with abundant sandy point bars along the outer bends of the meanders (Figure 112). Coastal plain rivers have similar, but smaller scale, features in their flood plains.

9 CONTINENTAL SHELF

Though not a place you are apt to visit during your naturalist explorations on the Georgia Coast, unless you own or hire an ocean-going vessel, the **continental shelf** represents the end member of the coastal zone continuum. The shelf off Georgia is quite wide, extending an average of about 70 miles off the beach. A distinct increase in slope at water depths of around 600 feet marks the edge of the shelf and the beginning of the **continental slope**. Figure 116 shows some details of the bottom topography of the shelf. It is clear that the continental shelf is not a boring flat surface, but has an abundance of topographic features, including:

1) As can be seen in Figure 116, there is a triangular bulge projecting off the continental slope margin that is located directly offshore from the mouth of the Santee River. That promontory is most likely what geologists call a **lowstand delta**. During one of the major glaciations, sea level was low enough for the mouth of the river to spill its sediment over the edge of the slope forming the prominent protuberance at the shelf edge. This particular delta has not been studied, but several ones at the edge of the shelf in the Gulf Coast region have (e.g., Sydow and Roberts, 1994).

2) Several shoal complexes exist on the shelf closer in to shore in the vicinity of the present mouth of the Santee River (Figure 116). According to research carried out by one of our former colleagues, Walter J. Sexton, the shoal directly off the river mouth is an ancient delta deposit formed during a stillstand episode when sea level paused during the rise after the last glaciation (Wisconsin).

3) The bathymetric data of Figure 116 also shows the presence of gigantic ebb-tidal deltas, as for example, the one directly off the entrance to Port Royal Sound, South Carolina. Several large ebb-tidal deltas also occur along the Georgia Coast (some of which are illustrated in Figure 53).

These features indicate that the post-glaciation rise of sea level was anything but steady, with a number of pauses, or stillstands, taking place from time to time.

Another consequence of the sea-level fluctuations during the Ice Ages was the carving of valleys (**lowstand valleys**) across the continental shelf when sea level was at its lowest levels. Henry and Idris (1992), reporting on a minerals assessment study of the Georgia continental shelf, stated that *the seismic records illustrate numerous incised valleys* but they did not reconstruct *Quaternary paleodrainage networks*. Henry et al. (1981), reporting on an earlier shelf survey, stated that *subsurface data show evidence of two possible fluvial erosion surfaces inferred to have been cut during sea-level lowstands in the Pleistocene and Pliocene. Large paleochannels incise to depths of -40 m* (131 feet) *MSL*.

A cross-section of the underlying geology of the continental shelf oriented directly off St. Catherines Sound and continued out to slightly beyond the shelf break was presented by Popenoe (1992). This cross-section, given in Figure 117, is based on a seismic-stratigraphic study that used existing U.S. Geological Survey high-resolution seismic-reflection profiles correlated with a number of core holes drilled into the shelf sediments and rocks. The oldest rocks shown on this section are Paleocene rocks located in the subsurface off the edge of the continental slope. Rocks of Eocene, Oligocene, Miocene, and Pliocene age (of the Paleogene and Neogene periods; Table 1) are mostly flat lying except where they dip offshore near the shelf edge. Quaternary period deposits are also present, especially near the shelf edge and on the shelf not too far from shore. Varying levels of the sea during the 60 million years or so represented by these rocks and sediments have resulted in some erosional events that have deleted representative sediment of certain intervals of time, as noted earlier, but, in general, this has been a relatively stable trailing edge of the North American Plate with relatively continuous deposition on the shelf and nearshore zone during much of this time.

Another detailed study sponsored by the U.S. Geological Survey showed that seven major ancient lowstand valleys occur on the inner shelf in the area from

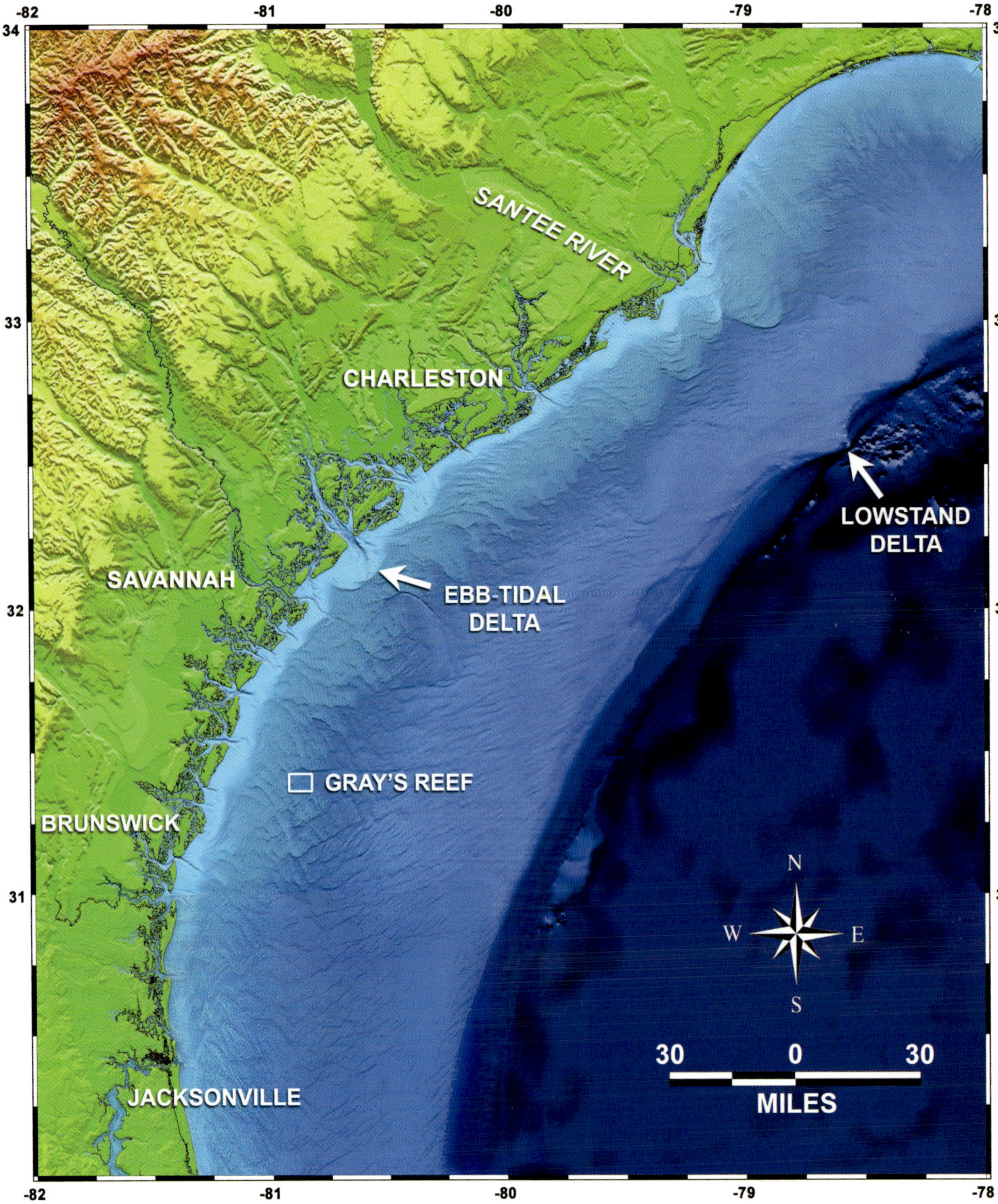

FIGURE 116. Bathymetry of the continental shelf off the South Carolina and Georgia shorelines. Note the lowstand delta at the edge of the continental shelf (arrow), as well as the numerous relict deltas further landward on the shelf, off the present mouths of the Santee and Pee Dee Rivers (South Carolina). The massive ebb-tidal delta off Port Royal Sound, South Carolina (arrow) is also clearly shown. From the National Geophysical Data Center.

the present mouth of the Pee Dee/Waccamaw Rivers to the North Carolina/South Carolina border, a distance of approximately 56 miles (Baldwin et al., 2004; 2006). When sea level was rising, these valleys were filled with sandy sediments in places as the streams built their flood plains within the valleys. Such sandy deposits have been dredged and used in a number of beach nourishment projects in the Myrtle Beach, South Carolina area. According to Clark Alexander (pers. comm.) inlet fill material has not been used for renourishment projects on the Georgia Coast.

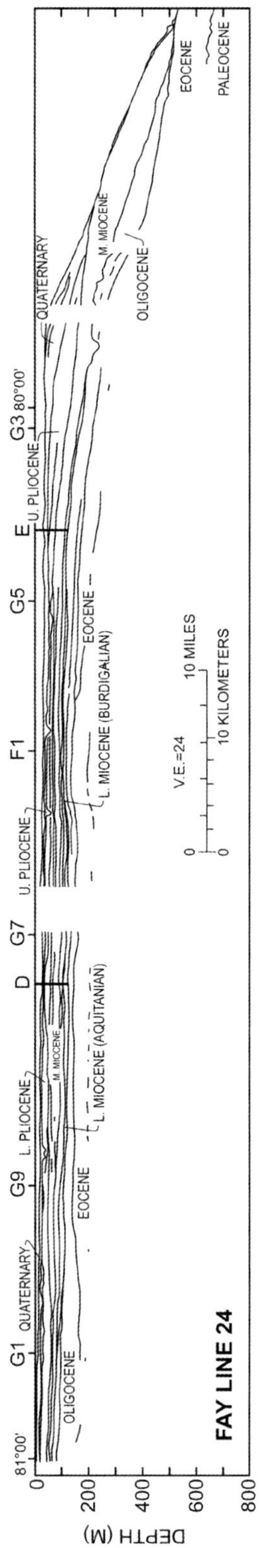

FIGURE 117. Geologic cross-section of the continental shelf of Georgia. Modified after Popenoe (1992). Line oriented ESE straight off St. Catherines Sound.

SECTION II
Coastal Ecology Overview

Caretta caretta (loggerhead sea turtle). Photo taken by Richard Daniel on 17 April 2006.

10 COASTAL ECOLOGY OF GEORGIA

In this section, we discuss the dominant marine and coastal animals that make the Georgia Coast such a great place to visit and enjoy. The wetland systems were discussed in Section I. In this Section, we briefly describe the life history, habitats, and conservation issues for marine mammals, sea turtles, and birds. In Section III, we point out the key species associated with the different regions of the coast.

MARINE MAMMALS

Though there are many marine mammals in the waters offshore of Georgia, the marine mammals of particular interest include North Atlantic right whales and bottlenose dolphins. The North Atlantic right whale (*Eubalaena glacialis*) is a large baleen whale, meaning they have hairy plates for filtering food from water, rather than having teeth. It is the rarest of all large whale species and among the rarest of all marine mammal species, being almost wiped out by commercial whaling (International Whaling Commission, 2001). The North Atlantic right whales were listed as endangered under the Endangered Species Act in 1970. The worldwide population is not known, but the western North Atlantic population is about 300-400 individuals.

The only known calving grounds for this critically endangered species are the coastal waters off northeastern Florida and Georgia. As best we know, three-quarters of all known right whale births occur in this area (Kraus et al., 1993)! Because of the importance of this area to the survival of the species, it has been designated as "critical habitat" under the Endangered Species Act, meaning that there can be special management considerations or protections to make sure that the animals and their habitat are not affected. Because they are slow moving and spend most of their time near the surface and close to shore during migration, they are particularly vulnerable to ship collisions. To protect the remaining population, NOAA, that has the responsibility for managing North Atlantic right whales, has taken both regulatory and non-regulatory actions to reduce the threat of ship collisions, such as mandatory and voluntary vessel speed restrictions in certain places and times, aircraft surveys to alert the shipping industry about when whales are in certain shipping lanes, and mandatory reporting of ship strikes.

North Atlantic right whales spend most of their time in their feeding grounds in the Gulf of Maine, the Bay of Fundy, and further north. However, starting in November, the adult female whales migrate south to the relatively calm nearshore waters off Georgia and northeastern Florida to give birth and nurse their young calves (Knowlton et al., 1994). The females do not eat during this period; instead they rely on the fat reserves built up during their time on the feeding grounds. Though mostly adult females make this migration, a few juveniles and adult males also occur in this area each winter. By March, females and their young calves head back north. It is during this migration that they are most vulnerable to ship strikes.

We could not find any information on "whale watching" tours along the Georgia Coast. However, you might be able to see these great animals on coastal fishing trips.

The most common marine mammal seen along the Georgia Coast and estuaries is the bottlenose dolphin (*Tursiops truncatus*). They are very social animals, often seen in groups of 2-15 nearshore, though there can be several hundred in offshore herds. They live a long time: 4-45 years for males and more than 50 years for females. They feed on a variety of prey, using echo-location to detect and capture food. We have seen them working together to herd fish onto a mud bank. It is quite exciting to see.

The biggest threats to the bottlenose dolphin are injury and mortality from fishing gear, such as gillnet, seine, trawl, and longline fishing, and viral outbreaks that might be linked to exposure to contaminants.

SEA TURTLES

The majority of sea turtles that nest on Georgia beaches are loggerhead sea turtles (*Caretta caretta*), the most abundant sea turtle occurring in U.S. waters. It was listed as threatened throughout its range in 1978 under the Endangered Species Act. Loggerheads mostly eat shellfish, such as horseshoe crabs, shrimps, clams, sea urchins, and mussels, which is why they have large, crushing jaws that is their namesake.

Loggerhead sea turtles have four life stages (Conant et al., 2009):

1) The hatchling stage begins when they emerge from their nests, although it takes them 4-7 days between when they first start to poke holes in their shells (called "pipping") until they emerge from their nests. The hatchlings usually emerge all at once and at night. They immediately head for the ocean, and begin a "swim frenzy" once they reach the ocean. Bright lights on land can confuse hatchlings, causing them to head toward land, thus the need to reduce artificial lighting on nesting beaches and adjacent land. They use magnetic cues and wave energy to head toward the North Atlantic gyre.

2) The oceanic juvenile stage lasts for 6.5-11.5 years, during which they feed while rotating along the gyre. Little is known about where they go and what they eat (scientists call these the "lost years"). However, it appears that patches of floating seaweed (*Sargassum*) are important.

3) The neritic juvenile stage lasts for ages 12-28 years during which they migrate to foraging grounds along the eastern U.S. continental shelf in water less than about 500 feet deep.

4) The adult stage begins at approximately age 28-30 years when they reach reproductive maturity. Tagging studies on Little Cumberland Island show that females can return to nest on the same beach for up to 25 years!

You will have the highest probability of seeing loggerheads when the females come ashore to nest. Nesting occurs mostly from mid May through mid August (Table 4). The number of eggs per nest, or clutch, ranges from 23-198, with a mean of 112 eggs. Females deposit clutches every 10-14 days; the same female will come ashore and deposit clutches 3-4 times per nesting season. The eggs take 50-70 days to hatch, and hatching occurs on Georgia's beaches between early July and mid October.

Most turtle nesting beaches in Georgia have programs to monitor for and protect the nests as soon as they are found. The Little Cumberland Turtle Project was started in 1964; it is the oldest loggerhead turtle project in the world. Little Cumberland is not currently considered a high-quality nesting beach because a much of the nesting habitat is highly eroded. The Blackbeard Island National Wildlife Refuge sea turtle project has the longest continuous running data set for sea turtles in Georgia, starting in 1966. The refuge has the highest density of nests per mile than any other island in Georgia. Cumberland Island has the highest number of nests because it is long, completely undeveloped, and only a small portion of the nesting habitat is highly eroded.

Table 5 shows that the number of sea turtles nests

TABLE 4. Sea turtle nesting numbers by beach for 2012. From Sea Turtle Nest Monitoring System, Georgia DNR Sea Turtle Conservation Program.

Beach	Unk.Date	Apr	May	Jun	July	Aug	Sept	Overall	
	Nests	Nests	Nests	Nests	Nests	Nests	Nests	Nests	Emergence Success (%)
Blackbeard	17	0	68	110	101	13	0	306	58.7
Cumberland	44	2	140	280	222	20	1	700	71.7
Jekyll	3	0	52	84	56	2	0	197	61.7
Little Cumberland	4	0	17	29	33	1	2	83	72.3
Little St. Simons	2	0	23	39	50	2	0	116	70.1
Little Tybee	0	0	2	4	9	0	0	15	57.1
Ossabaw	10	1	46	100	67	5	1	226	47
Sapelo	5	0	36	55	49	2	0	147	46
Sea Island	1	0	23	39	35	4	0	102	60.7
St. Catherines	28	0	41	65	46	9	7	185	51.4
St. Simons	0	0	3	3	0	0	0	6	64.3
Tybee Island	1	0	4	7	10	1	0	23	76.8
Wassaw	1	0	34	55	50	0	0	139	60.2
Total	115	2	489	870	728	59	11	2,245	61.9

in Georgia have been increasing for the last three years, which is great news. Surely the hard work of the numerous volunteers has paid off. However, nest counts are declining in many other areas and for the regional population as a whole (Conant et al., 2009). There are many on-going threats to the long-term recovery of sea turtles in Georgia: 1) Nest predation by hogs, raccoons, and dogs; 2) Nest failure due to tidal inundation and beach erosion; and 3) Drowning of juveniles and adults trapped in longline and shrimping nets. In Table 5, note that about half of the nests are "relocated." That means trained staff dig up the eggs shortly after they are laid in areas at risk of high predation, inundation, or erosion, then move them to protected locations where they have a better chance of survival. Other management techniques include covering newly identified nests with netting or even above-ground cages to reduce losses from predation.

Even though the number of emerged hatchlings on Georgia's beaches looks high, the survival rate from hatchling to adult is about 1 adult per 1,000 hatchlings (Frazer, 1986). The annual mortality rate for adult sea turtles that nest in Georgia is estimated to be about 20% per year. So, you can see why they are threatened!

MARINE AND COASTAL BIRDS

The Georgia Coast is a great place to observe birds. Much of this information was gleaned from a chapter in a literature synthesis written by Patrick Jodice, a professor at Clemson University (published in Michel, 2013). Figure 118 shows the National Audubon Society Important Bird Areas (IBA) and coastal marine protected areas (MPA) on the Georgia Coast.

If you are interested in birdwatching during your visit to Georgia, you should check out the Georgia Colonial Coast Birding Trail (http://www.georgiawildlife.com/ColonialCoastBirdingTrail). More than 300 species of birds (75 percent of the total species of birds seen in Georgia) have been spotted at the 18 sites along the birding trail.

Birds are discussed as four guilds: seabirds, shorebirds, wading birds, and ducks/divers. You may be able to see some pelagic seabirds from the shoreline, such as magnificent frigatebird and northern gannet; however, you will mostly need to go on a fishing charter to see offshore birds. As of 2012, there were no regular pelagic birdwatching offshore charter boats being advertised; however, groups do organize some trips.

Seabirds

Seabirds are species that forage in the marine environment over open water. The following seabirds nest in Georgia (see also Table 6):

Brown pelicans – The first record of brown pelicans nesting in Georgia is from 1988 along the north end of St. Catherines Island, and a second nesting site was located in 1990 in the mouth of the Satilla River (Wilkinson et al., 1994). Nests also occur on Little Egg Island Bar and the Brunswick Harbor Dredge Island. Brown pelicans mostly nest on islands to avoid predators, such as raccoon and feral cats, because they usually nest on the ground in dense colonies. Georgia's breeding population was estimated to be about 200 pairs in 1988, 600 pairs in 1991, and 3,600 pairs by 1999; however, in the past few years, the population has declined to <1,000 pairs, partly due to raccoons on Little Egg Island Bar (Michel, 2013). Breeding season is from April to September. Brown pelicans were listed as endangered under the Endangered Species Act in 1970 (mainly due to exposure to pesticides that led to egg thinning and breeding failure). With the ban on many of these pesticides, brown pelicans have recovered to the point that they were delisted throughout the Atlantic Coast in 1985.

Laughing gulls – These birds are common and occur year round. Laughing gulls can be seen in dense flocks following shrimp trawlers as they discard bycatch (the fish they catch in their nets but throw back).

Royal terns – These crested terns are present year round and breed in dense nesting colonies. The only royal tern nesting site in Georgia (on a dredge spoil island in

TABLE 5. Results of the Sea Turtle Nest Monitoring Program, managed by the Georgia DNR Sea Turtle Conservation Program for the period 2009-2011.

Year	Number of Unique Females	Total Number of Nests	Nests Relocated	Number Emerged Hatchlings	Mean Emergence (%)
2011	496	2003	833 (42%)	106,067	58
2010	547	1774	904 (51%)	115,255	62
2009	345	1005	490 (49%)	61,028	55

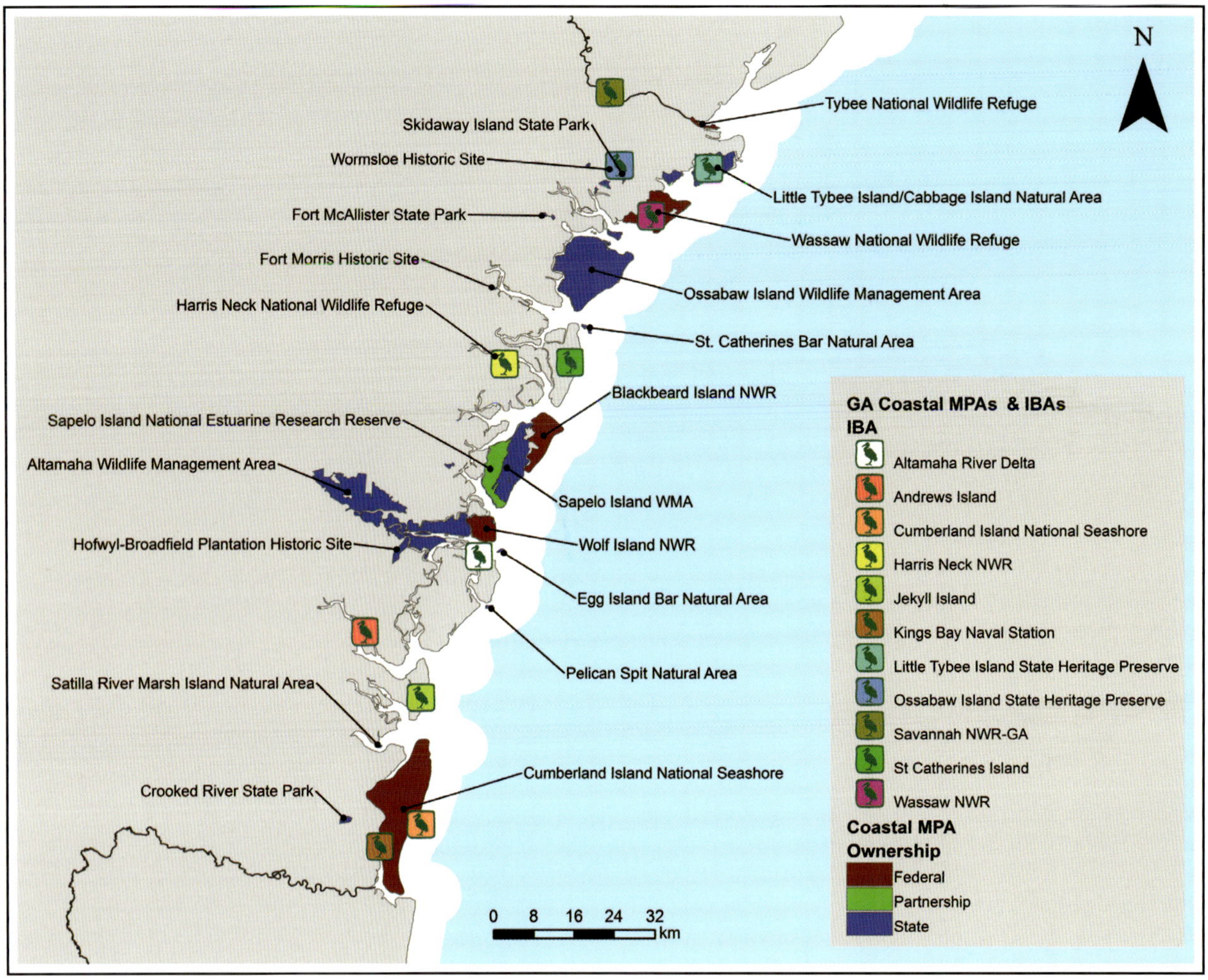

FIGURE 118. National Audubon Society Important Bird Areas (IBA) and coastal marine protected areas (MPA) on the Georgia Coast.

Brunswick Harbor) had 6,650 nests. Breeding season is from April to September, and each pair lays a single egg; the pair will re-lay if the hatchling is lost. You may also see sandwich terns, but they do not nest in Georgia.

Least terns – The eastern least tern is listed as "rare" in Georgia. It nests on beaches just above the highest tides but also on rooftops, and pairs usually lay two eggs. You will see them feeding in the surf zone along most beaches; they fly parallel to the beach, then hover and dive for small fish. If you are lucky, you will see the male bringing several fish to the female during mating and nesting, and to the chicks after they are born. The chicks are "semi-precocial," meaning that they are hatched with eyes open, covered with down, and capable of leaving the nest soon after hatching. However, they stay at the nest and are fed by parents.

Gull-billed terns – These birds are widely distributed but nest at few sites in Georgia, often in association with black skimmers. They are interesting in that they primarily feed on insects taken in flight, but they also eat fiddler crabs and fish.

Black skimmers –These birds are very distinctive, being black on top and having long bills that are red at the base and black at the tip. They feed by flying right above the water surface, with the lower mandible dragged in the water to detect and grab small fish. They are present year round and breed in only two locations in Georgia (Table 6).

Shorebirds

Nesting shorebirds in Georgia include American oystercatcher, black-necked stilt, willet, and Wilson's

TABLE 6. Major seabird nesting colonies in Georgia, nest counts in 2010.

Name	Brown Pelican	Royal Tern	Least Tern	Gull-billed Tern	Black Skimmer
Ossabaw	-	-	12	-	-
Little St. Simons Island	-	-	12	12	?
Little Egg Island Bar	150	-	-	-	-
Andrews Island	-	-	150	30	-
Brunswick Harbor Dredge	37	6,650	200	15	98
Satilla Marsh	441	-	-	-	-
Cumberland N. Beach	-	-	100+	-	-

plover. In 2010, the Georgia Department of Natural Resources conducted a comprehensive survey of the entire Georgia Coast to count territorial pairs of two species of particular concern: American oystercatchers (116) and Wilson's plovers (359), a significant increase since the last survey in 2000. They set out signs in areas with high numbers of birds and potential for human disturbance, so watch for those signs as you walk the beaches. These birds nest from right above the high-tide line up to the dunes from April to July. Wilson's plovers along the Atlantic coast are listed as threatened, meaning that the population would continue to decline if not protected. American oystercatchers are listed as a Species of High Concern in shorebird conservation plans because of its overall low population (11,000 birds) and widespread habitat loss.

There are many migratory and overwintering shorebirds in Georgia, including sandpipers, dowitchers, dunlin, yellowlegs, turnstones, plovers, whimbrels, and willets. Along the Atlantic coast, migratory shorebirds are more common during the fall migration; during spring, the central flyway is a more likely route. Georgia hosts the single largest concentration of wintering semipalmated plovers in the U.S.; it is also an important concentration area along the Atlantic coast for wintering piping plovers (listed as threatened throughout their wintering range) that come to this area from all three breeding populations (Atlantic Coast, the Northern Great Plains, and the Great Lakes).

In Georgia important migratory/wintering sites for coastal birds include Tybee and Little Tybee, St. Catherines, Sapelo, St. Simons, Jekyll, and Cumberland Islands.

Wading Birds

Wading birds, such as egrets, herons, ibises, and wood storks, are colonial nesters that usually nest in trees adjacent to water. As large birds, they are easily seen feeding in shallow water along tidal channels, tidal flats, ponds, roadside ditches, and even on beaches. Wood storks are of particular interest because they were listed as federally endangered under the Endangered Species Act in 1984. Wood storks were first reported in Georgia in 1965, and now about 20 percent of the U.S. nesting population (about 10,000 breeding pairs) nests in Georgia in 28 colonies. In Section 3, we point out some good places to see wood storks. They are impressive; at about 3 feet tall, they are the largest wading bird that breeds in North America. The colonies do best during wet periods, when there is water below the nesting trees so that raccoons can't climb the trees and prey on the eggs.

Ducks and Divers

This group of birds includes dabblers, bay ducks, sea ducks, loons, and grebes. You will see lots of them mostly in winter because many breed in northern areas. Those that breed in Georgia include pied-billed grebe, wood duck, mottled duck, and common moorhen. The wood duck is the most common duck as well as breeding duck in Georgia. In the fall and winter, you may see American widgeon, black duck, bufflehead, canvasback, gadwall, green and blue winged teal, hooded merganser, northern pintail, northern shoveler, redhead, and ring-necked duck in interior wetlands. Along the open coast, you will see scaups, scoters, double-crested cormorants, loons, and grebes. Georgia is toward the southern limit of the wintering areas along the Atlantic coast, thus the number of birds varies significantly each year. Sometimes, the rafts of seaducks in winter can number in the thousands, particularly during colder winters.

SECTION III
Major Compartments

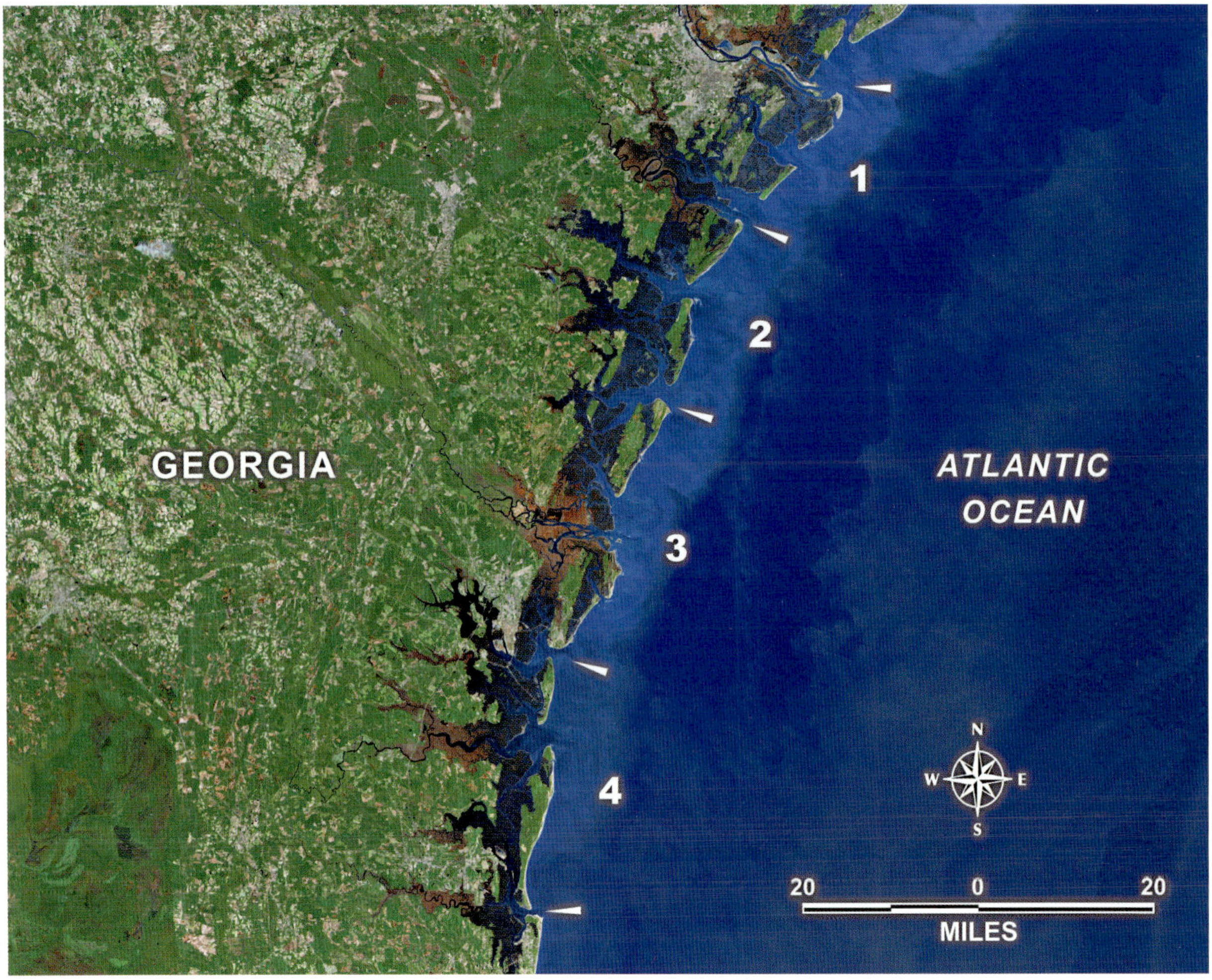

Landsat image mosaic acquired in 2000, courtesy of the NASA Applied Science & Technology Project Office.

11 COMPARTMENT 1 - SAVANNAH RIVER TO OSSABAW ISLAND

INTRODUCTION

For purposes of discussion, we have subdivided this compartment into five subcompartments (Figure 119):

1) **Savannah River Flood Plain and Delta** – The second largest river delta in the Georgia Bight.

2) **Tybee Islands** – The highly developed and continuously eroding Tybee Island plus a disorganized complex of salt marshes and isolated beach ridges called Little Tybee Island (on some maps).

3) **Wassaw Sound** – A two-mile-wide sound with very little freshwater influx.

4) **Wassaw Island** – The only large, isolated, and purely Holocene prograding barrier island on the Georgia Coast. However, there are three smaller ones – Tybee, Little St. Simons, and Sea Island.

5) **Ossabaw Sound** – The widest of the Georgia sounds, which also has the largest coastal plain (blackwater) river in the state emptying into it, the Ogeechee River.

SAVANNAH RIVER FLOOD PLAIN AND DELTA

Introduction

The Savannah River historically has delivered the second largest sediment load to the coast of any of the rivers in the Georgia Bight, second only to the combined output of the Santee and Pee Dee Rivers (Figure 120). Table 3, based on data collected between 1930/32 and 2008, gives a slightly lower flow rate for the Savannah River (11,454 ft^3/sec) than for the Altamaha River (13,323 ft^3/sec), which has no major dams along its course. However, the flow rates for the Savannah River at the present time are controlled to some extent by three major dams:

1) The dam that created the **Clarks Hill Lake** (if you live in Georgia) or the **J. Strom Thurmond Lake** (if you live in South Carolina) was constructed near the fall line about 18 miles upstream from Augusta, Georgia and filled with water in 1951/52 as part of a flood control, hydropower, and navigation project. This lake is the second largest man-built lake east of the Mississippi River. The largest is Kentucky Lake on the Tennessee River.

2) The **Hartwell Dam and Lake**, also a very large one, was created near Anderson, South Carolina in the foothills of the Blue Ridge Mountains between 1955 and 1963 for the same reasons given for the Clarks Hill Lake.

3) **Richard B. Russell Lake**, built by the U.S. Army Corps of Engineers (as were the other two) about half way between the two lakes just described, was filled with water in 1983/1984 for the purposes of flood control, hydroelectricity, recreation, additional stream flow regulation, water supply, and fish and wildlife management.

The lakes created by these three dams are visible on Figure 1. How well these dams have carried out their assigned tasks is not known to us in detail, except for a good effort in controlling floods. However, rest assured, they have done an exemplary job of keeping any sand grains heading for the coast from their source in the mountains or piedmont from reaching there. The data in Figure 120 shows that the sediment load in the Savannah River for the period 1985 to 2005 has been reduced to 14% of the sediment load before the dams were constructed.

It is probably a safe assumption that you would start your visit to this subcompartment in the city of Savannah. For a good introductory overview of the area you might visit, it would be worthwhile to study the map in Figure 119 and the satellite image in Figure 121 in order to make the following observations:

- The abundance of old rice ponds in the upper

FIGURE 119. Compartment 1, Savannah River to the north end of Ossabaw Island. Red arrows designate the boundaries of the compartment. Source: ESRI StreetMap USA.

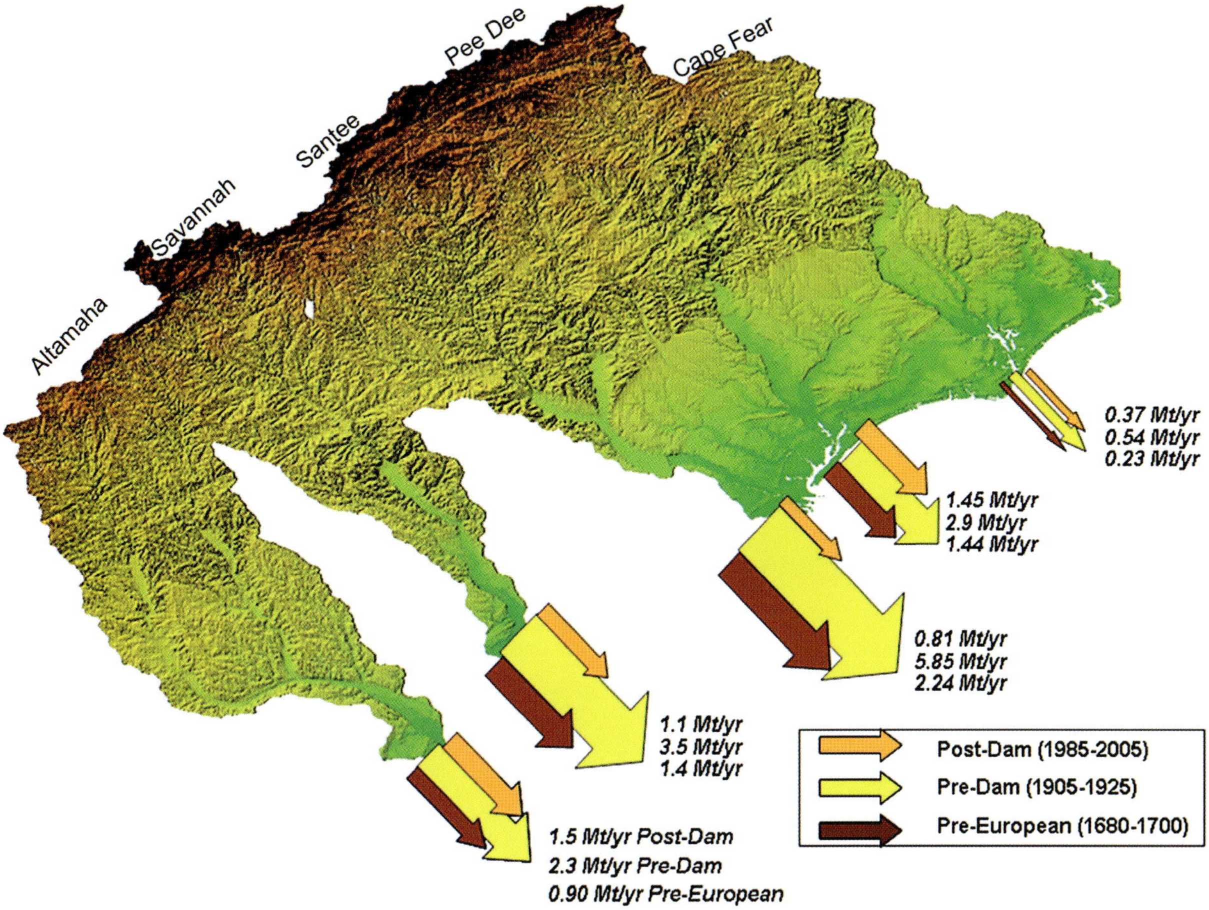

FIGURE 120. Sediment load (in millions of tons per year [Mt/yr]) from major rivers along the Georgia Bight (modified from McCarney-Castle et al., 2010). The three periods represent Pre-European (red), before deforestation occurred; Pre-Dam (yellow) when sediment loads greatly increased due to deforestation and agriculture; and Post-Dam (orange) when sediments became trapped behind the dams.

delta plain on the South Carolina side of the river (rectangular patterns in Figure 121), where the freshwater supplied by the Savannah River was used to support extensive rice plantations (now home of the Savannah National Wildlife Refuge). More than 10,000 acres were planted in "tide-flow" rice in the two decades before the War Between the States. An additional 6,500 acres were planted elsewhere on the Georgia Coast and islands (House, 1954).

- The city of Savannah, where you may want to spend some time. Great food and historic architecture, and who wouldn't want to visit the setting for the book *Midnight in the Garden of Good and Evil* by John Berendt.
- A series of protruding barrier islands along the delta front to the south (Tybee, Little Tybee complex, and Wassaw), building out as a result of the original input of sediment (including sand) from the river and the significant longshore southerly sand transport.
- Extensive salt and brackish marshes on both the northeast side of the main channel of the river in the lower delta plain (arrow D on Figure 121) and in a wide area behind the barrier islands south of the river mouth.

Morphology and Sediments

Going upstream from Savannah by boat, within about 10 miles one would reach the "head of tides" and the landward limit of the historic rice fields built and maintained by slave labor, an agricultural practice which started around the 1750s in Georgia and underwent an

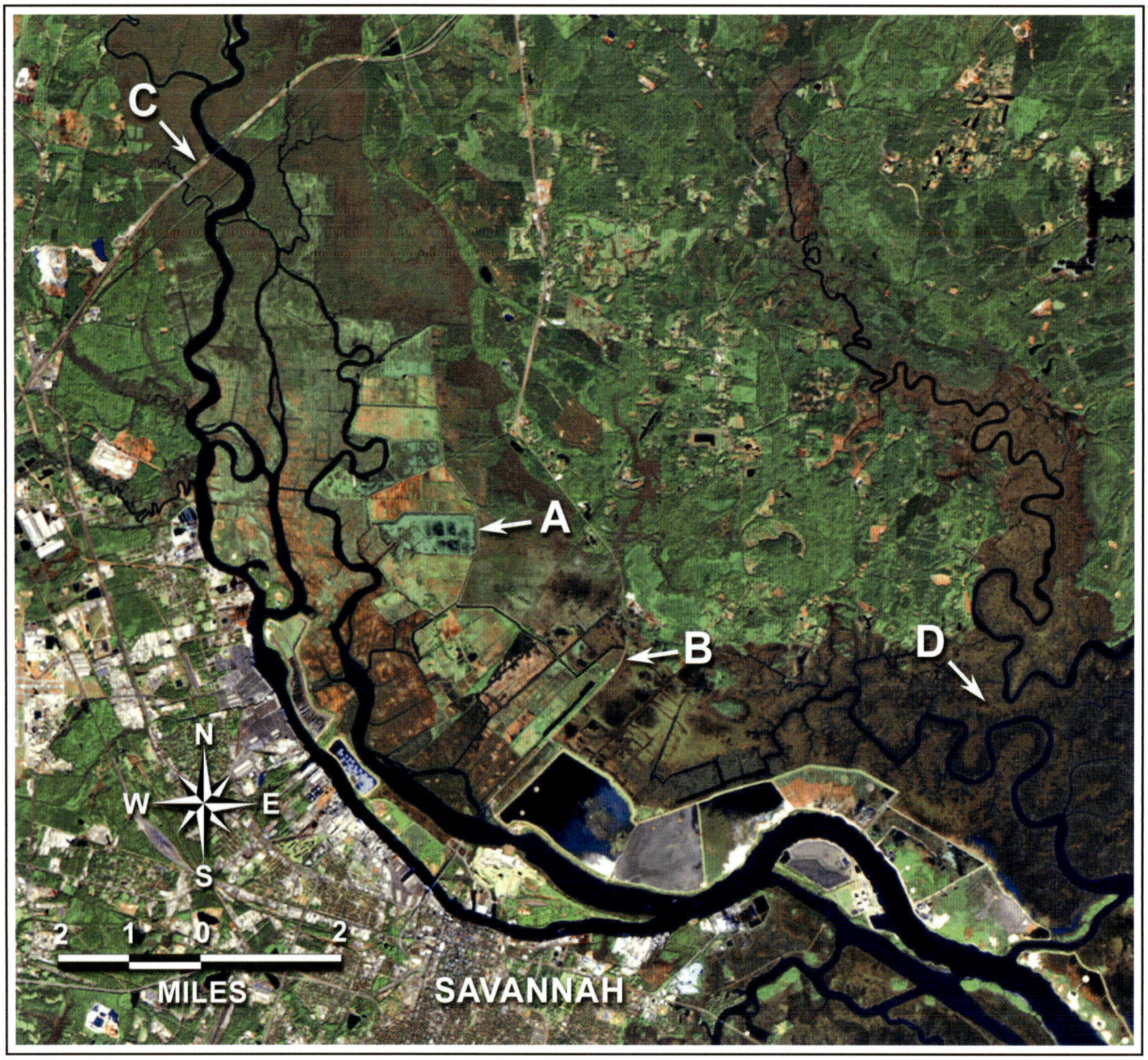

FIGURE 121. Upper/lower delta plain area of the Savannah River Delta, showing the location of the city of Savannah, the Savannah National Wildlife Refuge, and the marshes on the South Carolina side of the lower delta plain. Arrows: A – Savannah Wildlife Refuge; B – US 17; C – I-95; D – marshes of the lower delta plain (South Carolina side). Landsat image mosaic acquired in 2000, courtesy of the NASA Applied Science & Technology Project Office.

abrupt slowdown about 100 years later (discussed in considerably more detail later). Upstream of where the I-95 bridge crosses the river, a flood plain that averages around 3-4 miles in width is contained within a well-defined valley carved (at least partly) during the last lowstand of sea level. The morphology of the river contained within this valley is remarkably similar to the general model in Figure 112. As shown in Figure 7, the wide valley terminates at the fall line, on which the city of Augusta, Georgia is situated. In historic times, the fall line was where commercial boat traffic on the rivers coming from the coast ceased because of the rapids. Other major cites throughout the east and southeast located on the fall line include – Baltimore, Maryland; Washington, D.C.; Fredericksburg and Richmond, Virginia; Roanoke Falls, North Carolina; and Columbia, South Carolina.

As shown in Figure 120, before the dams were built, the Savannah River delivered a large volume of sediment to the coast, resulting in the second largest river delta in the Georgia Bight after the Santee/Pee Dee Delta in

South Carolina. It is also the second largest river delta on the east coast of North America.

We are not aware of any detailed studies of the geology of the Savannah River Delta similar to the ones our group has conducted on the Santee/Pee Dee Delta, South Carolina (Hayes et al., 1993; Hayes and Michel, 2008). A major difference is that the Santee/Pee Dee Delta is composed mostly of preserved natural areas, whereas the Savannah River Delta is in a highly populated area not so conducive to studies of natural habitats. It is likely that the two delta systems have a very similar geological origin and makeup.

In our surveys of the Santee/Pee Dee Delta, we subdivided it into three components – upper delta plain, lower delta plain, and delta front. The characteristics of these same delta subdivisions of the Savannah River Delta are given below:

- The **upper delta plain**, the landward limit of which coincides roughly with the "head of the tides," contains the slightly meandering main channel of the Savannah River as well as a number of side channels. The upper delta plane extends from approximately the U.S. 17 bridges (near the center of the city of Savannah) to the head of tides (near I-95). The Savannah National Wildlife Refuge is located within this part of the delta (see Figure 122B). All of the thus-defined upper delta plain and part of the lower delta plain are illustrated in the vertical image in Figure 121; an oblique aerial view of the upper delta plain is shown in Figure 122A.
- The **lower delta plain** of the Savannah River Delta extends from approximately the U.S. 17 bridges (near the center of the city of Savannah) to the landward side of the barrier islands. Because of the dominant southerly transport of sand historically delivered to the delta by the Savannah River, the delta is very asymmetric, with the land south of the river mouth projecting about 4.5 miles out into the ocean relative to land on the northern side (Figure 119). The lower delta plain of the recessed northern side of the delta contains highly meandering channels in the salt marsh (see examples in Figures 104B and 121). The southern side of the lower delta plain (south of the Savannah River) is a wide salt-marsh and estuary/sound complex that also averages around 4.5 miles in width. This part of the delta is essentially unattached to the upper delta plain, with interceding Pleistocene barrier island sequences of Silver Bluff and Princess Anne age (Figures 9 and 119). The "islands" named Skidaway and Wilmington on Figure 119 are composed of remnants of the Silver Bluff Pleistocene barrier island system formed 40,000-50,000 years ago. Clark Alexander (pers. comm.) thinks the Silver Bluff may be considerably older than this (as mentioned earlier), but in our discussions in this book, we are following the age given by Henry (2009) and Muhs et al. (2004).
- The **delta front** of the Savannah River Delta contains three main barrier island systems, the Tybee, Little Tybee complex, and Wassaw Island, and two major sounds, Wassaw Sound and Ossabaw Sound. Both of the sounds have massive ebb-tidal shoals and no flood-tidal deltas. Ossabaw Sound also contains the outlet of the Ogeechee River, the largest coastal plain (black-water) river in Georgia (Table 3).

Places to Visit

The lower reaches of the Savannah River Delta in South Carolina are mostly inaccessible by car, but the upper delta plain contains the **Savannah National Wildlife Refuge**, a site no visiting naturalist would want to miss, particularly those with a major interest in birds. If you are traveling south on I-95, take exit 5 off I-95 and go south on U.S. 17. In the vicinity of Limehouse, you pass through extensive pine forests that provide some of the pulp for the paper-mill industry in the area. A couple of miles south of Limehouse, turn right on State Road 170 and you are almost immediately on the upper delta plain and in the refuge.

If coming from Savannah, take U.S. 17 north across the Savannah River. From the mid point of the bridge, it is 6.7 miles to State Road 170 on which you take a left, almost immediately being inside the refuge. The valley wall is very subtle in this area. Going west, you pass by a swamp that was logged not too long ago. After leaving the swamp, you will begin driving across restored rice ponds with abundant birds, such as the glossy ibis and red-tailed hawk, plus immeasurable numbers of ducks (this road you will be driving on bisects the oblique aerial photograph of the refuge in Figure 122B).

To enter the refuge proper, go almost to the river and then take a left to begin a 4-mile driving tour on levees that extend throughout the restored rice fields. The last time we did this drive was on Thanksgiving weekend in 2006. As we began the drive, we looked to the right

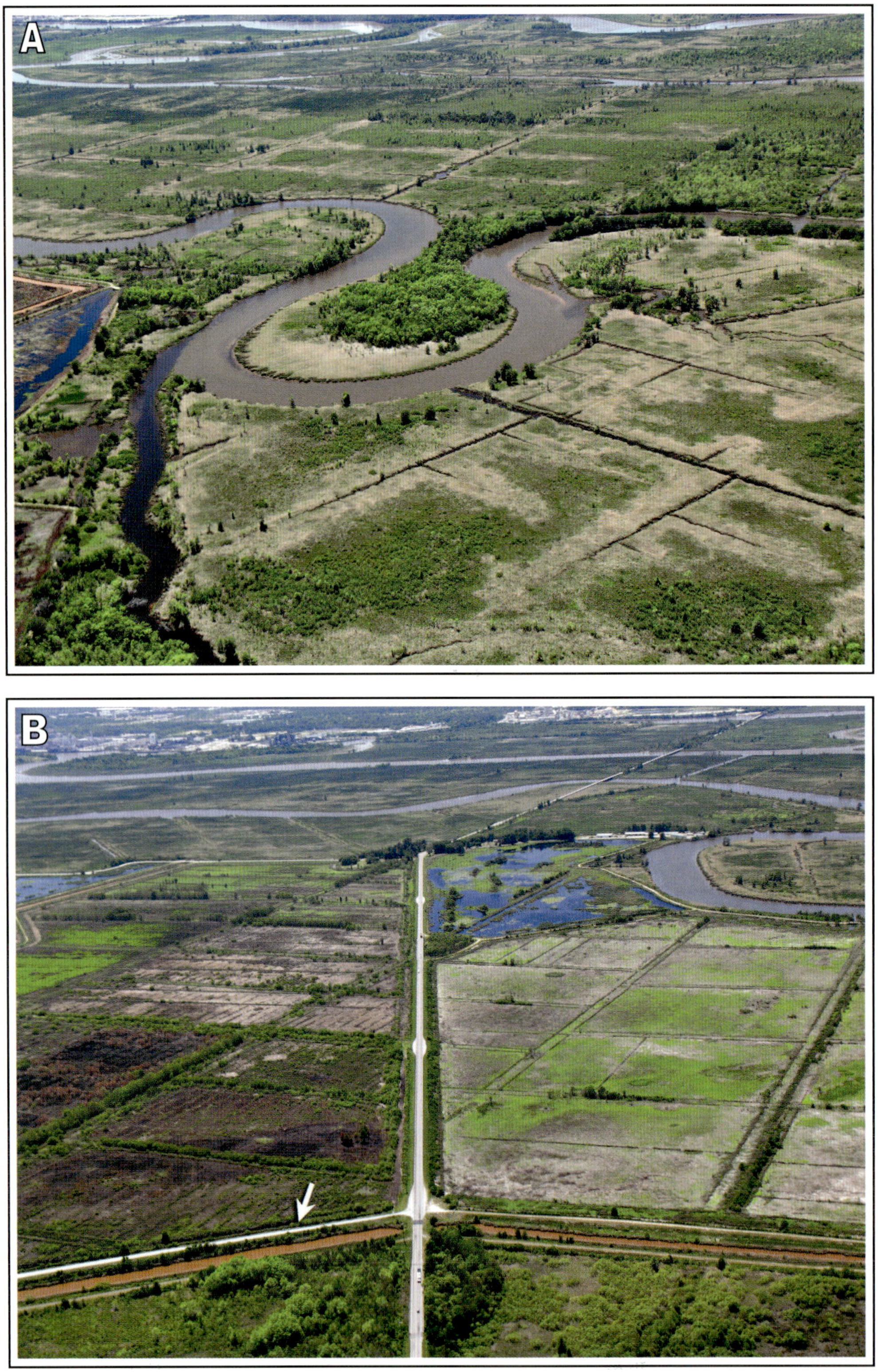

FIGURE 122. Upper delta plain, Savannah River Delta. (A) Near the upper limit of the upper delta plain. Rectangular patterns outline some of the old rice fields. (B) Center of the Savannah National Wildlife Refuge. State Road 107 bisects the photograph. The arrow points to the driving route described in the text. Both photographs taken on 28 April 2010.

where the port facilities for Savannah were conspicuous, as well as a paper mill.

Prescribed burning is one of the tools used by the refuge managers to eliminate the basis for exotic perennials, such as cattail, cutgrass species, and rattle bush. A sign by the drive explains the value of this process:

> *Annual plants soon colonize the bare soil and yield seeds of higher value to wildlife. Burning removes tons of dry, dead vegetation that could fuel a destructive wild fire. The high value species are wild millet, giant foxtail, and fall panicum.*

Just beyond mile 1, the road goes through an upland, island-like feature covered with trees, where another road sign informed us that:

> *This small island of trees was a slave community on a recessed plantation which bordered Laurel Hill Plantation called a* ***hammock****, a small area of high ground in a sea of rice fields. The round brick structure just ahead in the woods was a cistern that stored drinking water and perishable foods. .... Look for Chinese parasail trees with pale green trunks and 5-lobed leaves. The Asian trees were planted for shade on low country plantations. Notice periwinkle, a purple-flowered vine once cultivated as ground cover. Watch and listen for birds.*

If you follow that advice to watch for birds, you will not be disappointed, because they are everywhere!

At 1.5 miles, we spooked up a big flock of blue-winged teals as well as some hooded mergansers and coots. We didn't see any gators on this drive, but we remembered a similar drive in the late winter several years earlier when we came up on the largest gator we had ever seen while walking along one of the side trails. It was at least 13 feet long and covered with mud as it made its first "awakening" of the year.

At 3.5 miles, we stopped to admire what Hayes said was "The most ducks I have seen this side of Louisiana and Texas." There were thousands of them, ring-billed ducks, widgeons, and so on. We also saw flocks of ibises, abundant shorebirds, and several marsh hawks. In the early morning, this place has spectacular birding. If you are birding in coastal Georgia and South Carolina, you definitely do not want to miss the Savannah National Wildlife Refuge!

To see the lower delta plain, drive southeast on U.S. 80 to Tybee Island. Along the way, this highway, shown in Figure 123A, crosses over the marshes of the lower delta plain. At approximately 10 miles, the entrance to Fort Pulaski National Monument is on the left; it contains a well-preserved 19th century fort that was used during the War Between The States. It also has an excellent visitors center and one of the three walking trails is easily accessible to wheelchairs; the two others are earthen. The Fort is open 9:00 a.m.-5:00 p.m. daily, but the surrounding grounds are accessible during all daylight hours.

As noted by Georgia Wildlife, DNR (2002), in their web site titled *Georgia's Colonial Coast Birding Trail*, the birding is excellent here. This 5,600-acre National Monument, located near the mouth of the Savannah River, consists of McQueen's Island, Cockspur Island, and the adjacent salt marshes (Figure 123B). These diverse habitats are home to 200 species of birds, including songbirds, shorebirds, wading birds, and waterfowl, all of which, except for the waterfowl, are abundant year round.

Of special interest to birders are the **painted buntings** (Figure 124), which are *present from spring through summer along the edges of woodlands on Cockspur Island. This is also an excellent place to spot migratory songbirds in spring and late summer through fall. Clapper rails, seaside sparrows and marsh wrens can be seen and/or heard in the marshes around the fort throughout the year. Shorebirds can best be seen along the shoreline at low tide* (Georgia Wildlife, DNR, 2002).

TYBEE ISLANDS

Introduction

The delta front of the northern third of the Savannah River Delta (Figure 39B) is a headland that protrudes a little over 4 miles out into the ocean. This headland is composed of a densely populated barrier island called Tybee Island, which is approximately 3 miles long and three quarters of a mile wide (Figure 125), and an unpopulated, complex mixture of sand ridges and salt marsh to the south of it that is somewhat larger, with a length of 5.5 miles and a width of about 2 miles, the whole mass of which is labeled Little Tybee Island on some maps (e.g., Figure 119). This second area contains two sandy island features called Little Tybee Island (on some other maps) and Williamson Island (Figure 126). These two areas are separated by a moderate-sized tidal

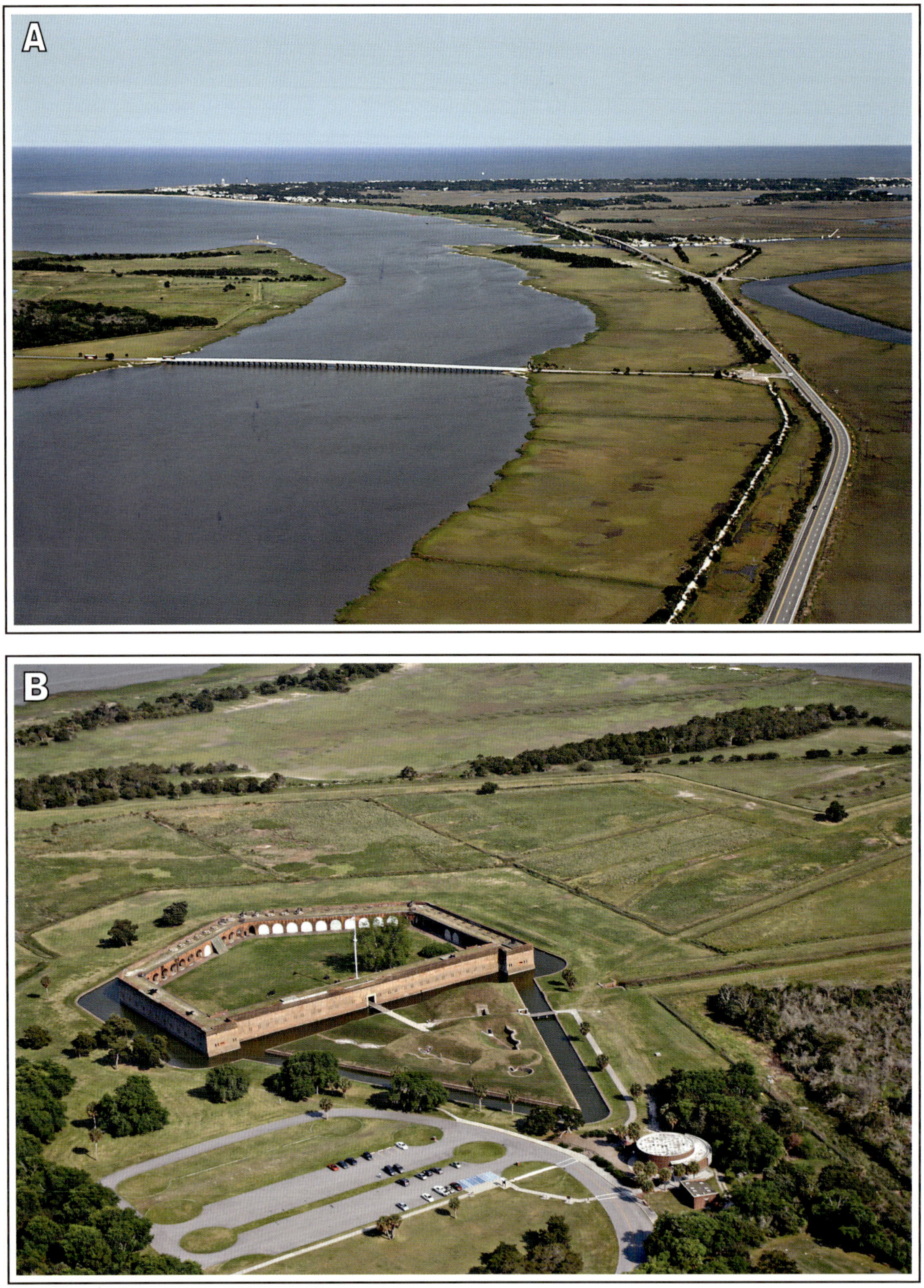

FIGURE 123. Lower delta plain of the Savannah River Delta. (A) Overview of the lower delta plain. Tybee Island is in the distance. The highway in the lower right is US 80, and the bridge in the left center leads to the Fort Pulaski National Monument. (B) Fort Pulaski National Monument. Both photographs taken on 28 April 2010.

FIGURE 124. Painted bunting. Photograph courtesy of the Friends of Hunting Island, South Carolina.

inlet at the outlet of Tybee Creek (Figure 125). These names are also a little odd. The main creek draining the marsh landward of the inlet is called Tybee Creek. For some strange reason, where the outlet channel runs alongside the south end of the island, it is called Back River. According to the nautical chart of the area, the inlet itself is called Tybee Inlet.

Geological and Human History

Before the dams were built on the Savannah River, the first major one being constructed in 1951/52, the Savannah River supplied enough sand to the shore so that three classical prograding barrier islands were formed by the waves along the delta's front as it slowly built in a seaward direction–Tybee, Wassaw, and Ossabaw Islands. This process was also severely interrupted with the dredging of the entrance channel into the Port of Savannah that started in the late 1800s. When sea level approached near its present elevation about 4,500 years ago, the high-water line was located along the front of the two elevated sandy "islands" called Skidaway Island and Wilmington Island (Howard and DePratter, 1980; DePratter and Thompson, 2013), that are now encased in marsh (Figure 127). As noted earlier, parts of these two "islands" are composed of the Silver Bluff barrier island system, that was formed around 40,000-50,000 years ago (Figures 9 and 10).

The northernmost of the modern prograding barrier islands, now called Tybee Island, is located to the immediate south of the mouth of the Savannah River. The mixed sandy beach ridge and salt marsh complex south of the mouth of Back Creek (Tybee Inlet) contains only one notable preserved prograding barrier island sequence, referred to as Little Tybee Island on some maps. It is 1.8 miles long and averages around 0.4 miles wide (Arrow B on Figure 126). This is a fragment of an older prograding barrier island that contains about 20 parallel beach ridges. The southwestern end of this second mass, referred to as Williamson Island on some maps, presently contains a recurved spit and a clearly defined transgressive barrier island (Figure 126).

As far as human habitation is concerned, a beach ridge buried beneath the prograded barrier island segment called Little Tybee Island (Figure 126) contains evidence of habitation as long ago as 3,100 years before the present (BP). This evidence was encountered in dragline operations. Howard and DePratter (1980) reported that *all pottery sherds recovered from the dredge spoil date to the Refuge (3,100-2,400 BP) or early Deptford (2,400-1,700 BP) periods, apparently originating on a site situated on the now-submerged surface.*

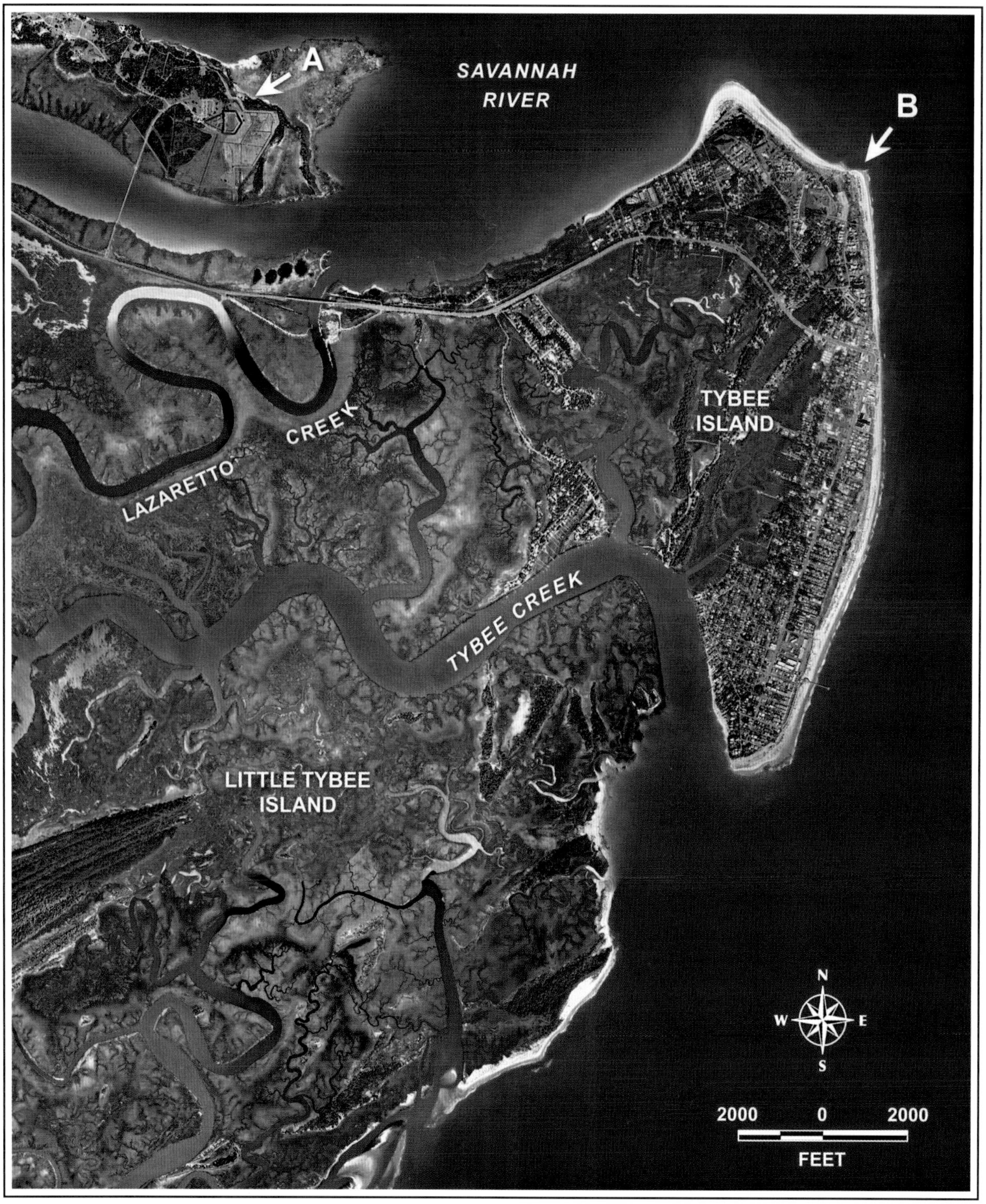

FIGURE 125. Vertical image of the highly developed Tybee Island and the undeveloped north end of a Holocene complex composed of sand ridges and salt marsh to the south (referred to as Little Tybee Island on some maps). Arrow A points to Fort Pulaski (Figure 123B) and arrow B points to the north terminal groin, which is also pictured in Figure 87A. Digital orthophoto acquired in March 2008, courtesy of USGS.

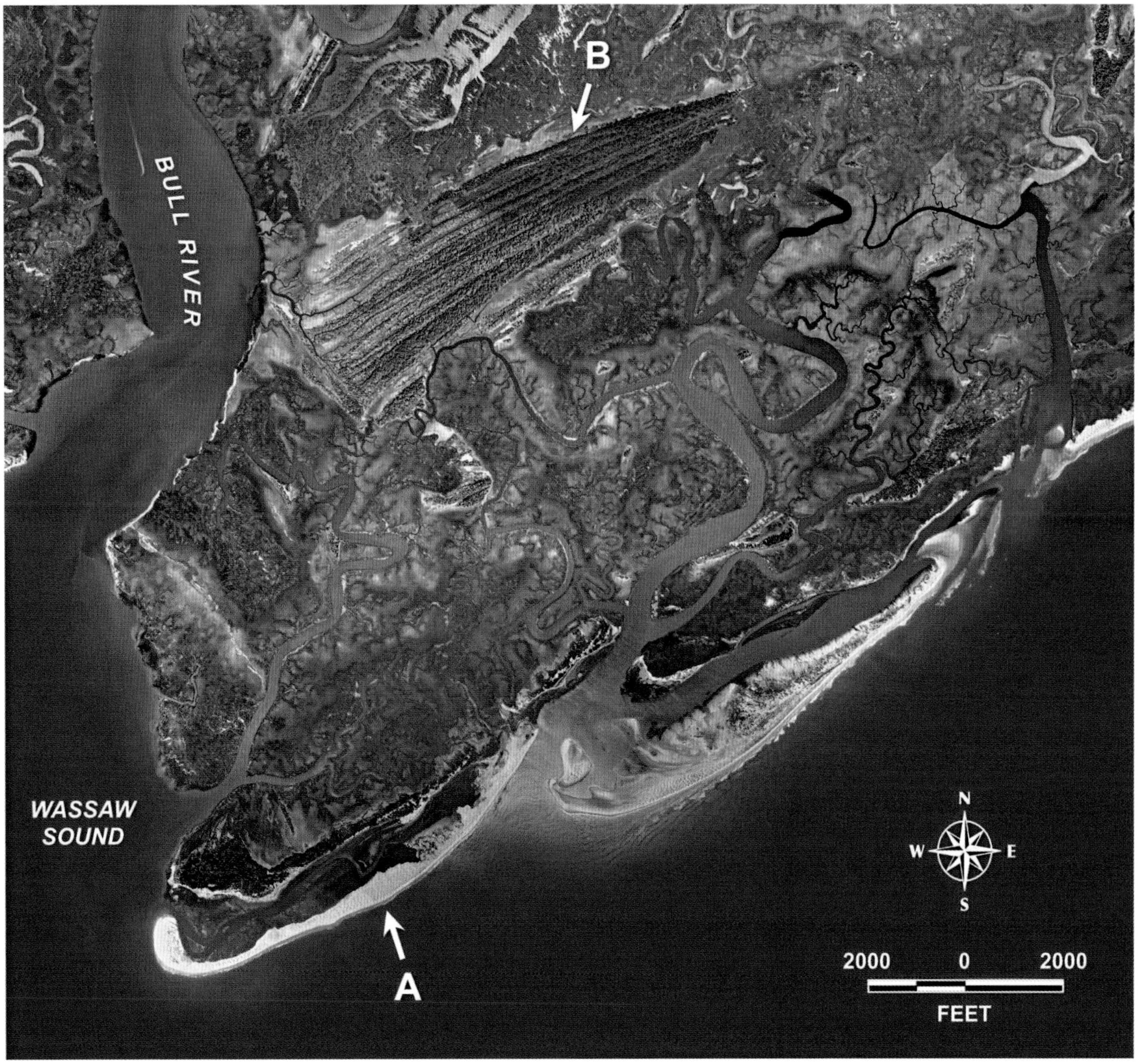

FIGURE 126. The central and southern portions of the Holocene complex of sand ridges and salt marsh located to the south of Tybee Island. Arrow A points to a classic example of a transgressive (landward migrating) barrier island (now known as Williamson Island). According to the 1978 USGS Wassaw Sound topographic map, the two prominent, unvegetated islands on the outer shore appear to have been at that time joined as one island named Williamson Island. By 1995 the NOAA nautical chart No. 11512 shows a new inlet had formed splitting the Island into two. Compare this island with the other examples of transgressive barrier islands in Georgia pictured in Figures 35A and 35B, as well as the general model given in Figure 36 (top row). Arrow B points to a remnant of a prograding barrier island formed between 1-3,000 years ago, which was referred to as Little Tybee Island by Howard and DePratter (1980), as well as on the USGS topographic map. Digital orthophoto acquired in March 2008, courtesy of USGS.

[NOTE: It should be pointed out again that some of the first detailed work correlating sea-level changes with archaeological data was carried out right here on the Georgia Coast. As was noted in Section I, DePratter and Howard (1980) reported that this was partly possible because *Indian populations in the coastal area produced copious amounts of ceramics, beginning about 4,500 years ago*, the characteristics of which changed rapidly as new methods of tempering and decoration were developed. A Carbon-14 chronology tied to this ceramic development

FIGURE 127. Infrared image of the northern portion of Compartment 1 showing Holocene time lines (in years before present), according to DePratter and Thompson (2013) and Turck and Alexander (2013). These data clearly revealed that this part of the coastline prograded eastward over time. The 2,900-year line is based on subsurface information; all others are from surface data. Landsat image acquired in October 2005, courtesy of USGS.

gave these scientists a convenient time frame. They also pointed out that until they did their work, previous workers *assumed that sea level on this coast remained nearly stable and that coastal morphology underwent little change over the past 5,000 years.* However, their combination of archaeology and coastal geology clearly indicated a more complicated picture, *shown by lateral and vertical shifts in human occupation sites and by changes in food resources.* Such changes in time and space permitted them *to determine relative and absolute rates of shoreline progradation and sea level fluctuations.* A **sherd** is a piece of broken pottery, especially one found in an archaeological dig.]

A lot changed when Tybee Island, along with the Savannah area, was colonized by General James Oglethorpe in 1733. The importance of the island with regard to the safety of the port of Savannah was recognized early, with a major lighthouse being established in 1773. Some more elements in the human development of the island include the following (Tybee Island Beach Management Plan, 2005):

1) By the mid 1760s, immigrant South Carolinians and Georgians were operating sizable (and profitable) rice plantations along the Savannah and Ogeechee Rivers that presumably made for a major export product leaving the port of Savannah.

2) Following the War of 1812, President James Madison ordered a new system of coastal fortifications to protect the U.S. ports against foreign invasion. Construction of a fort to protect the port of Savannah began in 1829 under the direction of Major General Babcock, and later Second Lieutenant Robert E. Lee, a recent graduate of West Point. The new fort, pictured in Figure 123B, was located on Cockspur Island at the mouth of the Savannah River. *In 1833, the new fort was named Fort Pulaski in honor of Kazimierz Pulaski, a Polish soldier and military commander who fought in the American Revolution* (a.k.a., The First Civil War) *under the command of George Washington.*

3) Some development of Tybee Island continued in the mid 1800s, when the island was still only accessible by water (by steamboat).

4) During the War Between The States, the Union Army placed siege batteries along the north coast of Tybee Island that aided in their successful bombardment and capture of Fort Pulaski on 10-11 April 1862. This was the first significant use of rifled cannons against masonry fortifications and demonstrated that masonry fortifications were obsolete.

5) With the establishment of a rail line with Savannah in 1887, the island became a true tourist destination, and by the 1890s, over 400 beach cottages and other buildings had been established on the island for summer residents.

6) U.S. Highway 80 was completed in 1923, linking Tybee Island with Thunderbolt, Georgia and, by that time, Tybee Island was one of the busiest seaside resorts in the southeast.

7) At the present time, Tybee Island is one of only four barrier islands in Georgia with public beach communities, the other three being Jekyll, St. Simons, and Sea Island.

Beach Erosion

The Tybee Island Beach Management Plan (2005) gave a detailed account of the erosion issues the island has undergone up to 2005, some of which follow:

1) A key part of this story is the dredging activities to keep a shipping channel open to the port. These dredging activities got underway in the late 1800s. At that time, a "point of land" extended for a half mile out into the ocean off the northeast side of the island. *By 1915, this land had disappeared and, by 1925, another 600 feet disappeared from the front beach.*

2) The dredging of the channel entrance to the port apparently impacted the migration of sand to the south around the shoals off the river mouth. The Management Plan states that *the channel was deep enough to stop most sand and the island had to be protected by the sea wall built in 1938.* The Management Plan states further that *over 130 groins helped hold some of the sand, but continuing deficits to the incoming sand flow due to the channel have caused increasing shoreline erosion beyond that expected from natural causes.* This statement is probably true, but, we have not found any *bona fide* scientific evidence to verify it. This evidence is possibly located in consultants' reports not available to us. According to Clark Alexander (pers. comm.), *the shipping channel has been intercepting all the sand that would be coming to Tybee since at least 1900.*

3) The Management Plan also states that *many older residents have memories of waves breaking against the sea wall from the Back River all the way around to the front beach.* Coastal geomorphologist George Oertel told us that, in the 1970s, he visited this beach many times and remembers standing on the beach at low tide between

14th and 18th Streets and looking landward at the 8-ft-high seawall with no dry sand being present between where he was standing and the seawall at high tide. Therefore, no doubt the waves did break against the seawall at high tide.

4) Two of the reviewers of this book who were early pioneering coastal geologists on the Georgia Coast, Orrin Pilkey and George Oertel, gave the following accounts regarding the erosion of the barrier islands in Georgia. This is Pilkey's account: *The riprap on Jekyll and St Simons has an important history. After a hurricane crossed the shore a little north of St. Augustine, Florida on 10 September 1964, President Lyndon Johnson flew along the coast of Georgia in a helicopter and expressed concern for the poor people of GA (who really had suffered little) but whose beaches showed a lot of washover and erosion, etc. and promptly and proudly arranged to have a seawall (rock revetment) built along effectively the entire developed shoreline of GA (St Simons and Jekyll) except Tybee. While I lived on Sapelo, it was always called the LBJ seawall by us and I believe the media. I guess that designation has long since gone by. I wasn't into seawalls at that time, but I did notice that within a year or two parts of it were completely covered by sand as the beach recovered on Jekyll.* According to Oertel, a revetment was also built along Tybee Island at that time and it was called "Johnson's rocks" by the locals.

5) Back to the Beach Management Plan, which states that after more erosion in 1963, the U.S. Army Corps of Engineers initiated a Federal Beach Control Project for Tybee Island, which presumably resulted in the construction of the north groin (pictured in Figures 87A and 128) in the mid 1970s, as well as the placement of 2.3 million cubic yards of sand on the front beach. This sand *was excavated from a source just south of the island.* This source was the ebb-tidal delta off the Tybee Tidal Inlet (Clayton et al., 1992), which is pictured in Figure 97. The feature located along the shore to the south, that is now known as Williamson Island, is composed of sand derived from this and later nourishment projects.

6) Following the mid 1970s nourishment project, *it was found that the new sand started migrating southward and disappearing off the end of the island* (Beach Management Plan, 2005). George Oertel stated that, actually, quite a bit of this sand was transported to the north. He surveyed 10-15 acres of new sand at the north end of the island after this particular project. This new spit-like protuberance of sand is indicated by arrow C in Figure 128. Then, in 1987, the south groin was established off 18th Street [presumably the one pictured in Figure 87C and 98A (arrow A)], and another million cubic yards of sand were placed on the *Back River, South Tip, and front beaches.*

7) In 1994, the entrance channel to the port was deepened by dredging, and some of the dredged material was placed on the north end of the island. However, this material was not pure sand and much of it soon washed away.

8) In 1995, three groins were built on the south tip of the island and 335,000 cubic yards of sand were placed there and northward to 14th Street.

9) In 2000, another 1.5 million cubic yards of sand, as well as *eight rock groins*, were added along the south end of the island. Oertel pointed out that some of these short, low-profile groins that were placed at mid tide elevations are functionally limited by the large Georgia tidal ranges (mean range 6.8 feet). Short low-profile groins are effective at mid tides when they intercept waves and drifting sand. At low tide, the groins are ineffective because they are out of the water. At high tide, waves pass over the groin tops without dissipating any energy.

10) Another nourishment project was carried out in late 2008. A little over 1.2 million cubic yards of sand were added to two locations on the oceanfront beach and to the Back River beach area nourished in 2000. This sand was dredged using a hydraulic dredge at a site located a little more than a mile southeast of the southern tip of Tybee Island. This sand was transferred via a pipeline to the south shore of the island, where it was distributed to match the natural system as much as possible (U.S. Army Corps of Engineers, 2009). In their report on this project, the Corps stated that *the next seven-year renourishment is scheduled to occur in 2015.*

Obviously, this has been quite a battle to keep a viable beach on Tybee Island. When we did a ground visit to the island in March 2010 and an overflight a month later, the beach looked to be in pretty good shape, as the discussion under places to visit will point out (also see the ground photographs in Figure 129). Williamson Island to the south (Figure 126) is obviously in an erosional mode at present, or, at least, the barrier island is migrating landward. At first, we thought that this was a little surprising, considering how much sand has been added to Tybee Island over the years, with Williamson Island being only a little over 4 miles downdrift from it. However, Clark Alexander (pers. comm.) noted that this island is there presently because of the dredging.

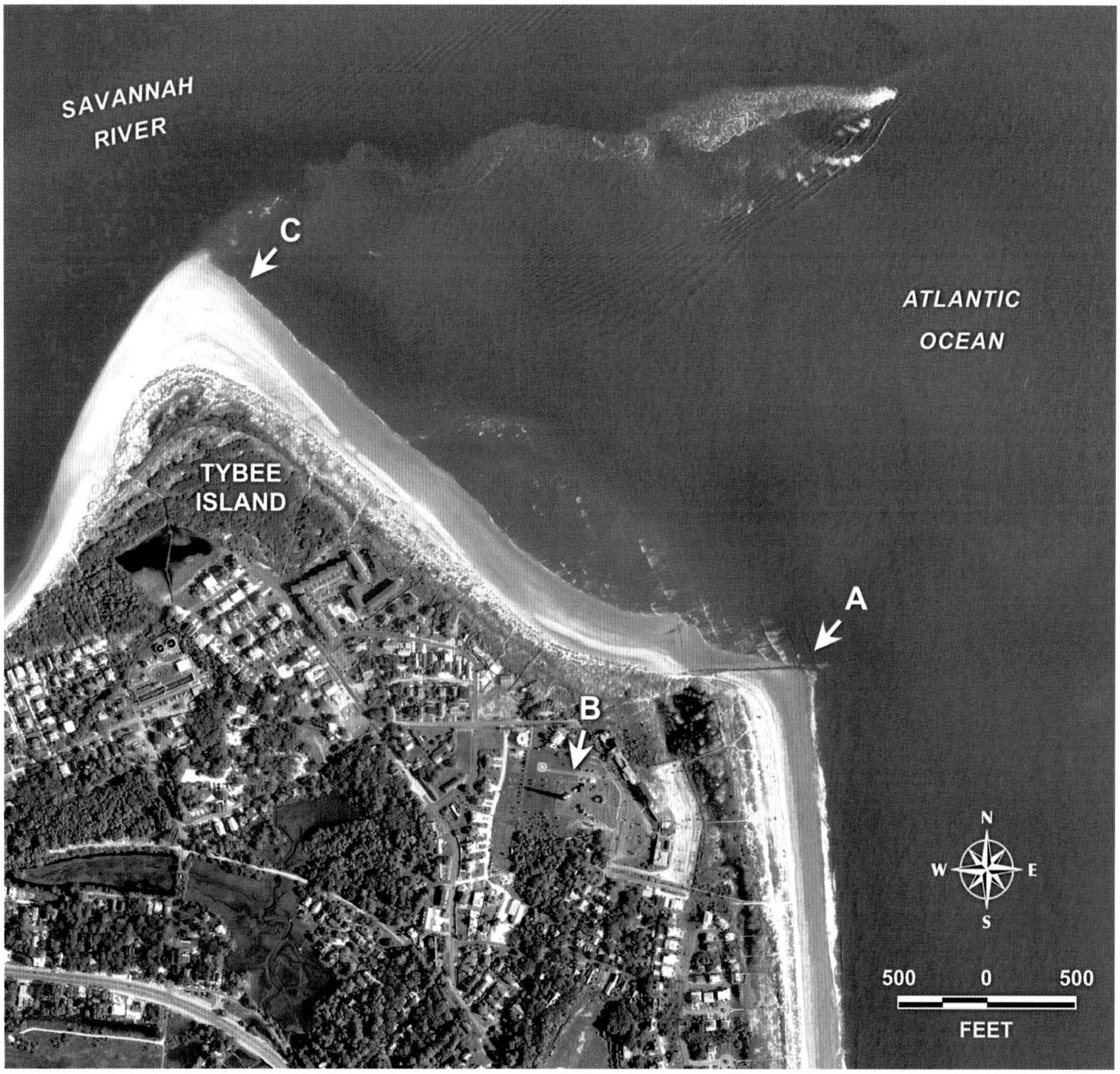

FIGURE 128. Vertical image of the north end of Tybee Island. Arrow A points to the terminal groin, which is also illustrated in Figure 87A, and arrow B points to the lighthouse. Arrow C points to a bulge of sand deposited by northerly longshore sediment transport after a major beach nourishment project in the 1970s. Digital orthophoto acquired in September 2007, courtesy of USGS.

More details on the erosional history of Tybee Island is included in the book, *Living With the Georgia Shore* (Clayton et al., 1992), a book all permanent residents and frequent visitors to the Georgia Coast should have.

Places to Visit

To get to Tybee Island, leave Savannah on U.S. 80 and pass through Thunderbolt and cross through Wilmington Island. You can also get to Wilmington Island via the Islands Expressway. Once you leave this "island" and cross over Bull River, there are 5.7 miles of almost uninterrupted viewing of a wide marsh on the right side of the road (Figure 123A). On the left side of the highway, there are many miles of a hiking trail along the old rail line, a nice place to take a walk and look at the marsh and for birds.

After 4 miles you pass by the bridge to Fort Pulaski

FIGURE 129. Tybee Island. (A) Ground view of the terminal groin (arrow) at the north end of Tybee Island. This groin is shown in the oblique aerial view in Figure 87A and the vertical image in Figure 128. (B) Intertidal bar located just south of the terminal groin at the north end of the Island. Arrow points to zone where the shell-bearing sample shown in Figure 61A was collected and where the beach surface shown in C was located. (C) Close-up photograph of the shell-hash rich sediment located on the side of the shallow intertidal trough shown in B. Scale is ~6 inches. All photographs taken during a low spring tide on 28 March 2010.

(discussed earlier). There are still several more hundred yards of marsh after you pass the first development. It is amazing how far that little sliver of Holocene (Tybee Island) sticks out there in the ocean. There was a time going back as far as 4,500 years ago that the Savannah River pumped a lot of sediment out to the coast, but not any more (refer back to Figure 120 to see how little sediment is coming down the river now). Looking off to the right as you drive along you can see the formerly prograding barrier island, Wassaw Island. You can also see the Fort off to the left, which is pretty auspicious over there. Then you go up over another bridge at Lazaretto Creek (Figure 125). There are high-tide bare spots beside the road, where presumably wrack deposited during high spring tides has killed the marsh grass that is mostly smooth cordgrass (*Spartina alterniflora).* You can see both short and tall *Spartina* as you drive along, the origin of which was discussed in Section I (see also the Glossary). Then you cross another creek that is full of *Spartina* marsh.

Follow the signs to the lighthouse on the northeastern end of the island; there is parking close to the lighthouse and a museum that provides access to the beach. The lighthouse can be clearly seen in the vertical image in Figure 128, as well as in the oblique aerial photograph in Figure 87A. So can the conspicuous "north terminal groin" that is located at the point where the beachfront takes an abrupt turn to the northwest.

As you walk around the area by the lighthouse, you will be within the former grounds of Fort Screven, built by the federal government in 1855 to provide modern seacoast defense (Explore Tybee.com, 2009). *Six poured-concrete, low-profile gun batteries (named for war heroes) and a minefield were ordered for Tybee along with hundreds of other military buildings. From 1897 to 1947, the fort was an integral part of America's Coastal Defense System. Troops stood guard on Tybee through the Spanish American War of 1898, World War I and World War II. In 1947, the fort was closed and sold to the Town of Tybee and tourism returned as a major part of Tybee's history. By the 1950s, many of the fort's buildings had been converted for use by private owners.*

In 1961, Battery Garland, the former gun battery and magazine for a 12-inch long-range gun, became the Tybee Island Museum. A room that once stored six hundred pound projectiles and two hundred pound bags of gunpowder now holds the collections and exhibits of more than four hundred years of Tybee Island history.

While there, you may want to check out Battery Brumby, the first six-gun emplacements at Fort Screven erected in 1897-1898.

On 28 March 2010, we took a walk from the parking area out onto the beach, which was very wide at that time, with lines of dunes back of the high-tide line. The intertidal beach itself was highly accretional, with multiple berms.

The photograph in Figure 129A shows the top of the high berm in the foreground, with its random patches of wrack. The berm crest was slightly erosional on that day. We concluded that this north terminal groin had been quite effective for many years, with the northward-moving sand being trapped south of the groin. The photograph in Figure 129B shows a subtle intertidal bar located a short distance south of the terminal groin. This bar contains a lot of shell hash material, shown in Figures 129C and 61A. Abundant shell material such as this commonly is derived from offshore shoal systems during beach nourishment operations. We suspect that this is also the situation here (confirmed by Clark Alexander).

Walking along the path through the dunes as you leave the beach, if you look to the north you can see some myrtle bushes that bend more-or-less permanently away from the beach and toward the land. This phenomenon is commonly referred to as **salt pruning**; an excellent example of this is shown in Figure 130.

(NOTE: Another term for salt pruning is **flagged**, as the trees or shrubs bend away from a persistent prevailing wind, with the growth only occurring on the side of the tree protected from the salt spray so that the limbs extend downwind like flags blowing in the wind. This presumably has something to do with the dehydrating effects of salt spray, but there may be other reasons. Whatever the reason, this is a very common phenomenon all around the world.)

According to Georgia Wildlife, DNR (2002), Tybee Island offers great birding opportunities, with one of the best birding spots being located along the beach at the north end of the island. The most common types of birds are shorebirds, present fall, winter, and spring, and gull-like birds that are present year round. They note further that *this is the best place in Georgia to find purple sandpipers. Look for these birds in winter along the rocks and near the beach at high tide. Look for northern gannets, gulls and terns flying just offshore. Sandpipers, gulls and terns gather in flocks along the beach at high tide.*

While in this area, you may want to visit the Tybee Island Marine Science Center located on the beach at 1510 Strand Avenue. This center has aquariums containing native marine species and exhibits of shells, sharks, etc.

Another excellent place to visit the beach is at the south end of the island. To get there, drive down U.S. 80 (Butler Avenue) to the end, where there are walkways to

FIGURE 130. Excellent example of salt pruning of shrubs located behind the beach about a mile and a half north of the south end of Jekyll Island. Arrow points in the prevailing wind direction. Photograph taken on 28 April 2010.

the shore. As you walk out to the beach, you will pass by several lines of vegetated sand dunes before reaching the high-tide line. One of the most conspicuous features on this beach is the somewhat submerged riprap groin pictured in Figure 87C and three combination sea wall/riprap features, such as the one shown in Figure 89. There are also four groins on the shoreline of Back River that appear to be trapping some sand on their southeastern sides.

In addition, this area provides a somewhat unique opportunity to walk out onto a wide intertidal sand flat (see ground and aerial photographs in Figure 98, and the vertical image in Figure 97), where shorebirds are abundant at times. In fact, if you are doing your tour of the Georgia Coast by driving, this may be your only chance to visit such a flat. As another added attraction, if you walk around near the high-tide line on the beach you may encounter some bubble sand, such as that pictured in Figure 81A, that, in fact, came from this location.

During our visit to the beaches of Tybee Island in March 2010, we didn't see much serious beach erosion. We did see those abandoned shore protection structures down on the southern end of the island, which have a lot of sand out in front of them. As discussed earlier, since the mid 1970s, at least six major nourishment projects have put over five million cubic yards of sand on the beaches of this small island. Therefore, it may be that the island is so small that the continued addition of large volumes of sand has kept the beaches in a state of quasi-equilibrium. The U.S. Army Corps of Engineers' next scheduled major nourishment project will take place in the year 2015.

WASSAW SOUND

Introduction

Wassaw Sound is the first of the eight major sounds on the Georgia Coast (going south). As can be seen in Figures 38, 119, 127, and 131, this two-mile-wide sound is formed by the confluence of two main streams, Bull River and Wilmington River. These two tidal rivers, together with ten additional creeks of some size, drain a zone of salt and brackish marshes that covers an area of about 16 square miles (Gross and Werner, 1997).

General Morphology and Coastal Processes

As shown by the bathymetric map in Figure 131, there is a very large ebb-tidal delta offshore of the entrance to Wassaw Sound, built by exceptionally strong ebb-tidal currents that have more focus and concentration at lower water levels and flow for a shorter amount of time during a single tidal cycle than do the corresponding flood currents (discussed in Section I). With a tidal range that commonly exceeds 9 feet during spring tides and accompanying current velocities of 3-4 knots, the entire sound is a very dynamic setting that is made even more dynamic during major storms. The bathymetry is complex with numerous channels, both intertidal and subtidal, with extremely strong tidal currents in some of them. Tidal flats many tens of feet wide are also fairly common.

(NOTE: A **knot** is a measure typically used to determine how fast tidal currents flow. A velocity of one knot means that a particle of water is moving at one nautical mile per hour. A nautical mile is 6,076 feet.)

Places to Visit

Assuming you are leaving Tybee Island after a visit there and you are planning to head south, your first encounter with the waters of Wassaw Sound will be where the U.S. 80 highway bridge crosses over Bull River, one of the two main tidal rivers that empty into the sound. This crossing is also over a wide expanse of salt marsh that contains abundant smooth cordgrass. Within a short distance, the road bends to the right up onto a forested upland called Talahi Island. Continuing northwest, you will soon cross the bridge over Turner Creek, after which you enter another somewhat subdued upland called Whitemarsh Island (see Figure 119). It is a little more than a mile from the Turner Creek bridge to the intersection of U.S. 80 with the Island Expressway. About 1 mile further along, if you stay on U.S. 80 on your way south, is the intersection with Johnny Mercer Boulevard, a major connecting road to Wilmington Island. From there, you will soon start crossing a wide salt marsh and about a quarter of a mile further from the marsh edge you will cross the bridge over the main channel of the spectacular tree-shaped tidal creek system shown in Figure 132. Along this part of the drive, there are big patches of black needlerush (*Juncus roemarianous*) on both sides of the road.

Once you cross the Intracoastal Waterway and enter Thunderbolt, we recommend that you next visit the **Wormsloe State Historic Site**, which is located

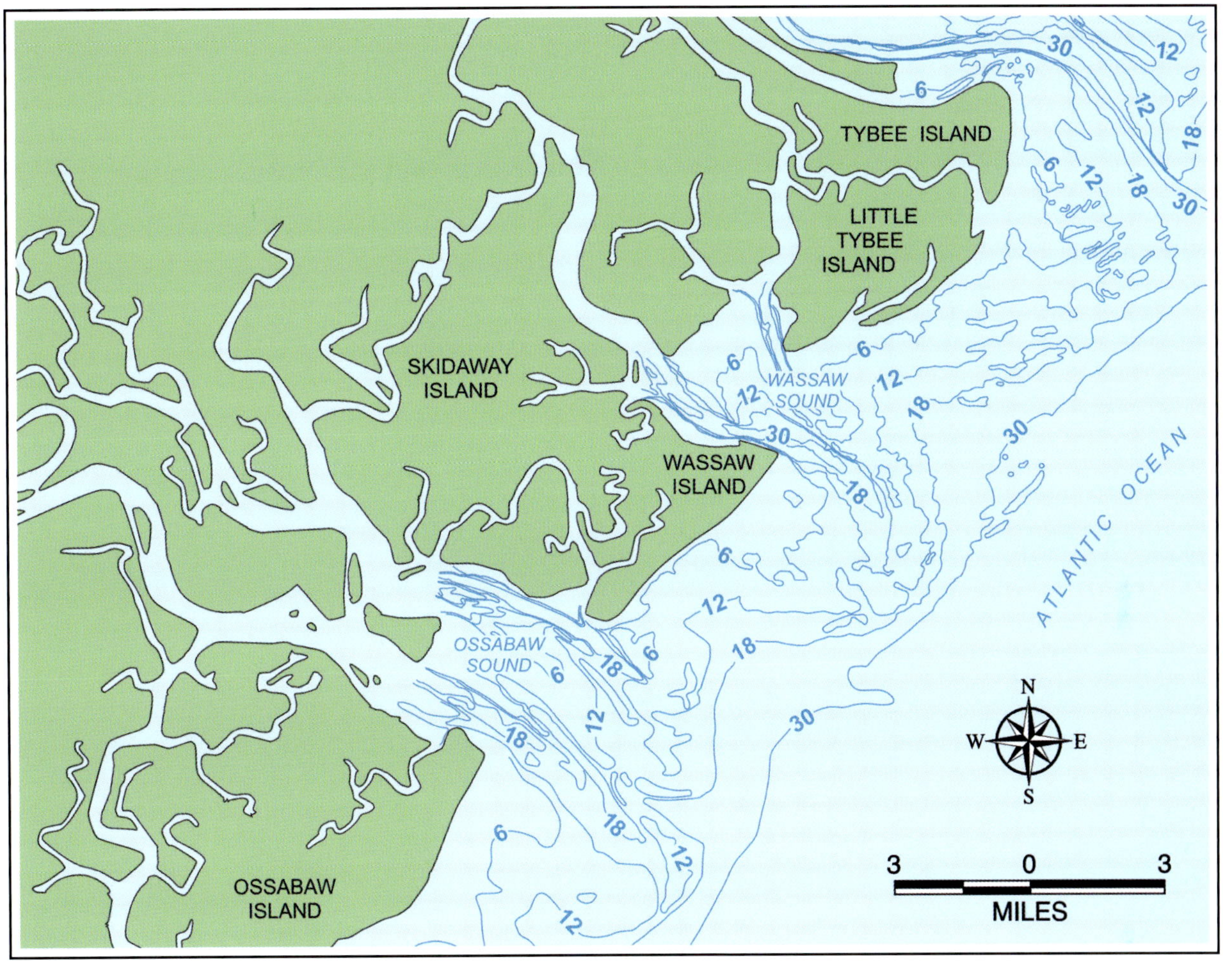

FIGURE 131. Bathymetry of Wassaw Sound and Ossabaw Sound. Note large shoals (ebb-tidal deltas) at entrances to the sounds. Slightly modified after Kellam (1986). Numbers give water depths in feet.

on the relatively high ground of the Princess Anne Pleistocene barrier island system. To get there, go south to the community of Isle of Hope and follow the signs to Wormsloe. There are a visitor's center and museum at Wormsloe that are worth seeing, but the main attraction is the fabulous avenue of oaks. It is about 1.2 miles long, with 112 live oak trees (Figure 133), and leads to the tabby ruins of Wormsloe. Resurrection fern and Spanish moss are of interest if you inspect the trees up close (see Glossary for detailed definitions of these plants).

Noble Jones, one of Georgia's first European settlers, oversaw the construction of the colonial estate of Wormsloe. Jones was an English physician and carpenter who carved out an even wider career in the colonial wilderness (Georgia State Parks & Historic Sites, 2006). He came to Savannah with James Oglethorpe in 1733 and commanded a company of marines charged with Georgia's coastal defense. The construction of the Jones fort house began in 1739 as hostilities between Britain and Spain escalated.

Jones had a son, Nobel W. Jones, who was a leader in the Georgia revolutionary movement. He was captured by the British during the war. Afterwards he served briefly in the Continental Congress. A medical doctor, in 1804 he became the first President of the Georgia Medical Society (New Georgia Encyclopedia, 2002). Following generations of the Joneses lived at Wormsloe for two more centuries.

A kiosk at the Wormsloe State Historic Site states that: *Wormsloe was also known for its horticultural efforts.*

FIGURE 132. Vertical image of a tree-shaped tidal creek system located about a mile east of Thunderbolt on the southeastern fringe of Savannah (Figure 119). The highway in the lower part of the image is US 80. Digital orthophoto acquired in September 2007, courtesy of USGS.

Indigo, rice, and silk were all cultivated here by Jones, his indentured servants, and later slaves. .... The tabby ruins of the original "Wormslow" home survive as the last architectural remnant of Savannah of the Oglethorpe era (1733-1743).

[NOTE: **Tabby** is a concrete-like substance *made of lime, water, powdered oyster shells, and crushed (but chunky) oyster shells* (Livingston, 2010). This material was one of the primary substances used to build dwellings in the early days of European settlement of this part of the world.]

WASSAW ISLAND

Introduction

Known as the most pristine of the Georgia barrier islands, Wassaw Island is a classic example of a prograding barrier island, of the type illustrated in Figures 35 and 36. This island, which is clearly illustrated by the infrared image in Figure 38, is 5.3 miles long and averages about 0.6 miles in width.

As mentioned earlier but probably worth repeating

FIGURE 133. Avenue of oaks at the Wormsloe State Historic Site. It is 1.2 miles long and contains 212 live oak trees. Photograph taken on 28 March 2010.

here, DePratter (1977), and DePratter and Thompson (2013) documented shoreline positions within the deltaic Pleistocene-Holocene interbarrier area between Skidaway and Wassaw Islands, based on Indian occupation sites. These data revealed that this part of the coastline prograded eastward over time, with human inhabitants moving with it to access resources (Turck and Alexander, 2013). About 700 years ago or so, the first of the prograding beach ridges that now compose the island was deposited (see infrared aerial photograph in Figure 127). There are some remnants of older beach ridges located further landward that were no doubt formed when some of the higher blips on the sea level curve occurred (see curve in Figure 8).

With regard to beach erosion on this island, a study by Langley et al. (2003) that compared the shorelines in the 1850s with those in the 1920s, concluded that, in general, *most erosion is occurring on the northern end of the island, whereas the most accretion is occurring on the southern end of the island.*

Wassaw National Wildlife Refuge

The entire 10,053 acres of this island area, that has been only slightly modified by human activities, were made into a national wildlife refuge on 20 October 1969. The refuge includes beaches with dunes, maritime forest, and extensive salt marshes. *Salt marsh and tidal creeks separate the Refuge from the mainland and Skidaway Island to the west. Refuge visitors may enjoy recreational activities such as birdwatching, beachcombing, hiking and general nature studies. The 20 miles of dirt roads on Wassaw Island and seven miles of beach provide an ideal wildlife trail system for hikers. Birdwatching is particularly fruitful during the spring and fall migrations* (U.S. Fish and Wildlife Service, 2010a).

Rookeries for egrets and herons are present on the island, and many wading birds are present in the summer months. In summer, about 10% of the loggerhead sea turtles that nest on Georgia's sand beaches come to lay their eggs on Wassaw Island. There were 165 nests in 2011 and 139 nests in 2012; all nesting is monitored by a loggerhead sea turtle volunteer program (Georgia DNR Sea Turtle Conservation Program).

This refuge is accessible only by boat and, as noted on the web by the U.S. Fish and Wildlife Service (2010a), *both Wassaw and Pine Island are open to the public during daylight hours - other upland areas are closed. Transportation to the Refuge must be arranged by the visitor. Several local marinas in the Savannah area (Skidaway Island and Isle of Hope) and a public boat ramp adjacent to the Skidaway Island bridge can serve as launching sites for trips to Wassaw.*

Places to Visit

Access to Wassaw Island and the Wassaw National Wildlife Refuge was discussed earlier. If you are restricted to land travel by car, there is still one excellent place to visit within this subcompartment, **Skidaway Island State Park**. To get there from downtown Savannah, take State Road 204 south to the south side of Hunter Army Airfield, where you turn left onto GA Spur 204 (also called the Diamond Causeway), which leads all the way to the island (Figure 119). If, however, you have just completed a visit to Wormsloe State Historic Site with its majestic avenue of oaks (Figure 133), go to the community of Isle of Hope and drive south through the community of Bethesda, a beautiful drive by some striking live oaks. At the intersection with GA Spur 204, turn left and go for 2.2 miles where you will pass over the bridge across the Intracoastal Waterway. As you drive across the bridge, you can see the trees in the distance on Skidaway Island, about a 9-mile-long remnant of the Silver Bluff Pleistocene barrier island system, with maximum elevations of about 14-15 feet. The densely populated core of the island contains some patches of climax maritime forest.

Skidaway Island State Park contains 588 acres of preserved marshes, tidal creeks, and maritime forest. It is open from 7:00 a.m. until 10:00 p.m. daily. At the entrance, you will receive a free, very detailed map of the Park. Some words on the back of the map follow:

1) *Located near historic Savannah, this park borders Skidaway Narrows, a part of the Intracoastal Waterway, and provides scenic camping and picnicking areas. Trails wind through marshes, live oaks, cabbage-palmettos, and pines, allowing visitors to watch for deer, raccoon, shorebirds, and rare migrating birds such as the painted buntings. An observation tower provides another chance for visitors to search for wildlife on this beautiful barrier island.* After reading that, we thought to ourselves that, personally, we wouldn't call Skidaway a barrier island; it was one about 40,000 years ago, but not now. This Pleistocene remnant is almost completely surrounded by salt marsh.

2) There are four major hiking trails, with the 1-mile Sandpiper Trail Loop being the most popular one. It takes you over salt flats and tidal creeks, and through maritime forests on island hammocks.

When we were there on 28 March 2010, we walked out the Big Ferry Trail to an observation tower. This is one of the better places along the coast to see a variety of marsh plants within a short walk. As you head out to the platform that leads to the observation tower, the first thing you walk by is a little fringe of black needlerush. On this short walk you can also get close-up views of saltgrass (*Distichlis spicata*) and sea oxeye (*Borrichia frutescens*). The next zone out contains some glasswort (*Salicornia virginica*). From the tower, you have a panoramic view of the salt marsh (Figure 94A).

As noted by Georgia Wildlife, DNR (2002), this park is a great birding spot where songbirds, shorebirds, wading birds, and waterfowl can be observed. Specialties include ospreys, painted buntings, pileated woodpeckers, and bald eagles. *Warbler watching can be spectacular during spring and fall migrations. Look for nesting osprey in spring and summer. Listen and look for marsh wrens and clapper rails in salt marshes. Look for painted buntings during the spring and summer.* A photograph of the much sought-after painted bunting, the most beautiful of the songbirds in the southeastern United States, is given in Figure 124.

OSSABAW SOUND

Introduction

Ossabaw Sound is the second of the eight major sounds on the Georgia Coast (going south). As can be seen in Figures 119, 131, and 134, this 3.3-mile-wide sound is fed by two main streams, the Ogeechee River,

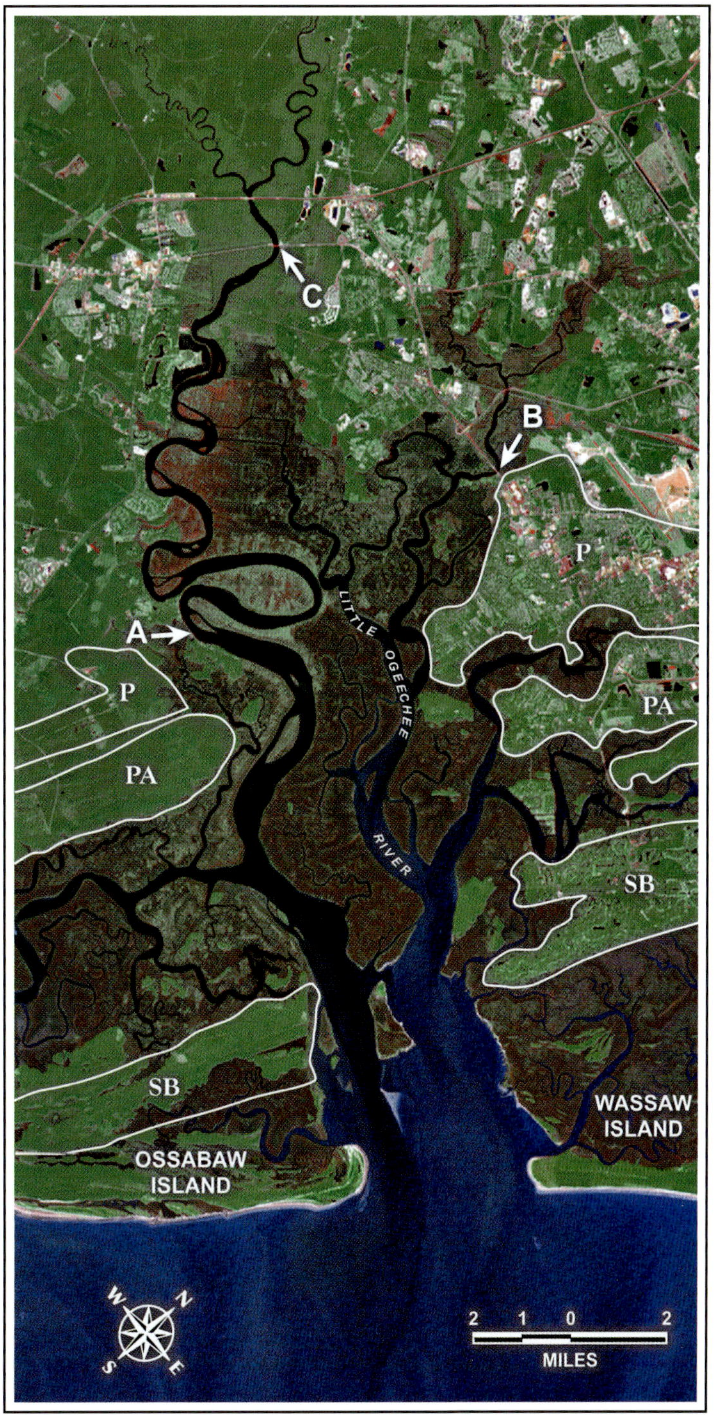

FIGURE 134. Vertical image of Ossabaw Sound. The highly meandering Ogeechee River (arrow A), the largest coastal plain (black-water) river in Georgia, adds a considerable amount of fresh water to this estuarine system (see Table 3). Arrow B points to where GA 204 crosses the Little Ogeechee River and arrow C points to where US 17 crosses the Ogeechee River. The approximate locations of some remnants of the Pleistocene Pamlico (P), Princess Anne (PA), and Silver Bluff (SB) barrier island systems are shown (after Huddlestun, 1986). The barrier island materials (with green color of plants on beach ridges) located seaward of the Silver Bluff (SB) are Holocene, or recent, in age. Landsat image acquired in October 2005, courtesy of USGS.

the largest coastal plain river in the state, to the south, and Little Ogeechee River and its numerous, mostly tidal, tributaries to the north. This multitude of streams entering the sound makes for a very complex and widespread estuarine system, the largest, in terms of area, on the Georgia Coast. This is made especially clear by inspection of the vertical image in Figure 134. Because of the large tides and generally depositional nature of this coastline, fairly wide tidal flats are common, the most notable one, perhaps, being the wide sandy tidal flat at the entrance to the sound, pictured in Figure 135.

General Morphology and Coastal Processes

As seen on the bathymetric map in Figure 131, there is a very large, complex ebb-tidal delta offshore of the entrance to the sound. The sound itself is made up of a heterogeneous set of intertidal and subtidal channels that are subject to very strong currents at times. With tides that commonly exceed 9 feet during full and new moons and accompanying current velocities of 3-4 knots, the entire sound is a very dynamic setting that is made even more dynamic during major storms.

Places to Visit

There are several places of interest to visit by driving around the margin of Ossabaw Sound. One place to start is where GA spur 204 (Diamond Causeway) crosses the Intracoastal Waterway just west of Skidaway Island. From there going west for a little over 1 mile to another bridge that crosses Moon River, the road goes right through extensive, odd-looking chopped up patches of marsh (mostly *Spartina alterniflora*) that is intersected by a number of small channels. The Frank O. Downing Fishing Pier is located right along side of the bridge over Moon River.

From the Moon River Bridge, continue in a north/northwest direction for 5 miles to where GA spur 204 intersects with GA 204 at the south side of Hunter Army Airfield. Along this drive, you will pass through some developments with big pine and live oak trees, many with extensive hangings of Spanish moss. You will also pass closely to several large zones of salt marsh.

At the intersection with GA 204, go left on this main artery and drive for about 4 miles to where the highway crosses the Little Ogeechee River (Figure 134; arrow B). This part of the road passes through a highly developed area. Beyond the bridge, however, the road crosses a mile-wide zone of marsh where the boundary between the saline smooth cordgrass marsh and the more brackish black needlerush marsh is conspicuous. Continuing on to the junction with U.S. 17, the topography is very flat. The reason for this is that from the vicinity of the bridge west, the underlying geological formation is the Pamlico backbarrier complex, which is over 15 miles wide along this area (Huddlestun, 1986). This backbarrier sequence is composed of muddy sediments deposited in the estuaries/sounds landward of the Pamlico barrier island system (125,000-130,000 years ago), so the topographic relief is minimal.

Another very interesting site to visit in this area is the **Savannah-Ogeechee Canal Museum & Nature Center**, which can be reached by continuing west on GA 204 past the intersection with U.S. 17. From the intersection with I-95 proceed straight ahead for another 2 miles to the museum/nature center, which has visiting hours from 9 a.m.-5 p.m. daily.

As noted by Georgia Wildlife, DNR (2002), *this site contains remnants of an extensive canal system that linked the Savannah and Ogeechee rivers during the 1800s. Today the area is a recreational facility that highlights the natural history of this rich floodplain forest while preserving the historic relics associated with a once-thriving artery of commerce.* There are more than 2 miles of hiking trails between the nature center and the Ogeechee River. This is an especially noteworthy birding spot with specialties such as prothonotary warblers, northern parula warblers, Swainson's warblers, wood ducks, Mississippi kites, and swallow-tailed kites. If you have not as yet seen these two kites, they are definitely worth the effort to observe them sailing gracefully overhead.

From there, we recommend that you go back to U.S. 17 and head southwest toward the Ogeechee River, the largest coastal plain (black-water) river in Georgia (Figure 110). This section of the river is highly meandering, as is shown on Figures 104A and 134.

Proceeding southwest on U.S. 17, soon you will come to the edge of the valley of the Ogeechee River carved by the river during the last lowstand of the sea. The valley averages about 2 miles in width at this point. Once in the valley, you will drive through what appears to be a "logged out" cypress/tupelo swamp (see Figures 114 and 115 for illustrations of this habitat). When you come to the river, before crossing the bridge turn left into a major boat ramp facility that contains a fishing platform,

FIGURE 135. Sandy tidal flat at the entrance to Ossabaw Sound. It can also be seen on the image in Figure 134. Arrow points to low-lying island called Raccoon Key. Photograph taken during a low spring tide on 28 April 2010.

an excellent stage for views across and up and down this beautiful coastal plain river. See the photograph in Figure 136, taken from this spot, that shows a fringe of giant cordgrass (*Spartina cynosuroides*) along the margin of some cypress trees on the southwest side of the river.

Back on U.S. 17 and heading southwest, it is 1.75 miles to the south bluff of the valley. On this drive, you will be heading across a big swamp (Lost Swamp?) that appears to be reclaiming the old rice ponds. It looks as if the cypress trees have been logged over at least once.

At the intersection with GA 144, turn left and proceed to the town of Richmond Hill. There you will have an opportunity to visit another one of the sites on the Colonial Coast Birding Trail called the **Richmond Hill J. F. Gregory Park**. To get there, take GA 144 into Richmond Hill and turn left on Cedar Street, which leads directly to the Park.

This unique Park is operated by the City of Richmond Hill. *The area is dominated by a 300-acre rice field that predates the* War Between The States. *A three-mile walking trail runs along the top of a dike that encompasses the field, which today is a wooded wetland* (Georgia Wildlife, DNR, 2002). Along the trail you can see songbirds and wading birds year round and numerous waterfowl in the winter. The bird specialties include prothonotary warblers, wood ducks, and barred owls. The checkerboard pattern of the old rice fields dominates the marshes south of the U.S. 17 bridge, as is clearly shown on the image in Figure 104A.

We recommend that you next drive southeast down GA 144 parallel to the Ogeechee River valley. At 4.9 miles, you should turn left on GA spur 144 and head toward **Fort McAllister State Historic Park**. At Rabbit Hill, where the west side of the river has cut into the

FIGURE 136. Ogeechee River where US 17 crosses it near Burroughs in Chatham County (located by arrow C on Figure 134). The tan-colored band at the water line is giant cordgrass (*Spartina cynosuroides*) and some of the trees in the background are bald cypress (*Taxodium distichum*). Photograph taken on 28 March 2010.

upland, there are over 60 houses with boat docks along the riverbank (see Figure 104A).

The Georgia Historical Society has erected a roadside kiosk on GA 144 at 4.5 miles before you reach the Historic Park that contains the following information: *East 4.5 miles, on the Great Ogeechee River, Fort McAllister was built 1861-62 to guard the "back door" to Savannah. During 1862-63 it repulsed 7 attacks by armored vessels, some mounting 15-inch guns. Dec. 13, 1864, its small garrison of 230 Georgians was overwhelmed by Hazen's Division, 15th Corps (U.S. flag), which had marched via this road. Its fall opened the Ogeechee to Union vessels, which, loaded with supplies for General Sherman's army, had been lying in Tybee Roads and Port Royal Sound. A wharf and depot were built at King's Bridge from which these supplies were distributed to invading forces*."

[NOTE: General **William Tecumseh Sherman** was the commanding officer during the famous "march to the sea" by the Union Army in 1864/65, which was accompanied by destruction of civilian properties along a wide swath. He and his army entered Savannah on 21 December 1864 and, for whatever reason, the city was spared, unlike the city where the authors of this book now live, Columbia, South Carolina, which was burned to the ground. One of the most famous of General Sherman's many notable quotes was spoken during his stay in Savannah in December 1864. When asked by a civilian about his opinion of General Grant, he is reported to have said – *General Grant is a great general. I know him well. He stood by me when I was crazy, and I stood by him when he was drunk; and now, sir, we stand by each other always.* Grant went on to be President. When the

Republicans asked Sherman to run for President in 1884, he is reputed to have said something like – *If nominated I will not run, and if elected, I will not serve.* This is now known as a *Shermanesque* statement.]

The park covers 1,725 acres and is open from 9:00 a.m.-5:00 p.m. Monday to Saturday, and from 2:00 p.m.-5:00 p.m. on Sunday. A pamphlet published by the Georgia State Parks gives the following information about this Park:

1) It is *nestled among giant live oaks and beautiful salt marsh* and is a good spot for fishing, birding, and other naturalist pursuits;
2) It is the *home of the best preserved earthwork fortification of the Confederacy*;
3) It has a museum dedicated to the War Between The States; and
4) Being bordered by Redbird Creek, it has a courtesy dock and boat ramp on it (on Savage Island, a hammock in the marsh) as well as on the Ogeechee River.

In your touring around the park, if you walk down off the natural levee, it is possible for you to encounter some extremely rich upper bottomland hardwoods, including huge loblolly pines (*Pinus taeda*) and the hardwoods mentioned on the cross-section in Figure 114, as well as some of those shown in the photograph in Figure 115A.

The river is very wide here (the fort is located on Figure 104A) and the eroded bluff is 12 feet high. The site is on the northern edge of a remnant of what is presumably the Pamlico barrier island system deposited about 125,000-130,000 years ago. The fort is at this spot because this is the first significantly high ground inland from the open ocean for boats traveling along the Ogeechee River. The highest elevation contour in the fort area is 14.4 feet. The community of Oak Level, located around 2.7 miles southwest of the fort, is positioned on the same upland and has a maximum elevation of 27 feet. That is the correct maximum elevation for the Pamlico system, so the barrier island remnant has probably been eroded down some along the river. As can be seen on Figure 134, the Princess Anne Pleistocene barrier island occurs southeast of the Pamlico system, but it does not extend to the west far enough to intercept the Ogeechee River. Therefore, the fort is located further landward than the Princess Anne barrier island system, which was deposited about 85,000 years ago. The bulk of the more seaward Silver Bluff Pleistocene barrier island system, deposited about 40,000-50,000 years ago, composes much of the northwestern portion of Ossabaw Island (Figures 9 and 134).

This Historic Park is also one of the sites described in the Georgia Wildlife, DNR (2002) discussion of Georgia's Colonial Coast Birding Trail. The specialties in the area that they list include painted buntings, wood ducks, northern harriers, bald eagles, and ospreys. They also made the following recommendations: *Painted buntings are most often seen in late spring through summer along the causeway. Look for migrating warblers during spring and fall migrations. Bald eagles are most often seen during the winter. Ospreys are most often seen in spring and summer. Look for northern harriers winging low over the marsh in winter.*

12 COMPARTMENT 2 - OSSABAW ISLAND TO BLACKBEARD ISLAND

INTRODUCTION

For purposes of discussion, we have subdivided this compartment into four subcompartments (Figure 137):

1) **Ossabaw Island** – A large "combination" island, the uplands of which are about half remnant of the Silver Bluff Pleistocene system and half Holocene sediments.

2) **St. Catherines Sound** – A huge sound with a relatively narrow entrance (1.5 miles) that has three major tidal river systems feeding into it.

3) **St. Catherines Island** – Also a large "combination" island with the Holocene segment being split in half by the ebb-dominated McQueen Inlet.

4) **Sapelo Sound** – Another huge sound also fed by three major tidal rivers with the Harris Neck National Wildlife Refuge located along the banks of the South Newport River.

OSSABAW ISLAND

Introduction

The more elevated elements of Ossabaw Island are composed primarily of two wide zones of prograding beach ridges deposited at two different times: 1) A seaward component deposited during the Holocene epoch, actually since when sea level reached near its present position about 4,500 years ago; and 2) A more landward component of approximately the same size deposited around 40,000-50,000 years ago (Silver Bluff barrier island system) that intersects the Holocene component at an acute angle like a chicken wing (see Figure 138). As the time lines presented on Figure 138 show, much of the Holocene component of the island is less than 1,000 years old (DePratter and Thompson, 2013). A wedge-shaped zone of salt marsh separates these two sandy, beach-ridge segments. Approximately 4 miles to the west of the Silver Bluff upland, another wooded elevated zone is present. This is a remnant of the Princess Anne barrier island system that was deposited around 85,000 years ago. The Princess Anne and Silver Bluff barrier island systems are separated by a spectacular complex of salt marshes and highly meandering tidal channels (see image in Figure 138).

General Geomorphology

The different geological components of Ossabaw Island, as well as the position of the Princess Anne and Pamlico barrier island sequences, are illustrated in Figure 138. The Holocene segment of the island composes the southernmost limb of the Savannah River Delta (Figure 39B). This Holocene limb of the island, which has been consistently progradational over the past few hundred years (at the least), is the largest such Holocene barrier-island deposit in the Savannah River Delta system, and the largest on the Georgia Coast, as well (Figure 39A). We speculate that this large size is the result of the volume of sand being delivered to the coast by the Ogeechee River, which, historically, has been added to that delivered by the Savannah River. Both of the outer ends of the island show evidence of notable progradation in recent times, as is illustrated for the northern end in Figure 34C, and also at the southern end in Figure 139B. Other aspects of the geomorphology of the Holocene component of the island are illustrated by the photographs of a small tidal inlet in the middle of the outer coast in Figure 139A, and of the spectacular sandy tidal flats adjacent to the southern end of the island that are pictured on the cover of this book.

Ownership History

According to Coastal Georgia Experience (2007), there is a long history of humans living on Ossabaw Island, including *Indians, Spanish missionaries, plantation owners, African slaves and millionaires. A Michigan*

FIGURE 137. Compartment 2, north end of Ossabaw Island to north end of Blackbeard Island. Red arrows mark boundaries of the compartment. Source: ESRI StreetMap USA.

FIGURE 138. Ossabaw Island and the adjoining mainland (located on Figure 137). The distributions of the four barrier island systems present in the area are indicated. The most easterly one (labeled A) is composed of sand deposited in prograding beach ridges since sea level reached near its present position around 4,500 years ago (Holocene time). However, note that the superimposed yellow time lines (after DePratter and Thompson, 2013) show that most of these Holocene beach ridges are <1,000 years old. This complex is attached on its landward side to an almost equally large set of prograding beach ridges deposited around 40,000-50,000 years ago (called the Silver Bluff barrier island system; labeled B). Further west at a higher elevation, two more prominent suites of prograding beach ridges, labeled C and D on the diagram, are present. These are the Princess Anne barrier island system deposited around 85,000 years ago and the Pamlico barrier island system, deposited around 130,000 years ago. The Princess Anne and Silver Bluff systems are separated by a wide zone of salt marshes and meandering tidal channels. Landsat image acquired on 23 October 1999, courtesy of Earth Science Data Interface at the Global Land Cover Facility.

FIGURE 139. Ossabaw Island, (A) Small tidal inlet located near the middle of the Island (seaward side). Note the large size of the ebb-tidal delta (900 feet wide) and the absence of a flood-tidal delta. (B) The prograding southerly end of the Island. Note the parallel bands of trees that are growing on the elevated portions of prograding beach ridges formed by the process illustrated in Figure 83. Both photographs were taken during a low spring tide on 28 April 2010.

physician acquired the island in 1923 as a winter residence. His daughter created the Ossabaw Foundation in 1961. In the late 1970s, the Nature Conservancy of Georgia purchased an option and transferred ownership to the state. Ossabaw became the first acquisition of the Heritage Trust Act of 1975, which protects it from development. The island is managed today by the Georgia Department of Natural Resource. The island has more than 100 miles of dirt roads and trails.

Places to Visit

This island can be visited only by boat. If you are so inclined, there are several important considerations to keep in mind, including the following (after Lenz, 2002b):

1) The beach is open sunrise to sunset, but the interior of the island is closed unless permission is granted by the Georgia Department of Natural Resources for a cultural mission or by special hunting permit, or by the Ossabaw Island Foundation;

2) During these daylight hours, only the following activities are allowed on the beach: hiking, bird watching, shelling, fishing, picnicking, and nature studies. Hunting is allowed on the island by special permit only; and

3) Two docks may be used by permission only. If you plan to go the south end of the island, Kilkenny Fish Camp, a full-service marina, provides the closest access. If traveling to the north end, many use the marinas at Fort McAllister, Coffee Bluff, and Isle of Hope. Dock access can be limited or not available.

ST. CATHERINES SOUND

Introduction

St. Catherines Sound, with an entrance about a 1.5 miles wide (Figure 140), is the third of the eight major sounds on the Georgia Coast (going south). Three major stream networks enter the sound: 1) Bear River, which exits the wetlands landward of Ossabaw Island; 2) Medway River, which drains an extensive swampland at Limerick and beyond in its headwaters; and 3) North Newport River, the outlet for an diverse network of tidal channels in the wetlands in the lee of St. Catherines Island.

General Geomorphology and Coastal Processes

As can be seen on the bathymetric map in Figure 141, a large ebb-tidal delta with an areal extent of 33 square miles (Figure 53), also shown in Figure 49A, is present off the mouth of the sound. This inlet also has a relatively small flood-tidal delta, pictured in Figures 49A and 142, a unique occurrence on the Georgia Coast. As with the other sounds, the large tidal range augments the formation of wide tidal flats and widespread salt marshes. The channel networks in the marshes are complex, as the vertical image of the headwaters of the Medway River in Figure 143 shows. In places where the major tidal channels intersect uplands, residences and their associated docks crowd the shoreline. The Belfast River, a tributary of the Medway River, at the development of Belfast, located on Figure 143, contains about 50 docks.

Places to Visit

If traveling south on I-95, the road between exits 87 and 78 crosses over the headwaters of the sound, specifically the landward tributaries of the Medway River, called Jones Creek and Jerico River. Driving across this area on the interstate, you will see an impressively wide marsh area. The streams in this area are highly meandering, probably for the same reason as the one given earlier for the streams in the headwaters of the Ossabaw Sound. That is, the underlying geological formation at this locality, the Pamlico backbarrier complex, is over 15 miles wide and is composed of muddy sediments (Huddlestun, 1986). This results in a very flat topography of relatively low relief, hence the intense meandering. The segment of the highway that crosses the Jerico River and Jones Creek is shown in Figure 143, where you can see that all this meandering of the channel has carved away at the adjacent uplands.

The community of Sunbury and the **Fort Morris State Historic Site**, places definitely worth the visit, are located on the banks of the Medway River, being positioned about a third of the way from I-95 to the mouth of the sound (located on Figures 137 and 140). To get there, take exit 76 off I-95 and head east on Islands Highway (GA 38; Figure 137). About a mile along, you will be passing a large sand mine on the left. Apparently, the whole area from Ft. Morris to the interstate is on the Pamlico barrier island system (Huddlestun, 1986; see Figure 140). A general elevation of 34 feet is fairly

FIGURE 140. Vertical image of St. Catherines Sound, St. Catherines Island, and Sapelo Sound. The approximate location of some of the remnants of the Pleistocene Pamlico (P), Princess Anne (PA), and Silver Bluff (SB) barrier island systems are shown (after Huddlestun, 1986). The barrier island materials (with green color of plants on beach ridges) located seaward of the Silver Bluff (SB) are Holocene, or recent, in age. Arrow A points to the Harris Neck National Wildlife Refuge and arrow B points to Shellman Bluff. Landsat image acquired on 23 October 1999, courtesy of Earth Science Data Interface at the Global Land Cover Facility.

common in this area. According to Huddlestun (1986), the Pamlico barrier island system ranges between 25-40 feet in elevation, and judging by the size of the huge holes in the ground in the sand mine, it must be pretty thick (30 feet or so?).

At the village of Seabrook, you should turn left on Trade Hill Road and drive 0.2 miles to the intersection with Fort Morris Road. Then turn left, and drive for 2 miles to the entrance to the fort. Continuing on past the fort entrance for a short distance, you will be in the

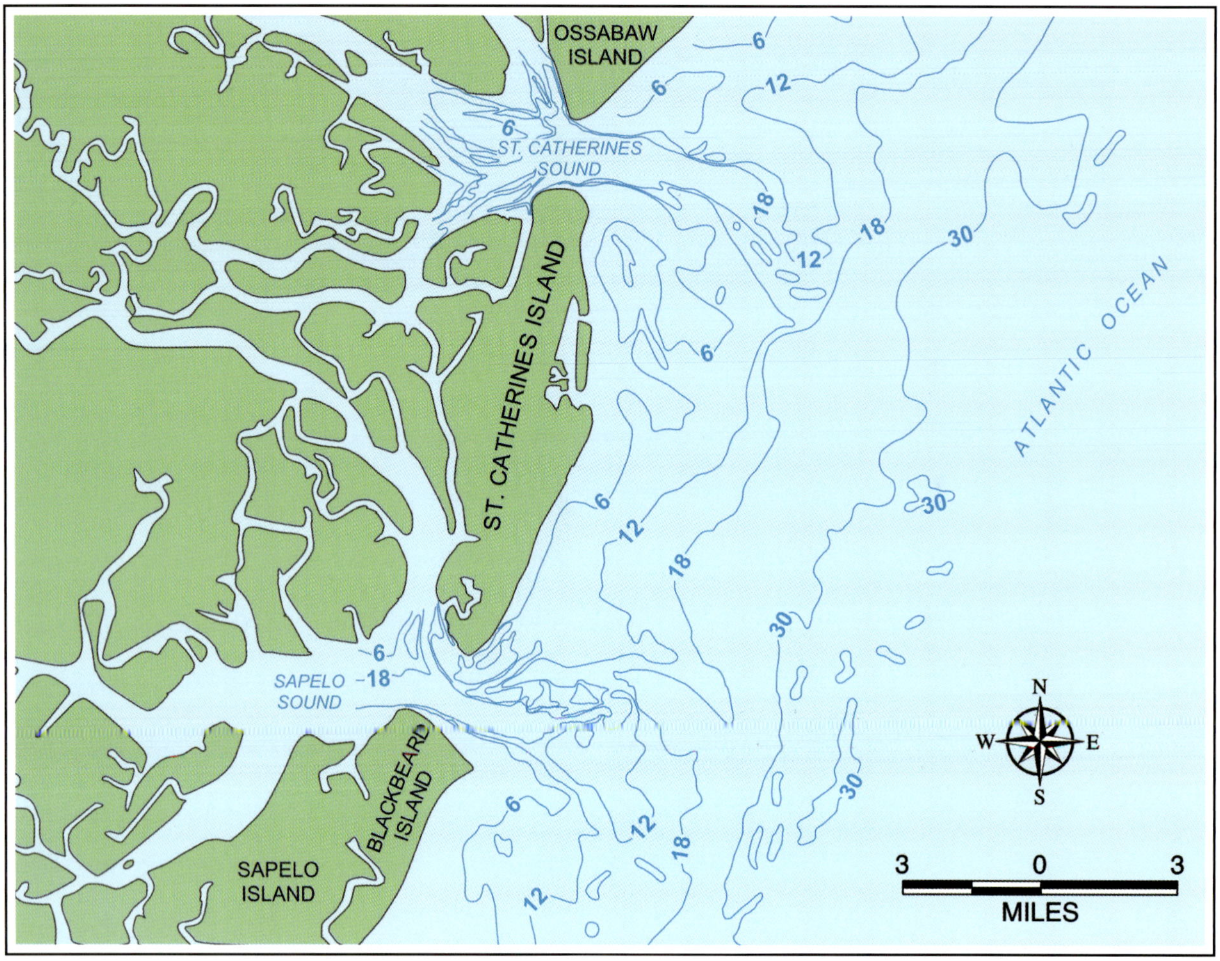

FIGURE 141. Bathymetry of St. Catherines Sound and Sapelo Sound. Note large shoals (ebb-tidal deltas) at entrances to the Sounds. Slightly modified after Kellam (1986). The numbers represent water depths in feet.

community of Sunbury, which has a very nice boat ramp on the Medway River with a platform you can walk out on for a good view of the river and the marsh.

Fort Morris was originally located on a remnant of the 125,000-130,000 years old Pamlico Pleistocene barrier island system, the first major elevated land surface available for ship landings upstream from the entrance to the Medway River. Similar to the situation at Fort McAllister, the remnant of the younger Princess Anne barrier island system (called Colonels Island at this location) does not extend far enough to the northeast to intersect the main channel of the Medway River, being separated from it by zone of salt marsh over 1,000 yards wide at the narrowest juncture (Figure 140).

According to Browns Guide to Georgia (2009a), *Fort Morris was built to defend the former town of Sunbury, once a bustling seaport second in Georgia only to Savannah. Fort Morris was used as a coastal fortification during the* The First Civil War. *The earthen works were reconstructed during the War of 1812 and were later used as a Civil War* (a.k.a. The War Between The States or The Second Civil War) *Encampment. The site's museum features displays of civilian and military life during Georgia's Colonial, Revolutionary and Antebellum past.* The hours for visiting the site are 9:00 a.m.-5:00 p.m. Tuesday to Saturday; 9:30 a.m.-5:30 p.m. Sunday. Closed Monday. In other words, don't show up to check out the facility on a Monday, which is what we did the last time we visited there.

The Fort Morris State Historic Site is also one of the localities included in the Georgia Wildlife, DNR (2002) discussion of Georgia's Colonial Coast Birding Trail. The most common birds are songbirds and wading birds, both of which are present year round, with the specialties

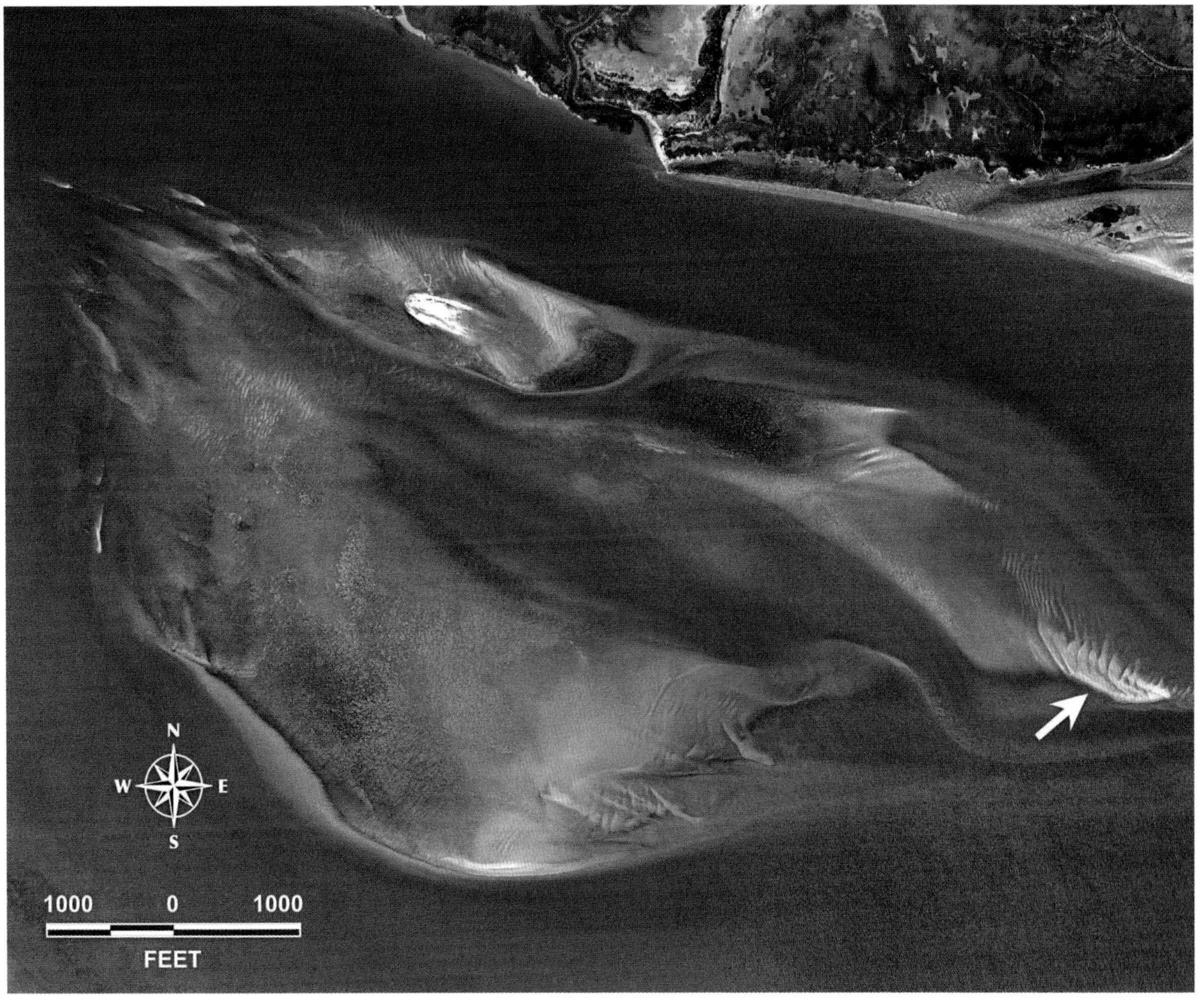

FIGURE 142. Flood-tidal delta at the entrance to St. Catherines Sound. Compare with Figure 49. Arrow points to flood-oriented sand waves that did not change their orientation when the tide dropped. Digital orthophoto acquired in June 2008, courtesy of USGS.

for this site being yellow-throated warbler, marsh wren, clapper rail, and, once again, the painted bunting. *The best times to view painted buntings are in spring and summer; the woodlands found on the site are havens for warblers during spring and fall migrations.*

ST. CATHERINES ISLAND

Introduction

This 10-mile-long barrier island is illustrated by the vertical image in Figure 144. It is bordered by two large sound/estuarine systems, St. Catherines Sound to the north and Sapelo Sound to the south, both of which are illustrated in Figures 140 and 141. A zone of salt marshes and tidal channels that averages a little over 5 miles in width separates the landward side of the island from the elevated, Princess Anne Pleistocene barrier island system to the west (Figure 140). According to Georgia Travel (2010), this barrier island is *comprised of 14,640 acres of marsh and wetlands and 6,700 acres of forest.*

A tremendous amount of information on this island is presented by Bishop et al. (2011a), a compendium of 16 papers on the geoarchaeology of St. Catherines Island, published in the proceedings of the Fourth Caldwell Conference, held on St. Catherines Island on March 27-29, 2009. This is great reference on a number of topics.

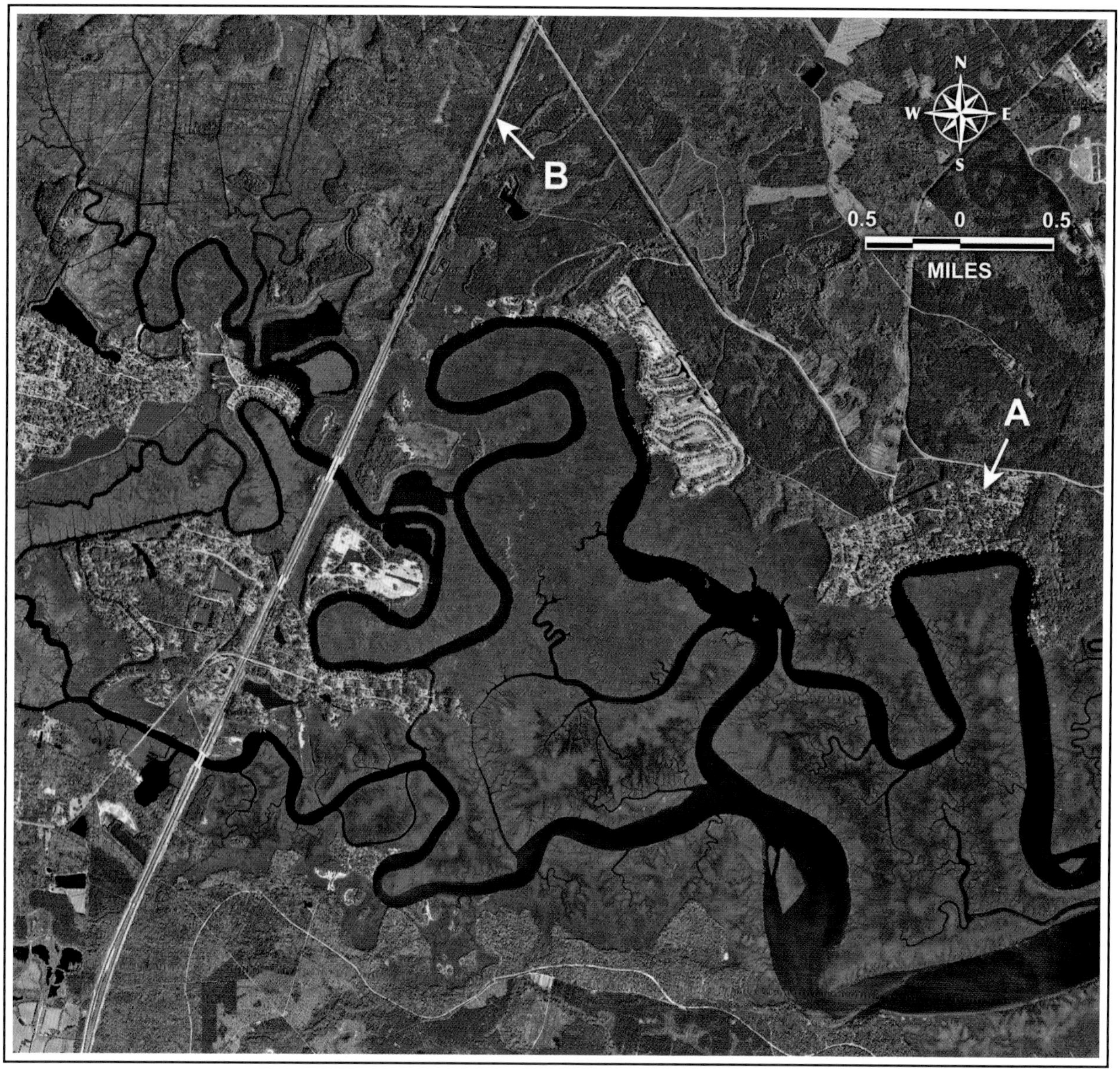

FIGURE 143. Complex meandering tidal channel systems at the head of St. Catherines Sound (located on Figure 137). Arrow A points to the community of Belfast, and arrow B points to I-95. The larger channel that I-95 crosses is called the Jerico River and the smaller one further southwest is called Jones Creek. Digital orthophoto acquired in March 2006, courtesy of USGS.

Beach Erosion

Two sources have succinctly summarized the beach erosion on this island, Bishop et al. (2011b) and Meyer, Bishop, and Vance (2011). A combination of their conclusions follow:

1) As was pointed out by Griffin and Henry (1984), a number of factors, including the interruption of the flow of sediment by impounding rivers to the north, dredging of navigable channels, and rising sea level, have accumulated to make St. Catherines one of Georgia's most erosional barrier islands; and

2) This erosion continues up until the present time, as it now occurs on virtually all margins of the island, with a net shoreline retreat or erosion rate of 14.2 ft/yr along the north-central portion of the island, 8.2 ft/yr along the south central portion of the island, and a significant erosion rate of 27.1 ft/yr along the Sapelo Sound margin.

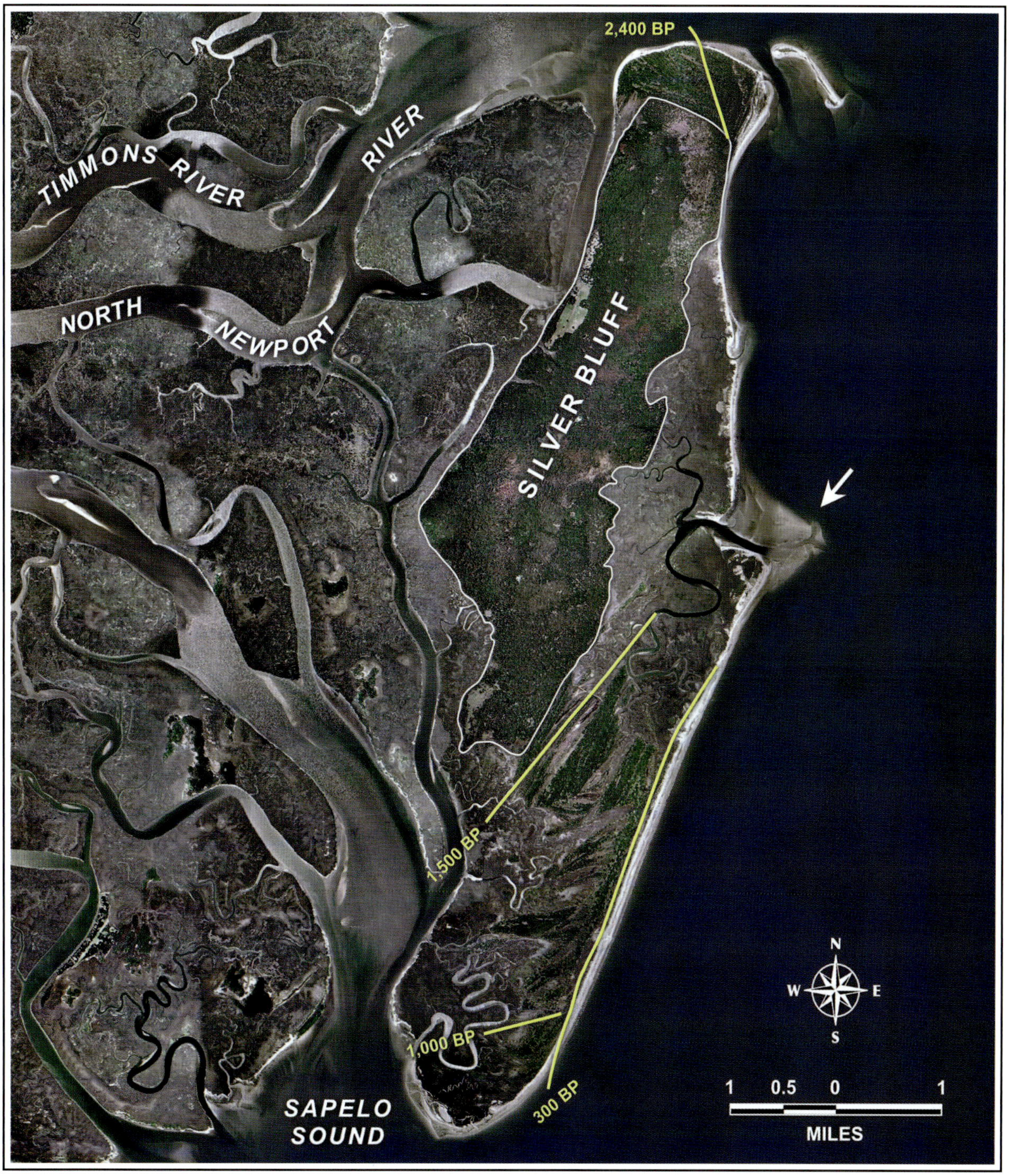

FIGURE 144. Vertical image of St. Catherines Island. The remnant of the 40,000-50,000-year-old Silver Bluff barrier island system is outlined (after Huddlestun, 1986). The barrier island sediments (some parts with green color of plants on beach ridges) located seaward of the Silver Bluff (SB) are Holocene, or recent, in age. Time lines for these Holocene beach ridges, based on data in DePratter and Thompson (2013), are also shown. Arrow points to the ebb-dominated McQueen Inlet, that is pictured in Figure 48. Digital orthophoto acquired in June 2008, courtesy of USGS.

General Geomorphology

As outlined on Figure 144, the landward side of the central and northeastern parts of the island is composed of a remnant of the Silver Bluff Pleistocene barrier island system, said by most to be 40,000-50,000-year-old (Henry, 2009; Muhs et al., 2004). A similar core of Silver Bluff sediments also occurs on Ossabaw Island, as well as on the remaining four major barrier islands to the south (Sapelo, St. Simons, Jekyll, and Cumberland Islands; Figure 39A). A zone of east-west oriented beach ridges a half-mile wide of Holocene age (Bishop et al., 2011b) is attached to the northeastern end of the island. The north end of this zone is presently highly erosional, with accumulations of large live oak trees on the beach. The presence of these trees on relatively low elevation beach ridges (maximum of 10 feet or so) indicates that the beach ridges at that location are probably as much as up to several hundred years old. There is also a zone of north-south oriented beach ridges (also Holocene in age) that measure about 1,000 feet wide attached to the eastern side of these beach ridges (see Figure 145A). Some Holocene time lines for this island determined by DePratter and Thompson (2013) are shown on Figure 144. More than half of the Holocene beach ridges on the north end of the island are more than 2,400 years old. On the other hand, the beach ridges on the southeast side of the island, many of which were deposited as recurved spits, are less than 1,500 years old.

As can be seen on Figure 39A, there is considerably less Holocene sand on St. Catherines Island than there is on Ossabaw and Wassaw Islands to the north. This is presumably due to the fact that St. Catherines Island is located beyond the south end of the Savannah River Delta (Figure 39B), thus it has a smaller source of Holocene sand than the islands further north.

A massive sand body that composes part of the southern flank of the large ebb-tidal delta of St. Catherines Sound is also attached to the northeastern end of the island, clearly shown in Figure 49A. A 500-feet-wide and 0.75 mile long inlet-like channel separates a large sandy shoal, that presently has a small island on it, from the end of the main island. Oblique aerial photographs of this inlet/channel and its associated sandy tidal flat are given in Figures 145B and 145C.

To the south along the beach, there is a zone about half a mile long that is composed of a scarp tens of feet high that has been eroded into the remnant of some Holocene beach ridges. From the southern end of this

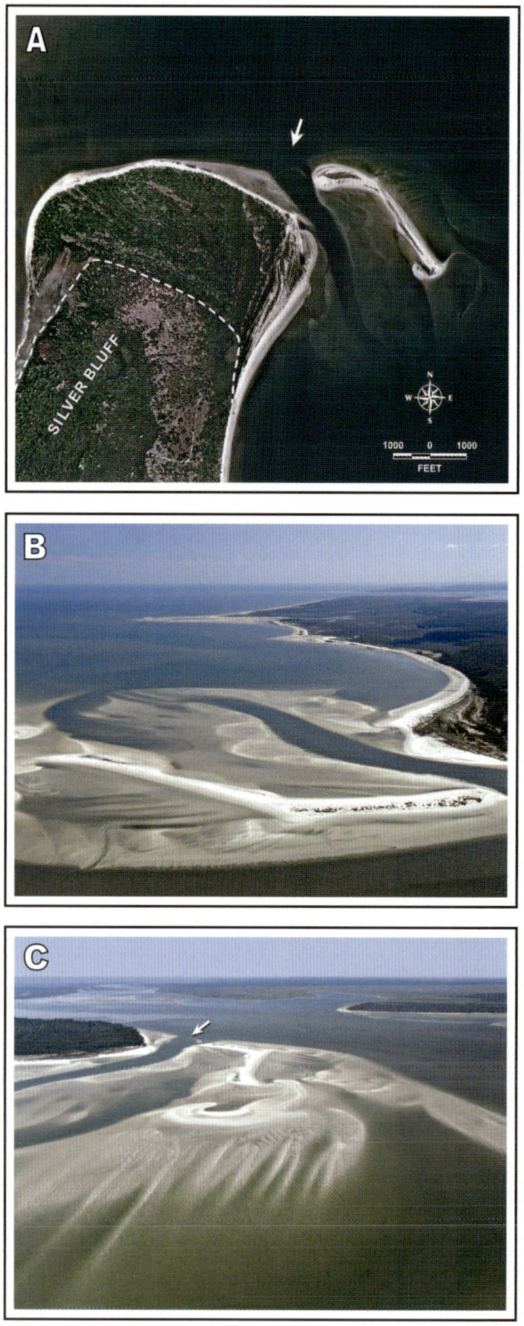

FIGURE 145. St. Catherines Island. (A) Vertical image of the north end of the Island, which is composed mostly of Holocene beach ridges (<4,500 years old). Arrow points to small tidal inlet/channel shown in B. Digital orthophoto acquired in June 2008, courtesy of USGS. (B) Tidal inlet/channel off the northeast end of St. Catherines Island (located by arrow in A). (C) Intertidal sand flat off the northeast end of the Island. Arrow points to inlet/channel pictured in B. Photographs in B and C were taken during a low spring tide on 28 April 2010.

clearly erosional zone, a transgressive barrier island is separated from the Silver Bluff remnant by a zone of salt marsh (Figure 144). This transgressive barrier island system, which is intersected by a small tidal inlet, continues south for about 2.5 miles to where it joins the classic, ebb-dominated McQueen Inlet, illustrated in Figure 48.

From McQueen Inlet to the southwestern end of the island is a zone 6 miles long and about 1 mile wide that is a purely Holocene depositional feature (Linsley, Bishop, and Rollins, 2008). It includes fragments of prograded beach ridges, including some recurved ridges, which are interspersed with salt marsh that has some small tidal channels in places. However, in recent times, this area has been notably erosional, as evidenced by still-standing trees in the intertidal zone (Figure 146A). The southwestern end of the island has a zone of classically displayed transgressive barrier islands, with marsh sediments exposed in the intertidal zone (Figure 146B).

As mentioned earlier, this large barrier island complex is backed by a zone of salt marshes around 5 miles wide, which contains an array of large tidal channels, two of which are illustrated by the vertical image in Figure 147.

Places to Visit

As noted by Georgia Travel (2010), St. Catherines Island was originally settled *by the Guale Indians, who were later joined by Spanish missionaries in the 1500s. Today, the island is owned by the St. Catherines Island Foundation and is used for conservation of its natural resources, protecting endangered species, and preserving historic sites.*

Accordingly, at the present time, *the shores of St. Catherines Island are reached by boat only. You can reserve your place at Halfmoon Marina on Colonels Island or Shellman Fish Camp in McIntosh County. You are welcome to spend the day at the island, but there is no place for overnight accommodations.*

We are limiting the rest of this discussion of places to visit in this subcompartment to locations you can drive to on the landward side of St. Catherines Island.

Browns Guides (2009b) gives a description of a driving tour in Liberty County called the Historic Liberty Trail. This drive is described as *a unique driving tour offering a diverse experience integrating history, culture and ecology. It covers 10 stops, including: Midway Museum and Historic District, Cay Creek Wetlands, Geechee Kunda Cultural Arts Center, LeConte-Woodmanston Botanical Gardens, Dorchester Academy and Museum, Fort Stewart Museum, Melon Bluff Nature and Heritage Preserve, Seabrook Village, Fort Morris State Historic Site, and Sunbury Cemetery.*

One of the sites on this list, **Melon Bluff Nature and Heritage Preserve**, sits on the bluff overlooking the North Newport River. As noted by Browns Guides (2009b), this site, located on lands that were once part of on old rice plantation, is *nestled amongst 3,000 unspoiled acres ... set amid gorgeous, moss-hung oaks at the river's edge.* It has 25 miles of grassy, forest trails for hiking, biking, picnicking, and riding.

To get there, take exit 76 off I-95 and travel east on Islands Highway (GA 38; Figure 137) for approximately 3.2 miles. The Nature Center is on the right. Hours are 9:00 a.m.-4:00 p.m., Tuesday to Sunday. There is an admission fee.

This site is also one of the localities included in the Georgia Wildlife, DNR (2002) discussion of Georgia's Colonial Coast Birding Trail. Its specialties are wild turkey, wood stork, clapper rail, roseate spoonbill, and painted bunting. The best seasons are for songbirds (all), wading birds (all), marsh birds (all), waterfowl (winter), and birds of prey (winter).

From this location, it is possible to continue southeast on Islands Highway for approximately 4 miles to Colonels Island. A village on this "island," Halfmoon Landing, is on a bluff that is being eroded into by a meander of the North Newport River. The piece of high land called Colonels Island is almost completely surrounded by marsh. The upland portion of the island, that has a maximum elevation of about 19 feet, is composed of Princess Anne barrier island sand deposited about 85,000 years ago (Figures 9, 10, and 140), according to Huddlestun (1986).

The **LeConte-Woodmanston Botanical Gardens** is another interesting site to visit, even though it is a ways off the beaten path. To get there from Colonels Island, go back to exit 76 on I-95 and go south on I-95. At about 3 miles, the highway crosses an incredible marsh system associated with the headwaters of the North Newport River (Figure 137). About a mile or so upstream of the bridge, the stacks of the large Interstate Paper Mill at Riceboro can be seen. At exit 67, turn right (north) on U.S. 17 and travel 3 miles to Sandy Run Road. At that intersection, turn left and follow this road to the stop

FIGURE 146. St. Catherines Island. (A) Highly erosional beach located approximately 2.4 miles northeast of the southernmost end of the Island (Figure 144). (B) Transgressive barrier island on the southernmost tip of the Island. Arrow points to marsh sediments exposed in the intertidal zone. Both photographs were taken during a low spring tide on 28 April 2010.

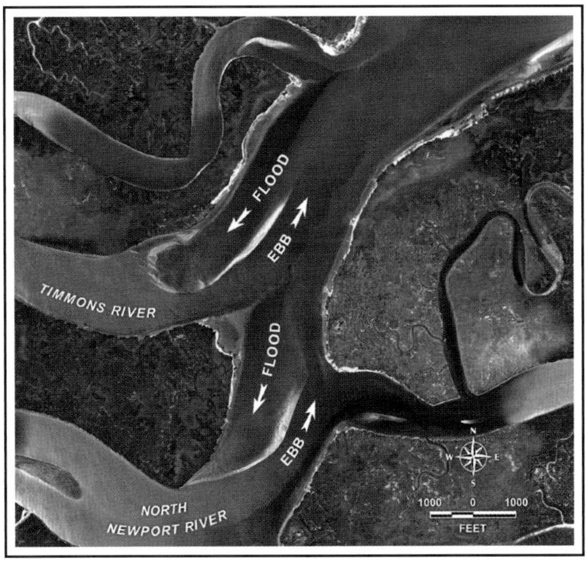

FIGURE 147. The intersection of two major tidal channels – the North Newport River and the Timmons River. This part of these channels is located about 2 miles west of the north end of St. Catherines Island (located on Figures 140 and 144). The trailing ebb spits of two well-developed tidal river point bars were exposed in the intertidal zone at low tide when this image was taken. The segregation of the dominant ebb- and flood-tidal current zones can also be clearly seen. Compare this image with the general model given in Figure 105B. Digital orthophoto acquired in June 2008, courtesy of USGS.

sign at Barrington Ferry Road, where you make a left turn and follow the signs to the gardens.

Browns Guides (2009b) gave the following description of this site:

> *LeConte-Woodmanston, formerly the home of Dr. Louis LeConte, flourished as one of Georgia's earliest inland swamp rice plantations and is now a nature preserve. Dr. LeConte achieved international fame in scientific circles as did his sons, John and Joseph. John was the first president of the University of California at Berkeley. Joseph and his friend, John Muir, co-founded the Sierra Club. Today, Louis LeConte's world-famous 18th Century botanical gardens are being recreated with a myriad of antique plants. Visit the cypress forest and walk the interpretative trail along the earthen rice dikes leading through the Bulltown Swamp black-water eco-system. Take a stroll along the Avenue of Oaks or bask in an 18th Century nature experience. They are all part of the Historic Bartram Trail.*

As fate would have it, for years the Geology Department at the University of South Carolina, where Michel earned her Ph.D. and Hayes served as a professor for twelve years, was located in a building on the Columbia campus named LeConte College. This building was named after one of Louis LeConte's sons, Joseph, who joined the faculty of South Carolina College in 1857. In 1869, he moved on to the faculty of the University of California, where he gained fame for his success as a professor of geology and physiology, publishing among numerous other books, the noted *Elements of Geology* (1877). He also published another famous book called *Evolution and its Relation to Religious Thought* in 1888, an effort to reconcile what was then called the theory of evolution with Christian beliefs.

SAPELO SOUND

Introduction

Sapelo Sound is the fourth of the eight major sounds on the Georgia Coast (going south). As can be seen in Figure 140, this 2-mile-wide sound is fed by three main streams: 1) South Newport River to the north, that drains the marsh complex landward of the southern half of St. Catherines Island; 2) Mud River to the south, that drains the marsh complex landward of the northern half of Sapelo Island; and 3) Sapelo River, that is situated directly in a landward direction in line with the long axis of the sound. As was also true for the three sounds already discussed, because of the large tides and generally depositional nature of this coastline, fairly wide tidal flats and extensive salt and brackish marshes are present. And, also like the others, a large ebb-tidal delta is located offshore of the entrance to the sound (Figure 141), in this case with an areal extent of 33 square miles (Figure 53).

Tidal Channels

In many respects, this sound is quite similar to the three sounds to the north of it (Wassaw, Ossabaw, and St. Catherines). However, presumably because of the flat gradients in some of the marsh areas, some of the smaller tidal channels show extreme sinuosity, as the two examples illustrated in Figure 148 show. The extreme upper end of the Sapelo River, shown in Figures 148B and 148C, gives the appearance of having "tied itself into a knot!"

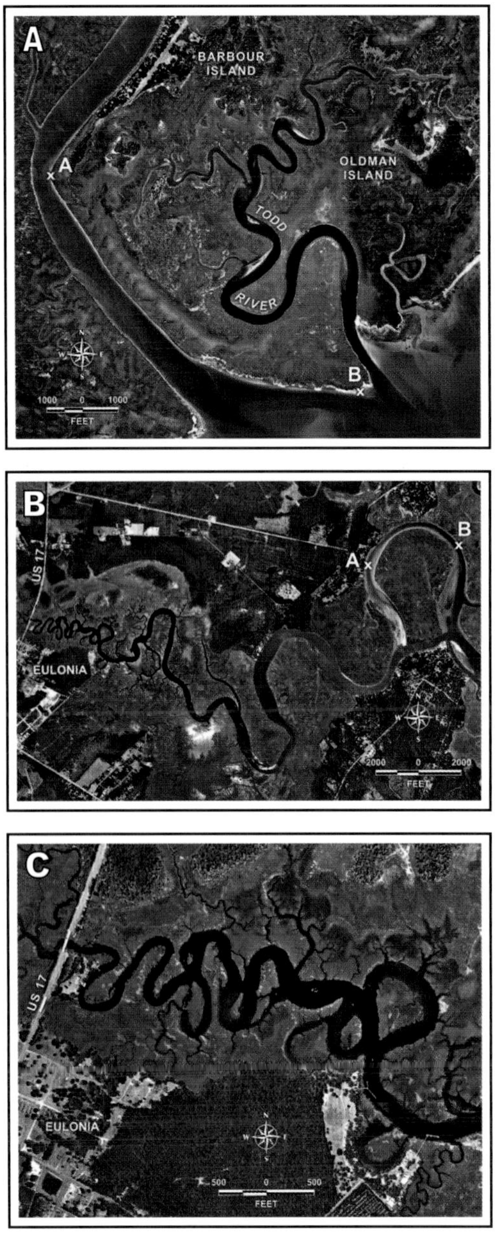

FIGURE 148. Examples of extreme sinuosity (meandering) of tidal channels in the tributaries of Sapelo Sound. (A) A highly meandering small tidal channel in a salt marsh near the entrance to the sound (located in the southwest corner of Figure 144). The straight-line distance between points A and B is four miles. (B) The extreme northern end of the Sapelo River near Eulonia in McIntosh County (Figure 137). The straight-line distance between A and B across the largest meander loop in the upper right is 0.9 miles. (C) Detail of the extreme end of the Sapelo River, one of the more complex meandering systems on the coast. Digital orthophotos acquired in June 2008, courtesy of USGS.

Places to Visit

The first place that we recommend for you to visit is **Harris Neck National Wildlife Refuge** (Figure 137). To get there, if you are heading south on I-95, take exit 67 and go south on U.S. 17. Almost immediately, you will be driving across a wide marsh associated with the highly meandering headwaters channel of the South Newport River that eventually empties into Sapelo Sound. After crossing the river, keep travelling south on U.S. 17 for approximately 0.4 miles to Harris Neck Road. Take a left on this road and drive 6.5 miles to the refuge entrance.

A kiosk at the refuge includes the following: *This Harris Neck Refuge is part of a chain of seven refuges comprised of the Savannah Coastal Refuges network managed by the U.S. Fish and Wildlife Service. It spans 100 miles, Pinckney Island Wildlife Refuge at the northern end and Wolf Island near Darien, Georgia as the southernmost one. Between these lie Savannah, Tybee, Wassaw, Harris Neck, and Blackbeard.* Of these refuges, only three are accessible by vehicle – Savannah, Pinckney Island, and this one, Harris Neck. With regard to the other refuges in Georgia, Wassaw and Blackbeard can be visited by boat, but Wolf Island Refuge is closed to the public. The refuges in this network are located on Figure 149.

A flyer available at the entrance gate points out the following:

1) The refuge covers 2,824 acres, has seven hiking/biking trails, and a wildlife drive that circles through the middle of the refuge;

2) This area started out as farming area in 1750, but the plantations closed down after The War Between The States;

3) During World War II, a pilot training airport was built. Remnants of the paved runways are still present;

4) On 25 May 1962, the forerunner of the U.S. Fish and Wildlife Service acquired this land, which includes 1,297 acres of salt marsh and 157 acres of man made freshwater ponds, for a migratory bird refuge. These habitats are now managed primarily for marsh and wading birds, waterfowl, resident wildlife, and wildlife-dependent recreation;

5) Management of six man-made freshwater impoundments enhances the refuge's importance as a wintering area for migratory birds. These ponds serve also as roosting and feeding areas for many wading birds and waterfowl. A photograph of the largest of the ponds, Woody Pond, is given in Figure 150; and

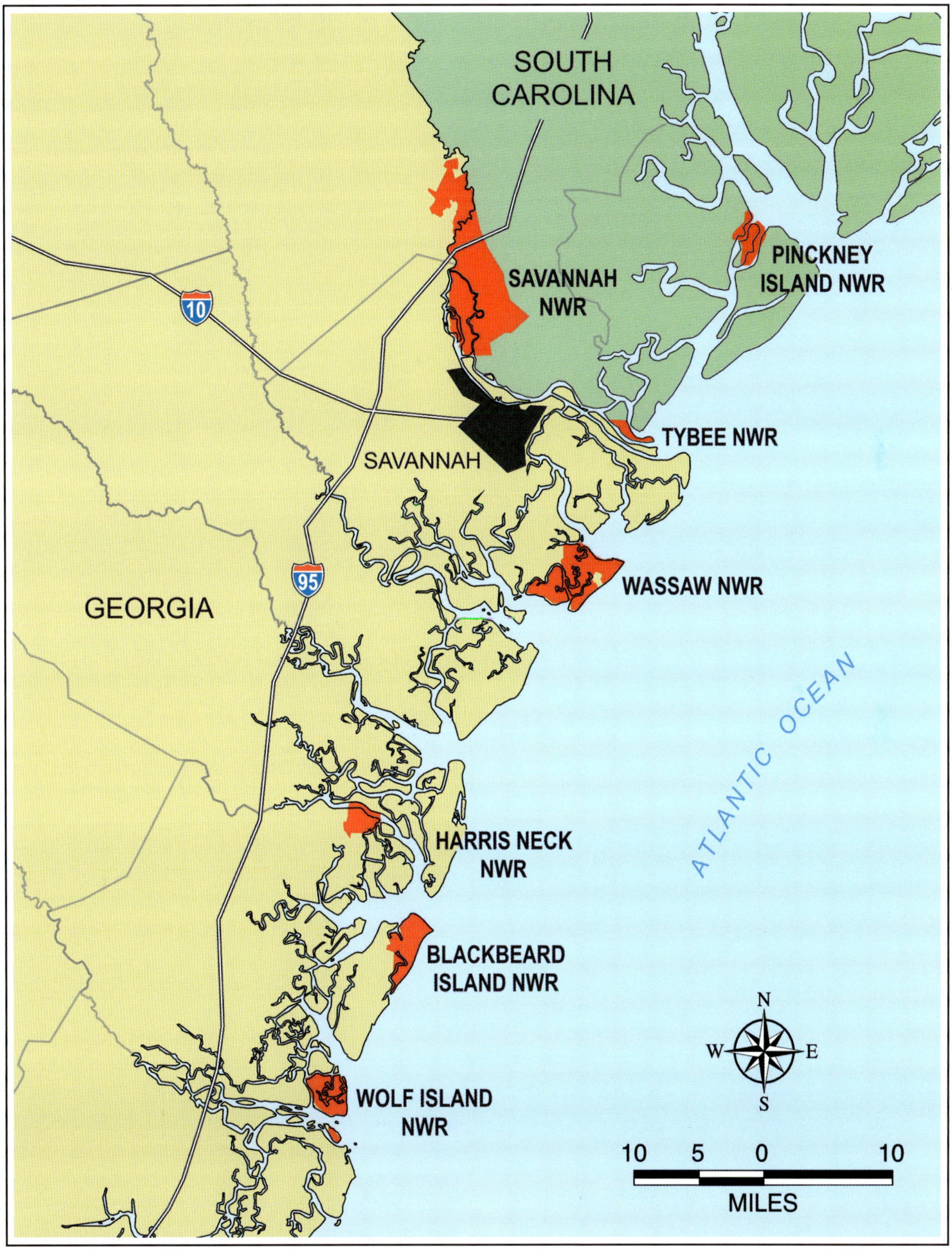

FIGURE 149. The National Wildlife Refuges (shown in red) that make up the Savannah Coastal Refuges network managed by the U.S. Fish and Wildlife Service.

FIGURE 150. Woody Pond nesting and roosting area on the Harris Neck National Wildlife Refuge. Photograph taken in late March 2010.

6) Hours are sunrise to sunset seven days a week. There is no entrance fee.

We visited this refuge on 27 March 2010 and took the 4-mile-long Wildlife Drive. Our impression was that the refuge is a little piece of high ground surrounded by wetlands. According to Huddlestun (1986), the refuge is located on a remnant of the Princess Anne Pleistocene barrier island system (Figure 140).

The first part of drive in the refuge is through live oaks and pines, hosts to lots of Spanish moss. While on this drive, we saw cormorants, wood storks, common egrets, turkey vultures, and coots.

This site is included in the Georgia Wildlife, DNR (2002) discussion of Georgia's Colonial Coast Birding Trail. The specialties they listed include wood stork, white ibis, and painted bunting. They also gave the following tips: *The best time to view nesting wood storks, great egrets, snowy egrets, black crowned night herons, anhingas and other waders is May and June. One of the best places in Georgia to view nesting wading birds, Harris Neck provides the only man-made nesting structures for wood storks. Nesting wading birds are best viewed with a spotting scope. Look for painted buntings in late spring through summer.*

The U.S. breeding population of wood storks was listed as endangered in 1984 under the Endangered Species Act. The population declined from an estimated 20,000 pairs in the 1930s to about 5,000 pairs in the late 1970s, mainly because of reductions in the quality of feeding habitats around their preferred breeding sites (they build nests in trees to avoid predators). They have specialized feeding methods–they grope around in shallow water with their bill partly open, then snap it shut when they touch a fish. Therefore, they need larger concentrations of fish in shallow waters that often dry up during droughts or as a result of water diversions. Ideal water conditions include periods of flooding, during which fish populations increase, followed by dryer periods, when the fish are more concentrated, coinciding with nesting season.

After completing our drive, we concluded "this is the place to see wood storks." Wood storks started coming to this Refuge in 1987. A write-up on a kiosk by Woody Pond records that, *in 1991, all of the available platforms were occupied.* Over 400 wood storks arrive annually, but *more than 800 have been recorded in some years. Two or more chicks are usually fledged from every nest.*

Another place to visit in this vicinity is the village of Shellman Bluff (Figure 137). To get there from the Harris Neck Wildlife Refuge, go west on Harris Neck Road for 1.9 miles, where you turn left on Young Man Road. Go 1.4 miles on this road, headed to the southwest, until you come to the first significant road on your left. Take this road and head southeast. After about 3 miles, you will be in Shellman Bluff. This village is on the elevated remnant of the 125,000-130,000 year old Pamlico Pleistocene barrier island system (Figure 140). The village is located on a couple of tributaries of the Sapelo River, the Broro River closest to the heart of town, and the larger Julienton River located a little further along to the northeast. These rivers, as well as the Sapelo River itself, are highly meandering with abundant tidal river point bars, such as the large one in the Julienton River very close to Shellman Bluff shown in Figure 107. The three tidal river point bars shown in Figure 106 are located in the Savannah Cut of the Sapelo River near another village on a bluff in the Pamlico remnant called Pine Harbor.

We have never spent much time in this area, so we will add the following description of it by Lenz (2002c):

> *Shellman Bluff is a peaceful, picturesque fish camp village that retains the distinctive charm that once was common on the Georgia Coast but has become harder to find. This is the real thing. Quaint, screened fishing cottages sit back among oaks festooned with Spanish moss. All the dirt roads of the quiet village seem to lead inevitably to the high bluff that overlooks the Broro and Julienton rivers. Winding along the edge of the high bluff is another sandy road that offers one of the best coastal views in Georgia. In the morning, the sun rises from behind Harris Neck, lighting the green marsh and dappling the tidal rivers. Birds roost in distant hammock islands. .... Today, Shellman and Sutherland Bluff have several excellent low-country seafood restaurants and is a great place to charter a boat for fishing, a nature outing, or a trip to Blackbeard Island National Wildlife Refuge.*

13 COMPARTMENT 3 - BLACKBEARD ISLAND TO JEKYLL ISLAND

INTRODUCTION

For purposes of discussion, we have subdivided this compartment into five subcompartments (Figure 151):

1) **Blackbeard and Sapelo Islands** – A large "combination" barrier island complex, with the drumstick-shaped Blackbeard Island of Holocene age as the northern flank. These islands are very famous in the geological literature.

2) **Doboy Sound** – A modest-sized sound with a 1-mile-wide entrance that is squeezed in between the southern end of Sapelo Island and the lower delta plain of the Altamaha River Delta.

3) **Altamaha River Delta** – Third largest river delta in the Georgia Bight, a spectacular geomorphological phenomenon with a fascinating human history. The Santee/Pee Dee Delta in South Carolina and the Savannah River Delta at the South Carolina/Georgia border are the two largest deltas.

4) **St. Simons and Sea Islands** – Another "combination" barrier island complex that contains the largest developed coastal area in Georgia.

5) **St. Simons Sound** – Although it has a narrow, 1-mile-wide entrance, this major sound has an abundance of tidal rivers, with the second largest city on the Georgia Coast, Brunswick, located along the banks of the Brunswick/Turtle Rivers.

BLACKBEARD AND SAPELO ISLANDS

Introduction

This welded complex of Holocene and Silver Bluff Pleistocene barrier islands is about 12 miles long, with an average width of about 3 miles (Figure 152). This barrier island system is the most noteworthy one in Georgia with regard to the scientific literature. As pointed out in the dedication of this book, early researchers at the University of Georgia Marine Institute on Sapelo Island made many original discoveries in the fields of coastal geology and ecology. This Institute, founded in 1953, is still very active and plays a major role in the culture of Sapelo Island, as is made clear by the following comments from UGAMI (2010):

The University of Georgia Marine Institute (UGAMI), located on this somewhat isolated barrier island, is *accessed from Meridian, Georgia by a passenger ferry which is run by the Georgia Department of Natural Resources 3 times per day (20-30 minutes between Meridian and Sapelo). UGAMI has a long-term lease for facilities totaling 1,500 acres on the south end of Sapelo Island within and bordering the NOAA Sapelo Island National Estuarine Research Reserve (SINERR). The facility is very unique in its situation of relatively pristine surroundings, as well as the designation of the surrounding marshes as a NOAA SINERR* (located on Figure 152). *This is a fortuitous and productive research situation. The remaining 2/3 of Sapelo Island are managed by the Georgia DNR as a Wildlife Management Area. There are approximately 100 full time residents on Sapelo Island, including UGAMI personnel, DNR personnel, and the small Hog Hammock community. Hog Hammock (500 acres) is the only privately owned property on Sapelo Island. Hog Hammock is a community of descendents of African-American plantation workers, and many members of the community are employed by UGAMI.*

The ***Sapelo Island National Estuarine Research Reserve*** *occupies just over one-third of Sapelo Island* (Figure 152), *the fourth largest Georgia barrier island and one of the most pristine. The reserve's 6,110 acres contain the Duplin River and its estuary, and several upland tracts. The reserve comprises 2,110 acres of upland maritime forest and hammock land and 4,000 acres of tidal salt marsh. The upland maritime forest of the reserve is composed of a mix of native hardwoods and about 90% of the reserve's marshland is covered by smooth cordgrass. The most conspicuous animals of the salt marsh are the*

FIGURE 151. Compartment 3, north end of Blackbeard Island to north end of Jekyll Island. The red arrows indicate the boundaries of the compartment. Source: ESRI StreetMap USA.

FIGURE 152. Vertical infrared image of Sapelo Island. The remnant of the Silver Bluff Pleistocene barrier island system that forms the western part of the Island is outlined (after Huddlestun, 1986). The barrier island sediments located seaward of the Silver Bluff are Holocene, or recent, in age. The time lines for these Holocene components of the complex, after Depratter and Thompson (2013), are also shown. Note the clearly defined beach ridges on Blackbeard Island. The boundary of the Sapelo Island Estuarine Research Reserve is outlined in yellow dashes. Landsat image mosaic acquired in 2000-2001, courtesy of USGS.

graceful egrets and herons, fiddler crabs, and raccoons; however, many other less visible creatures live within the reserve, including mollusks. Endangered and threatened species of Sapelo Island include Southern bald eagles, wood storks, American alligators, loggerhead sea turtles, the northern right whales and manatees. Not only is the island rich in natural history, but also in human history dating back 4,000 years (NERRS, 2012).

Coastal Processes and Geomorphology

This island has two very distinct geomorphic components: 1) The Holocene outer fringe, which was deposited in the last 4,500 years; and 2) The western core of the island composed of a remnant of the Silver Bluff Pleistocene barrier island system deposited 40,000-50,000 years ago. The developmental history of Sapelo Island and the associated Pleistocene barrier island systems is shown by the evolutionary sequence presented in Figure 45 (after Hoyt, Weimer, and Henry, 1964).

Three time lines for the Holocene portion of Blackbeard and Sapelo Islands are shown on Figure 152 (after DePratter and Thompson, 2013). The northeastern part of the Holocene component is the 7-mile-long Blackbeard Island, most of which is <1,400 years old. As is illustrated in Figure 36B, Blackbeard Island has a **classic drumstick shape**. The trailing **recurved spit** at its southwestern end also has a classic shape (Figure 30). As shown in Figure 54B, the inlet at the southwest end of the spit (Cabretta Inlet) migrates consistently toward the southwest under the influence of waves approaching the shore from the northeast, as is illustrated in Figure 31. Once the main channel of the inlet "loses its gradient advantage" due to over-extension to the southwest, the neck of the spit will be cut through, most commonly during a hurricane, and the process starts all over again (Figure 54B). The ebb-tidal delta of Cabretta Inlet is relatively large and conforms fairly well to the general model of ebb-tidal deltas in Figure 52. Figure 36C shows a large swash bar welding to the beach on the southwest edge of the ebb-tidal delta on 28 April 2010.

A narrow zone of Holocene sand, backed by a 0.5 mile of salt marsh, is located to the southwest of Cabretta Inlet, extending 1.6 miles to another rather small tidal inlet. From that inlet, it is another 3.7 miles to the extreme southwest end of the island, where the Holocene sand deposit is considerably wider, with the zone of curving beach ridges 500 yards wide shown in Figure 65B composing that end of the island (see also Figure 152). According to DePratter and Thompson (2013), this spit is <400 years old.

Many of the early studies on the beaches of this area described such features as the intertidal bars, like the ones also shown in Figure 65B (discussed in Section I; notable references include Hoyt, 1962; Hoyt, Weimer, and Henry, 1964). The nearshore zone was also studied in some detail (Frey and Howard, 1986).

From the standpoint of coastal processes and geomorphology, the remainder of the island, that is the Silver Bluff part, is somewhat overshadowed by the Holocene component. However, the climax maritime forest is extensive in places on the Silver Bluff sediments, containing a variety of birds and plant species, and the surrounding marsh of the Sapelo Island National Estuarine Research Reserve, located on Figure 152 and discussed earlier, is very noteworthy.

The salt marshes of the area were also studied in detail by the earlier workers. Their general model of the Georgia salt marshes based on the Sapelo Island studies, which was discussed in Section I, is given in Figure 58.

Blackbeard National Wildlife Refuge

U.S. Fish and Wildlife Service (2009a) reported the following interesting information about this national wildlife refuge that encompasses all of Blackbeard Island:

1) *The island was acquired by the U.S. Navy in 1800 as a source for live oak timber for ship building;*

2) *In 1924, the island was placed under the jurisdiction of the Bureau of Biological Survey to be maintained as a preserve and breeding ground for native wildlife and migratory birds;*

3) *In 1975, three thousand acres of the refuge were set aside as National Wilderness;*

4) As indicated in the images in Figures 30, 36B, and 152, *the island is comprised of interconnecting linear dunes thickly covered by oak/palmetto vegetation*;

5) *There are approximately 1,163 acres of open freshwater or freshwater marsh, 2,000 acres of regularly flooded salt marsh, 2,115 acres of maritime forest, and 340 acres of sandy beach*; and

6) At the present time, *the primary objectives of the refuge are to provide wintering habitat and protection for migratory birds; provide protection and habitat to promote resident and migratory wildlife diversity; and to provide protection and management for endangered and threatened species (loggerhead sea turtle, wood stork,*

piping plover). Notable concentrations of waterfowl, wading birds, shorebirds, songbirds, raptors, deer, and alligators can be seen at various times of the year.
(NOTE: *Blackbeard Island was named for Edward Teach, alias Blackbeard the Pirate. Rumors of Blackbeard's buried treasure still flourish, but no evidence of his fortune has ever been discovered.*)

Blackbeard Island is accessible only by boat. *Visitors are welcome to bring their own private watercraft, or arrangements for trips to the refuge can be made through local boat captains and marinas, as well as authorized tour guides. A public boat ramp on Harris Neck NWR (Barbour River Landing) may be used as a launching site for trips to the island.*

Places to Visit

The most obvious place to visit in this subcompartment would be Sapelo Island itself. However, you can only get there by boat. Whether you go to the island or not, we recommend that you visit the Sapelo Island Visitors Center, located about 6 miles north of Darien off GA 99 near Meridian (Figures 151 and 153). The Center, located beside the DNR boat launch for trips to the island, is open to the public. It has an abundance of displays on the wildlife, geology, and so on about the island.

The required reservations necessary to visit the island must be made through the Sapelo Island Visitor Center. Visitor Center hours are Tuesday-Friday 7:30 a.m.-5:30 p.m.; Saturday 8:00 a.m.-5:30 p.m.; Sunday

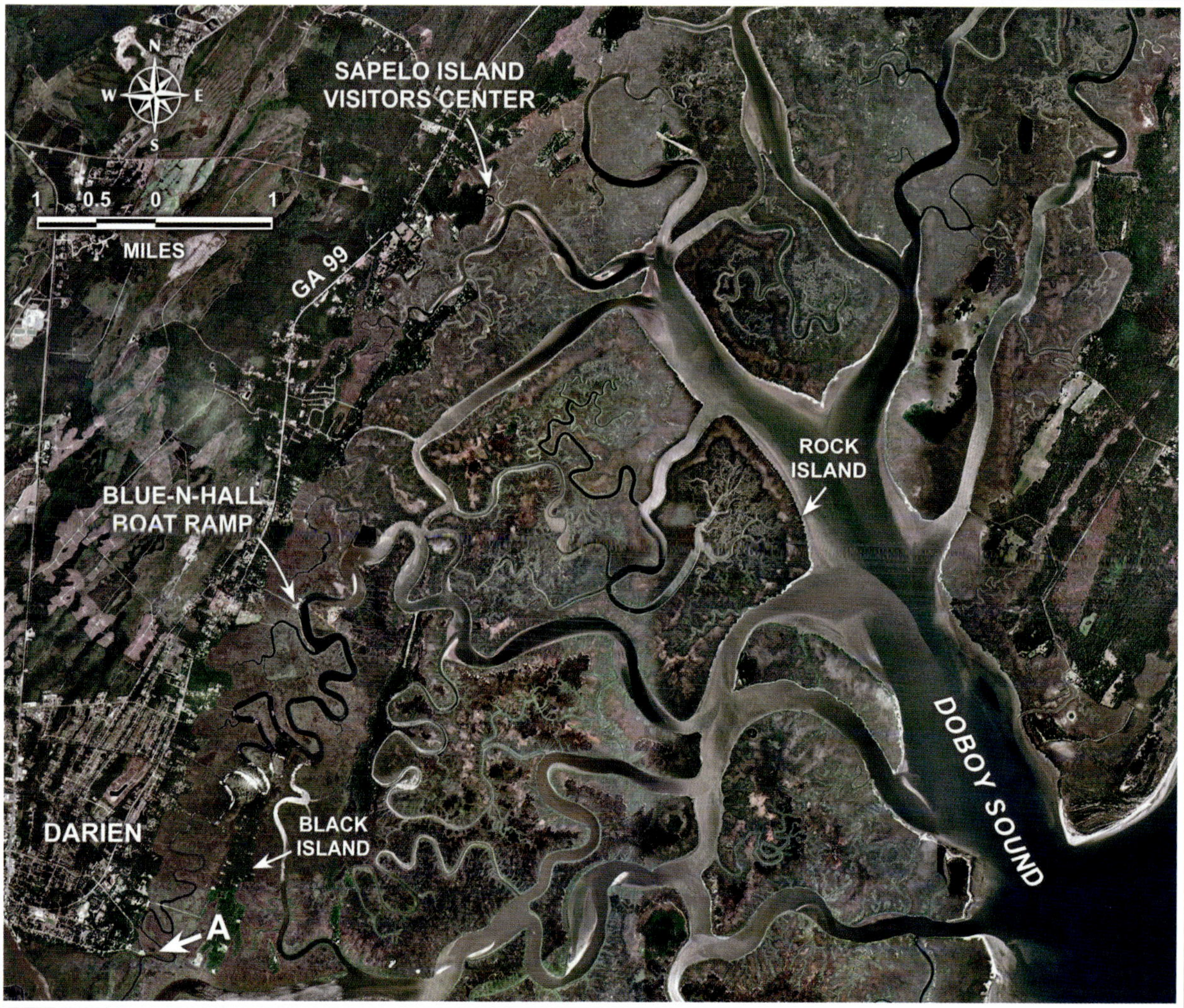

FIGURE 153. Vertical image of Doboy Sound. Note the high sinuosity of the streams in the area where the waters of the sound and the lower delta plain of the Altamaha River Delta meet (lower left-central portion of image). Arrow A points to the King George State Historic Site. Digital orthophoto acquired in June 2008, courtesy of USGS.

1:30-5:00 p.m.; Closed Monday.
(NOTE: A kiosk near the center presents the following details on the history of Sapelo Island: *Situated five miles to the east, Sapelo was home to Native Americans four thousand years ago and was the site of a sixteenth century Franciscan mission. Thomas Spalding was a leading planter of Sea Island cotton and sugar cane during his ownership from 1802 to 1851. Slave descendents* (presumably from the Spalding era) *still live on the island at Hog Hammock. Sapelo was owned from 1912 to 1934 by automotive pioneer Howard E. Coffin, followed by tobacco heir Richard J. Reynolds from 1934 to 1964. The University of Georgia Marine Institute opened on Sapelo in 1953. The state of Georgia bought the island in separate purchases in 1969 and 1976, almost 18,000 acres, from the Reynolds estate.*)

This island area is also home to Sapelo Island Wildlife Management Area (WMA), an area of 9,000 acres open to hunting according to certain conditions and regulations. For information about this WMA consult SIWMA (2010).

After you leave the Visitors Center, if you head north on GA 99, it is a pleasant drive of about 4 miles to the village of Crescent, located on a meander bend of the Crescent River, with about 20 docks projecting out into the river. On this trip, you will be driving on an elevated wooded area composed of a remnant of the Princess Anne Pleistocene barrier island system. Several excellent views off to the right of the salt marshes along the Mud River tributaries are also available on this drive.

Once you leave the village of Crescent heading west, you will no doubt notice that the road goes up to an upland even higher than the one you left behind on the Princess Anne remnant. Once on the higher elevation, you will be on the Pamlico Pleistocene barrier island system, a very conspicuous change in this area (see the southwest corner of Figure 140).

DOBOY SOUND

Introduction

Doboy Sound is the fifth of the eight major sounds on the Georgia Coast (going south). As can be seen in Figures 153 and 154, this one-mile-wide sound is somewhat restricted in its dimensions compared to the larger sounds to the north, being sandwiched between the lower delta plain of the Altamaha River Delta and the backside of Sapelo Island. Consequently, the ebb-tidal delta (Figure 154) is relatively small, only covering an areal extent of about 17 square miles, as compared to the 33 square miles covered by the ebb-tidal delta off Sapelo Sound (Figure 53). As was also true for the four sounds already discussed, because of the large tides and generally depositional nature of this coastline, fairly wide tidal flats and extensive salt and brackish marshes are present.

As can be seen on the vertical image in Figure 153, an amazing array of highly meandering, moderate-sized tidal channels is located to the northeast of Darien where the lower delta plain of the Altamaha River Delta merges with Doboy Sound. This extreme meandering is probably a result of the widening of the lower delta plain, with its exceptionally flat surface, as a result of the introduction of the abundant fine-grained sediments brought into the area by the Altamaha River. The tree-shaped channel network on Rock Island, located about 7 miles northeast of Darien (Figure 108C), is another interesting channel system in this same area.

Places to Visit

If you take the boat ride from the Sapelo Island Visitors Center out to the island, you will have ridden the boat right down the middle of Doboy Sound, a great chance to experience a large estuarine system, see large numbers of birds, and so on. Otherwise, a drive along the northwest rim of the sound along GA 99 is also a worthwhile experience.

To take this drive, we suggest that you go south from the road to the Sapelo Island Visitors Center (Figure 153) on GA 99, which is designated as a Georgia Scenic Byway in this area. Drive southwest for about 3.9 miles to the village of Ridgeville, where you turn left to go out to the Blue-N-Hall boat ramp (Figure 153). After a drive of about 300 yards you will "zoom down" off the Pleistocene upland, a remnant of the Princess Anne Pleistocene barrier island system in this case, and drive out across the marsh (Figure 155A). This part of the marsh is an extremely wide zone of black needlerush shown in Figure 155B.

The substantial Blue-N-Hall Boat Ramp/Dock complex is located about 600 yards out in the marsh on a big tidal channel called North River. This is a highly meandering channel, with some very well developed tidal river point bars. A short distance downriver, this channel expands abruptly into a much larger tidal

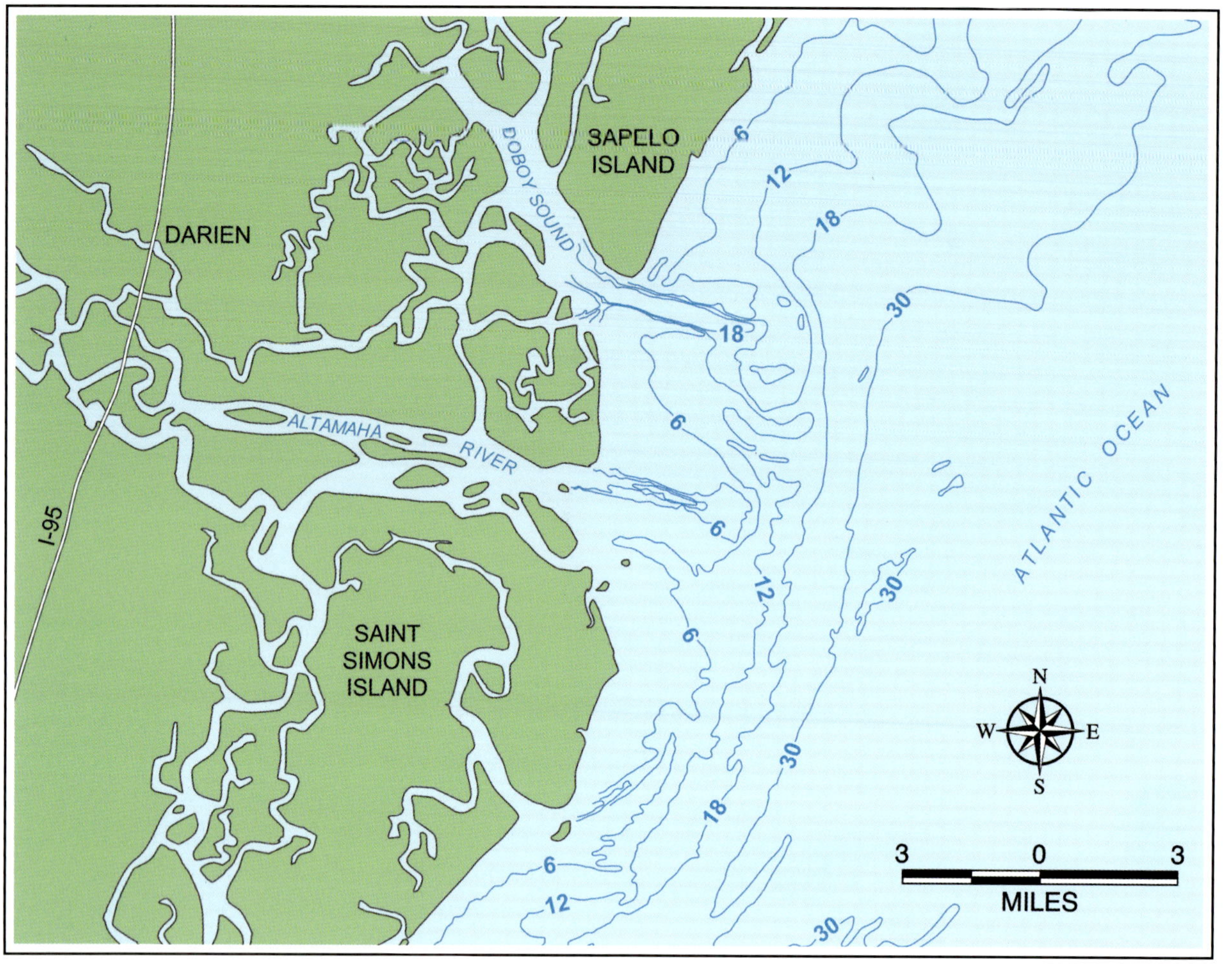

FIGURE 154. Bathymetry of the nearshore zone off Doboy Sound and the delta front of the Altamaha River Delta. Slightly modified after Kellam (1986). Numbers represent water depths in feet.

channel that feeds into Doboy Sound.

During a visit to this site on 27 March 2010, we looked offshore from the boat dock and surprisingly saw some trees. They are growing on an island called Hird Island, that is very similar to Black Island seaward of Fort King George (located in southwest corner of Figure 153). It is oriented the same way but sets offshore by about 0.8 miles (stairstep like). This island is about 1.8 miles long and averages about 300 yards wide. It has two large eroded scallops on its east side made by a highly sinuous tidal channel (Hird Island Creek) that averages about 125 feet in width (Figure 155A). There is also a landing strip on the island that has been there since 1980 (at least). There are well over 50 houses on this island, which appears to have a maximum height of about 10 feet. Black Island appears to be a little higher than that.

This observation leaves us with a question. Are these islands part of the remnant of either the Silver Bluff or Princess Anne Pleistocene barrier island systems, or were they formed during one of the earlier Holocene highstands? Interesting problem.

We thought this was a very interesting place to visit, with lots of birds and a great brackish marsh (Figure 155B).

ALTAMAHA RIVER DELTA

Introduction

As is clear from the map of the offshore bathymetry in Figure 154 and the vertical image in Figure 156, the sediments supplied by the Altamaha River's discharge rate of over 13,000 ft^3/second (Table 3) have created a conspicuous bulge in the shore properly defined as a

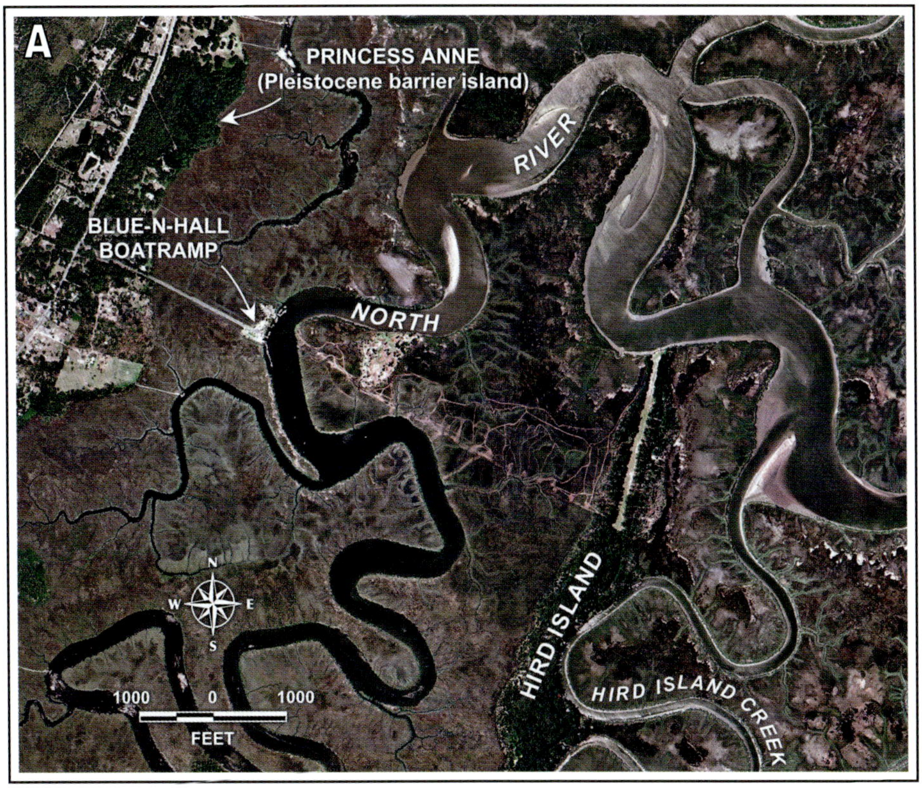

FIGURE 155. Doboy Sound, cont. (A) Vertical image of the northwest side of the sound. Note location of the Blue-N-Hall boat ramp and Hird Island. The outstanding tidal river point bars in the North River (to the northeast of the boat ramp) are also clearly visible in this image. Digital orthophoto acquired in June 2008, courtesy of USGS. (B) Black needlerush (*Juncus roemarianus*) marsh landward of the Blue-N-Hall Boat Ramp near Ridgeville.

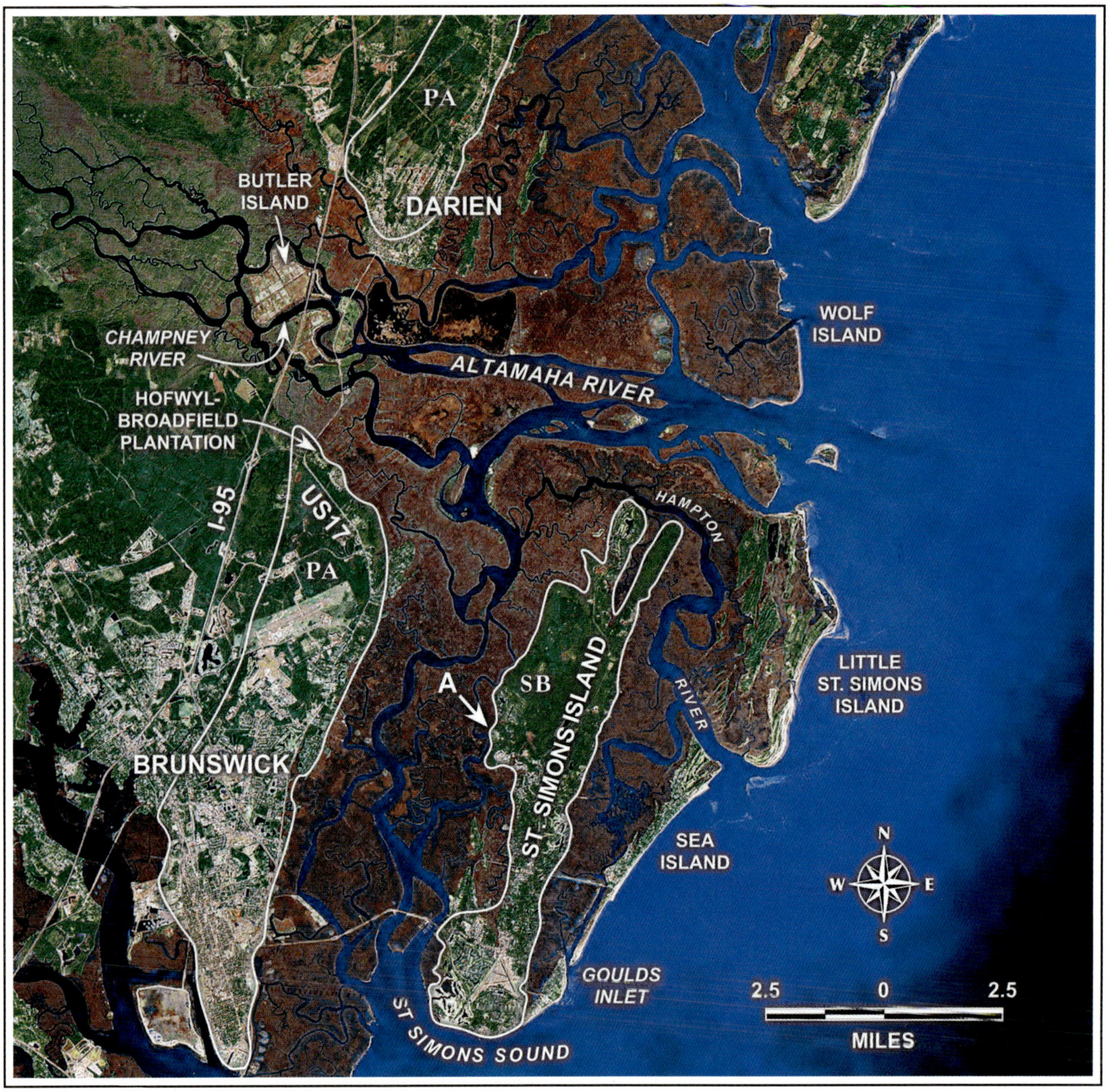

FIGURE 156. Vertical image of the Altamaha River Delta. Sea, Little St. Simons, and Wolf Islands compose the *delta front.* The approximate boundary of the Pleistocene Silver Bluff barrier island system on St. Simons Island is indicated (heavy white line; after Huddlestun, 1986). The barrier island sediments (with green color of plants on beach ridges) located seaward of the Silver Bluff (SB) are Holocene, or recent, in age. The city of Brunswick and the town of Darien are located on remnants of the Princess Anne barrier island system, which formed approximately 85,000 years ago (Huddlestun, 1986). Arrow A points to Fort Frederica. Landsat image mosaic acquired in 2000, courtesy of the NASA Applied Science & Technology Project Office.

river delta. With tributaries that drain the piedmont area in Georgia, this river is classified as a red-water, or piedmont, river. Therefore, this discharge of a relatively large sediment load (Figure 120) has allowed the river to build the third such prominent delta in the Georgia Bight (Figure 39B). Similar to the Savannah River Delta, this is a **mixed-energy delta**, with the obvious influence of the tides at the entrance, with three major tidal inlets being present (the river mouth itself, Hampton River outlet at the south end of Little St. Simons Island, and the outlet to Doboy Sound in the north), and the role of waves in building the barrier islands along the delta front. Note the approximate location of the Georgia deltas on the ternary diagram in Figure 59.

Some interesting notes on the human history along the river, provided by Sherpa Guides (1999), follow:

> *Hernando De Soto wrote about the river in 1539, and so did early naturalists who were fascinated by the New World flora and fauna found at "Georgia's Little Amazon." The name Altamaha, pronounced All′-ta-mahaw′, is from an immigrant Yamassee Indian group descended from an interior chiefdom originally known as Altamaha or Tama, located on the Oconee River just below Milledgeville, visited by Hernando de Soto in 1540. The Altamaha chiefdom was forced into slavery, but rebelled and eventually settled in St. Augustine until it evacuated with the Spanish to Cuba in 1763. The Spanish referred to it during the 1600s as Rio de Santa Isabel, referring to an early mission called Santa Isabel de Utinahica, established in the Timucuan chiefdom of Utinahica located at the forks of the river near present-day Lumber City. The ruins of more than 1,000 Indian sites along the river are evidence of how important the river was to Indians, who relied on it for food and transportation.*
>
> *The river has seen not only the dugout canoe, but also Spanish galleons, huge rafts of cotton and timber, and paddlewheel steamboats. Today, most watercrafts are motor-powered bass boats that belong to sport fishermen.*

Geology and Geomorphology

We are not aware of any detailed studies of the geology of the Altamaha River Delta similar to the ones our group has conducted on the Santee/Pee Dee Delta, South Carolina (Hayes et al., 1993; Hayes and Michel, 2008). We have made an effort, however, to apply the nomenclature we used for the Santee/Pee Dee Delta to the Altamaha River Delta, namely the delineation of three primary components – upper delta plain, lower delta plain, and delta front. The characteristics of these delta subdivisions of the Altamaha River Delta, as we have defined them, follow:

1) The **upper delta plain** of the Altamaha River Delta extends from above where the lowstand valley has cut through the Princess Anne barrier island system up to what we assume to be the "head of tides" located close to the north end of Lewis Island, one of a series of such features positioned between the bifurcated channels of the Altamaha River. These islands in the upper delta plain, including Lewis, Cambers, Wrights, and Butler Islands, are located within the Atamaha River Waterfowl Area. Lenz (2002d) referred to Lewis Island as *a naturalist's wonderland,* containing virgin cypress stands over 1,000 years old. He states further that *most of the other islands are dominated by a variety of marsh grasses that have flourished since antebellum times, when hundreds of slaves cleared the land of timber, dug canals, and built water-control dikes that were used to establish successful rice plantations* (also discussed later; see Figure 157). Thus, the upper delta plain contains two or three bifurcated, usually meandering, channels (see Figures 156 and 157) that are surrounded by swamps and old rice fields. The seaward boundary of the upper delta plain is near where U.S. 17 crosses the delta, and the landward boundary is about 7 miles inland from there. In this part the delta, which is confined mostly to within the valley carved by the river when sea level was low during glacial episodes, averages about 4 miles in width. The seaward half of the upper delta plain was the center for intense rice cultivation during the years preceding the War Between The States, as discussed in some detail later. The checkerboard pattern of the old rice fields can be clearly seen on the vertical image in Figure 157. From the landward end of the upper delta plain, the river extends upstream for about 75 miles before it breaks up into major tributaries. In this stretch, the flood plain of the river, which averages around 3-4 miles in width, is contained within the well-defined valley carved (at least partly) during the last lowstand of sea level. The morphology of the river present within this valley is remarkably similar to the general model given in Figure 112A, and, in fact, the actual example of this morphology in Figure 112B is an image of the flood plain of the Altamaha River near Baxley.

2) The **lower delta plain** of the Altamaha River Delta extends from just seaward of U.S. 17 for another 9 miles to the landward side of the outer islands that we define as the delta front. This somewhat ill-defined lower delta plain averages about 7 miles in width. On its northern end, it merges with the tidal channels and marshes on the southwestern side of Doboy Sound (Figure 153). Because of the dominant southerly transport of sand historically delivered to the delta by the Altamaha River, its shape is very asymmetric, with the land south of the river mouth projecting about 1.5 miles out into the ocean relative to land on the northern side (Figures 151 and 156). In fact, the Altamaha River Delta, as we have defined it, extends

FIGURE 157. Old rice fields. The lower portion of the upper delta plain of the Altamaha River Delta (compare with image in Figure 156). The fields on Butler Island were part of the holdings of the famous plantation owner, Major Pierce Butler, who is discussed in some detail in the text. Digital orthophoto acquired in March 2006, courtesy of USGS.

from the mouth of the Altamaha River to the southwest for 15 miles, all the way to the end of St. Simons Island (Figures 39B and 156). The landward end of the lower delta plain also contains some remnants of the old rice fields at the Hofyl-Broadfield State Historic Site. The lower delta plain of the recessed northern side of the complex contains highly meandering channels in the salt marsh, as can be seen in Figure 153. The southern side of what we are calling the lower delta plain (south of the Altamaha River channel that contains the sand waves illustrated in Figure 70A) is composed of three parts: a) A zone of salt marsh and tidal channels about 3 miles wide that merges with the waters of St. Simons Sound to the south about half way down the back side of St. Simons Island (Figure 156); b) The core of St. Simons Island, composed of a remnant of the 40,000-50,000 year old Silver Bluff Pleistocene barrier island system, a highly developed area that is 12 miles long and has a maximum width of a little over 2 miles; and c) Another salt marsh/tidal channel system about a 1.5 miles wide that separates Little St. Simons and Sea Island from the Silver Bluff sediments of St. Simons Island (Figure 156).

3) Because of the southerly transport of the sand supplied by the Altamaha River, the marsh-dominated islands on the north side of the **delta front** have a limited sand supply. Wolf Island, located immediately to the north of the mouth of the river, shows classic transgressive barrier island washover fans (Figure 34B). This barrier island system can also be seen clearly in Figure 158. Two small islands, Egg and Little Egg, occur right at the river mouth, and so does the small horseshoe-shaped shoal/island shown in Figure 159A. The southern side of the delta front, made up of Little St. Simons Island (Figure 159B), Sea Island, and the Holocene component of St. Simons Island, will be discussed later.

Wolf Island National Wildlife Refuge

The U.S. Fish and Wildlife Service (2009b) gave the following succinct description of this national wildlife refuge:

1) This refuge includes Egg Island and Little Egg Island, as well as all of Wolf Island itself (see Figure 158). It *was established on 3 April 1930 as a migratory bird sanctuary, and in 1975, it was designated a National Wilderness Area. A major purpose of the Wilderness Area is to provide protection for migratory birds and such endangered and threatened species as the loggerhead sea turtle and piping plover.* The NWR consists of *a long, narrow strip of oceanfront beach backed by a broad band of salt marsh. Over 75% of the refuge's 5,126 acres are composed of saltwater marshes;*

2) *Though the refuge's saltwaters are open to a variety of recreational activities, all beach, marsh, and upland areas are closed to the public; and*

3) *Visitors must make their own arrangements to reach the refuge. Marinas in the Darien, Georgia area may offer transportation to the refuge,* presumably for fishing, birding, etc. from a boat.

Altamaha Wildlife Management Area/Altamaha River Waterfowl Area

Lenz (2002d) gave the following information on the 27,078-acre Altamaha Wildlife Management Area/Altamaha River Waterfowl Area (ARWA), the second largest waterfowl area east of the Mississippi (the largest being the Chesapeake). This area *is visited by more than 30,000 ducks from mid October through mid April. A stop on the Colonial Coast Birding Trail, the area is equally popular with naturalists, bird watchers, duck hunters, and fishermen.*

The ARWA consists of several "islands" that are created by the Altamaha's meandering channels: Lewis, Cambers, Wrights, Butler, Champney, Broughton, Rhetts, Rockdedundy, and Dolbow islands.

Some of these canal-crossed islands, such as Butler and Champney, are former sites of plantations that today are important nesting and refuge sites managed for migratory waterfowl. Freshwater plants such as giant cutgrass, pickerelweed, wild rice, cattails, widgeon grass, and wild millet help support the duck population. Many species use the freshwater sites, including mallards, scaup, ring-necks, black ducks, pintails, canvasbacks, buffleheads, gadwalls, scaup, widgeons, mergansers, shovelers, green-winged teals, and occasionally the fulvous tree duck. A large population of snipe is found in January and February, and rails are also common in the marshes.

Mammals observed in the wildlife management area include deer, red and gray foxes, beavers, cottontail and marsh rabbits, feral hogs, otters, bobcats, minks, opossums, armadillos, raccoons, bats, and mice. In the tidal waters are dolphin and an occasional manatee. More than 26 species of snake have been reported here, including coral snakes, cottonmouths, copperheads, and pigmy and diamondback rattlesnakes.

Places to Visit

The land and marsh areas of the Wolf Island National Wildlife Refuge are closed to the public, with the surrounding marshes and tidal channels being accessible only by boat. There are, however, many areas of interest within the Altamaha River Delta system to visit by driving.

Starting on the north side of the lower delta plain, the **Ashantilly Center** is the site of the mainland home of Thomas Spalding, the early Georgia planter (and legislator) who had the large cotton plantation on Sapelo Island. The web site for Ashantilly Center (2010) points out that the *original home, circa 1820, burned in 1937, and the current house was an ongoing project of the Haynes family. Today, the Ashantilly Center is a non-profit educational and cultural historic site,*

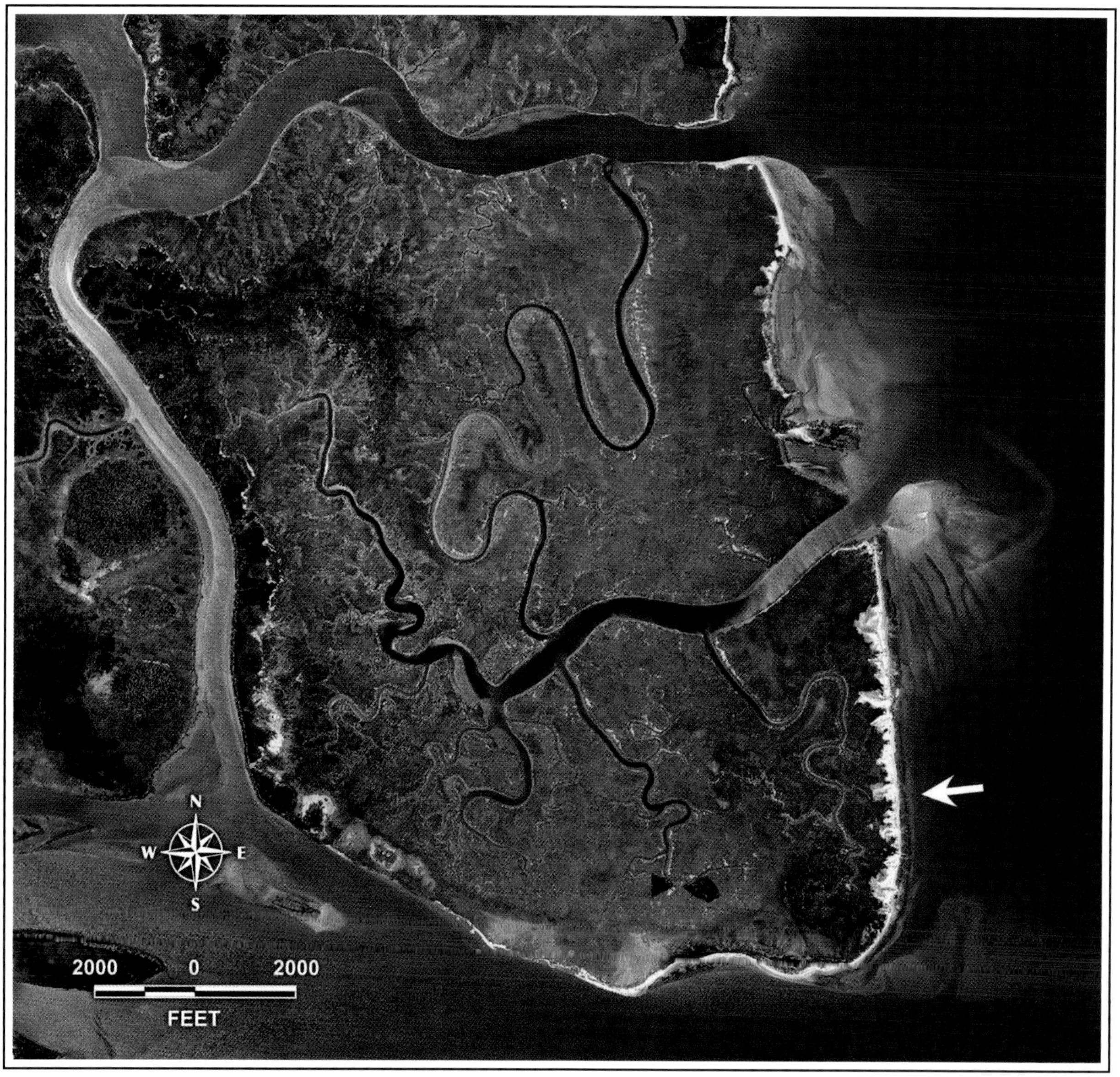

FIGURE 158. Wolf Island, a part of the northern limb of the delta front of the Altamaha River Delta (Figure 156). Arrow points to the transgressive barrier islands pictured in the oblique aerial photograph given in Figure 34B. This island is part of a National Wildlife Refuge and Wilderness Area with no land access permitted. However, boat traffic is allowed in the surrounding waters. Digital orthophoto acquired in June 2008, courtesy of USGS.

organized and founded by William G. Haynes, Jr., an artist, small letterpress printer (The Ashantilly Press), and environmentalist. His legacy, the Ashantilly Center, hosts cultural and historic events and workshops.

To get to Ashantilly, go about 2 miles south on GA 99 from the turn to Blue-N-Hall boat ramp (Figure 153), located a short distance to the north of the village of Ridgeville, to St. Andrews Cemetery Road, where you turn left. From the turn, it is about 350 yards to a dirt road on the left that leads to the entrance to Ashantilly.

The next place we recommend you visit is **Fort King George State Historic Site** (located on Figures 153 and 160A). This fort, one of the most frequently visited sites in the state, is an extremely interesting area from the standpoint of its history, some of which is given below (from Lenz, 1999; and a flyer by Georgia State Parks and Historic Sites):

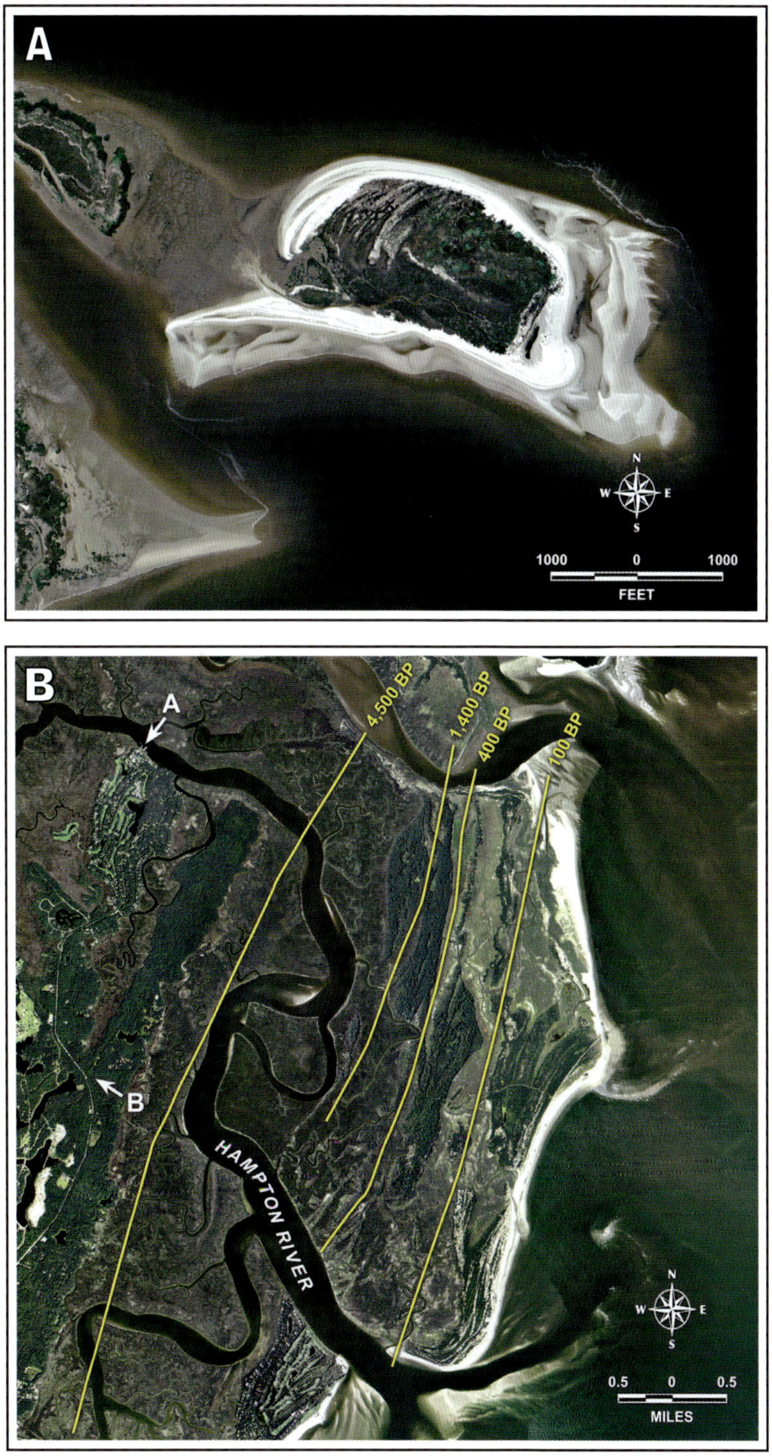

FIGURE 159. Delta front islands. (A) Small horseshoe-shaped shoal located directly off the mouth of the Altamaha River. Digital orthophoto acquired in March 2006, courtesy of USGS. (B) Little St. Simons Island, an entirely Holocene component of the delta front of the Altamaha River Delta. The superimposed time lines, after Depratter and Thompson (2013), show that most of Little St. Simons Island is <1,400 years old. See Figure 156 for specific location of these features relative to the rest of the delta. Arrow A points to the River Club Marina on the Hampton River and arrow B points to the road intersection discussed in the text. Digital orthophoto acquired in 2009, courtesy of USDA-FSA Aerial Photography Field Office.

FIGURE 160. Vicinity of Darien (located on Figure 156). (A) The town of Darien, in the northwest portion of this image, is located on a remnant of the 85,000-year-old Princess Anne Pleistocene barrier island system (Huddlestun, 1986). Arrow A points to the port of Darien on the Darien River, a branch of the Altamaha River system located near the boundary between the upper and lower delta plains of the Altamaha River Delta. Arrow B points to the Fort King George State Historic Site. The dashed red lines outline the boundaries of the Darien River channel in the mid 1800s, at about which time the Pico Cut was made to shorten the navigation route to the port. Since the cut was made, that part of the channel became a man-made oxbow lake that was eventually mostly filled up with sediment carried into the "lake" by tidal and river currents. Digital orthophoto acquired in March 2006, courtesy of USGS. (B) Historical nautical chart for 1871 showing the channel relatively soon after the Pico Cut was made.

1) In 1721, the fort was the first English settlement on the land that is now Georgia, also becoming the southernmost extension of the British Empire in North America.

2) *At this location, in what was then considered the colony of South Carolina, a cypress blockhouse, barracks, and palisaded earthen fort was constructed by Scotsmen led by British Colonel "Tuscarora Jack" Barnwell.* Located many days away from the nearest British town, Beaufort, South Carolina, the fort was intended to advance frontier settlements that would expand Great Britain's southern borders from the Savannah River to the Altamaha River. At that time, Britain was involved in struggles with France and Spain in attempts to control the southernmost areas, especially the Savannah River-Altamaha River region.

3) For seven years the officers and soldiers maintaining the fort endured severe hardships, including diseases such as malaria and dysentery. In addition, other problems developed, such as the periodic flooding by the northern limb of the diverging Altamaha River (now called Darien River) in the lower delta plain of the Altamaha River Delta. In all, 140 men died while at the fort. The graves of 17 of these British soldiers are located in a cemetery adjoining the fort.

4) *Although it seemed the fort was a failure, it did serve as a blueprint for General James Oglethorpe, founder of the colony of Georgia, who successfully set up a southern defense system on the site of the abandoned fort. In 1736, Oglethorpe brought in the Scots of Darien, led by General McIntosh, who rebuilt a fort on the site.*

5) *In the 1930s, historian Bessie Lewis uncovered old maps pinpointing the exact location of Fort King George and by 1940, Lewis, along with her group, The Fort King George Association, convinced the State of Georgia to purchase the land and preserve it as a historic site and park. In 1970, a museum was built at the site. Over the next 35 years, the fort was slowly reconstructed under the direction of the Georgia Department of Natural Resources.*

When you visit the site, you will see structures rebuilt based on old maps and designs of the fort, including a blockhouse, guardhouse, moat, and earthen palisades. The constantly growing museum features a variety of interesting books and maps, as well as a film, which covers the first inhabitants of the area, the Guale Indians, and a variety of other topics.

When we first started researching where the location of the old fort may have been, we concluded that the place where the fort had been reconstructed didn't seem correct, because it is located so far from the main channel of the Darien River. We thought that the actual fort may have been located on Black Island, a mysterious "island" in the marsh of the lower delta plain (located on Figures 153 and 160). This "island," which is about 1.6 miles long and averages about 400 yards wide, could be either a fragment of the Princess Anne barrier island system or of entirely Holocene origin. Therefore, Black Island probably has an origin similar to Hird Island located offshore of the Blue-N-Hall boat ramp that was discussed earlier.

If the fort were in that location on Black Island, it would have been about 200 yards across the marsh from the banks of what is now called the Darien River (assuming that this part of the river channel has not migrated much since then, which seems to be a good assumption). The north bank of the main channel of the Altamaha River proper is another 1.4 miles to the south of the south end of Black Island, which is now heavily forested.

However, upon further research, we found that in the mid 1800s, an artificial channel, called Pico Cut, was made through the narrow limb of a meander bend of the Darien River (Lenz, 1999), presumably to expedite boat traffic access to the port of Darien, a thriving port for the export of timber around that time. The port is located about 1.3 miles upstream of the cut (Figure 160A). As a result of the cut, the abandoned meander loop, then an artificially created "oxbow lake," began to fill up with sediment and, at present, only a small tidal creek flows out of the marsh-filled channel. The original channel as outlined on Figure 160A was based on the present marsh types. After this diagram was created, illustrator Joe Holmes discovered an historical nautical chart made in 1871 (see Figure 160B). This historic chart verified our earlier conclusion.

This means that, at the time the fort was first established and even on up until the mid 1800s, there was access to the main channel of the Darien River right along the border of the original fort. By the way, this location on the seaward edge of the well-elevated Princess Anne Pleistocene barrier island system is the first significant highland a large boat would have encountered when coming upstream along this northern limb of the Altamaha River Delta at that time (Figure 156). The main channel of the Altamaha River itself does not intersect a major upland area for many miles beyond the landward end of the upper delta plain. From the mouth of the present Darien River, it is 22 miles to the south, at the southern end of St. Simons Island, before another readily accessible highland area is available for large boats approaching the land from offshore. Thus, the locations of the fort and of the port of Darien on the bank of the Darien River at the edge of the Princess Anne upland (Figure 160A) are where they are because of the geological history and geomorphology of the area.

Figure 161 illustrates that the abandonment of meander loops on tidal rivers can also be a natural process. It shows one of these natural cut-offs, located further seaward of Darien on the Darien River. This cut-off looks very similar to the man-made abandonment that resulted from the excavation of Pico Cut, and the elimination of the original Fort King George area as a boat landing site (see Figure 160A).

Next, we recommend that you visit the town of Darien, which has a fascinating history and a number of historic churches. A map and walking tour of Darien, published by Great Southern Publishers (2010), is a helpful aid for exploration of the town, which, as noted earlier, is located on the prominent remnant of the Princess Anne barrier island system, formed during a highstand of the sea around 85,000 years ago. This upland reaches heights of over 20 feet.

The City of Darien was established in 1736 as a

FIGURE 161. A segment of the Darien River approximately 3.6 miles east of the Fort King George State Historic Site (see location in Figure 160A). Arrow A points to a sediment-filled abandoned oxbow lake off the main channel of the river. Unlike the one illustrated in Figure 160A, this "lake" was created and filled in by totally natural processes. Arrow B points to a classically shaped tidal river point bar. Compare it with the general model given in Figure 105B and the orthophoto in Figure 107. Digital orthophoto acquired in June 2008, courtesy of USGS.

military outpost by Scottish Highlanders under the command of General James Oglethorpe, making this historic town the second-oldest planned town in Georgia, after Savannah (Great Southern Publishers, 2010). These new Highlanders, now referred to as the Scots of Darien, were led by a General McIntosh.

This part of the country, including most notably the adjacent old rice fields of the Altamaha River Delta, played a major role in the issue of slavery, which eventually led to the calamitous War Between The States of 1861-1865. It is most interesting to point out that in 1739, eighteen of the most prominent members of the Darien colony signed the first petition against the introduction of slavery into Georgia. This was in response to pleas to Oglethorpe and the trustees by inhabitants of Savannah to lift their prohibition on slavery. The Highlanders' petition was successful at first and slavery was not introduced until ten years later in 1749 (101 years before the war started; Parker, 1997).

Rice farming, which took advantage of the freshwater streams and significant tides along the coast at the head of the Georgia Bight, was begun in South Carolina in the mid 1690s, the rice itself probably originating from Madagascar. As a result of this newfound mode of agriculture, *South Carolina and Georgia grew and amassed great wealth from the slave labor obtained from the Senegambia area of West Africa and from coastal Sierra Leone. At the port of Charleston, through which 40% of all American slave imports passed, slaves from this region of Africa brought the highest prices, in recognition of their prior knowledge of rice culture, which was put to use on the many rice plantations around Georgetown, Charleston, and Savannah.* (Wikipedia encyclopedia).

Being residents of Columbia, South Carolina, a city that was mostly burned to the ground by General Sherman's troops in 1865, it was not until we wrote this book that we were surprised to find out that the town of Darien suffered a similar fate during the War Between The States. The following account by Edling (2010) tells this stunning story:

> *Darien ... saw some of the first action of black troops in the Civil War when Federals looted and burned the town "to the ground" on June 11, 1863. Involved in the action was the 2nd South Carolina and the Massachusetts 54th, the U.S.C.T. regiment commanded by 25-year-old Col. Robert Gould Shaw, and made famous in the movie Glory. Darien was an important shipping port before the war, exporting cotton grown inland and rafted down the river. During the war, the port was used by blockade runners and consisted of 75 to 100 residences, including three churches, a market house, a courthouse, and an academy.*
>
> *Shaw's troops, stationed on St. Simons Island June 10-24, 1863, were ordered to Darien by Col. James Montgomery. The next day, the town was looted and burned, with the light from the fires visible from St. Simons Island, 15 miles away. Col. Shaw, the son of a famous Boston abolitionist, did not approve of burning the town and wrote letters to superiors about his feelings on the matter. Twenty-five days after leading troops to Darien, Shaw would be dead and buried in a mass grave with many of his men. He led the Massachusetts 54th on a suicidal assault of Battery Wagner, South Carolina, forever proving the battle courage of black troops in U.S. history.*

Darien recovered from that war to *become the second largest lumber shipping port on the southern coast from 1870-1910. When timber resources were depleted, Darien became what it is today, a fishing village.*

When we visited Darien, on 27 March 2010, we were impressed with the large number of shrimp boats in the port. We were also impressed how high the Princess Anne bluff appeared to be. This town is a beautiful setting with the usual large live oaks and pines, surely a place you wouldn't want to miss during a visit to this part of the coast.

If you depart from Darien on U.S. 17 and cross over the usually muddy Darien River, you will be heading into perhaps the most exciting area on the Georgia Coast from the standpoint of a naturalist. On this drive, U.S. 17 crosses four major river channels, the Darien, Butler, Champney, and South Altamaha Rivers. The Butler and Champney Rivers join a ways down from the highway to become the main Altamaha River (again). Meanwhile, the Darien River flows to the northeast and into the entrance to Doboy Sound. Also, the so-called South Altamaha River meanders on southeast for about 6 miles before it more-or-less loses its identity by mixing with the tidal waters of the lower delta plain. If you look south as you cross the four bridges, you will be looking toward a classic lower delta plain composed of branching distributary channels.

After crossing the Darien River, you will be driving for 0.8 miles across Generals Island, that, at the present time, is composed of the remnants of the old rice fields.

According to a roadside kiosk, this island was the property of General Lachlan McIntosh by virtue of a grant in 1758. Lachlan was the son of General John McIntosh, who, as noted earlier, was the leader of the Scots of Darien who were brought to Georgia by Oglethorpe in 1736 to re-establish Fort King George near Darien. Generals Island was the home of Lachlan McIntosh and his family up until the years of the First Civil War, during which time, as a champion of independence from the British, he became a General in the Continental Army, leading many campaigns. He was captured while helping to defend the city of Charleston, South Carolina when it was surrendered to the British on 12 May 1780, and held prisoner by the British until February 1782. When he returned to his home on Generals Island, it had been destroyed by the occupying British forces. The state of Georgia named the county in which Darien, Sapelo Island, and most of the Altamaha River Delta are located McIntosh County in honor of his family (HMdb.org, 2008).

The kiosk discussing Generals Island also informs us that *in 1808, a Canal, called General's Cut, was dug through the island to connect the Darien River with the middle branch of the Altamaha River* (the Butler River) *"for the convenience of the adjoining planters."*

As soon as you cross the bridge over Butler River, you will be on Butler Island, one of the most famous (notorious?) locations on the Georgia Coast, as it is the heart of Butler Plantation, originally owned by one **Major Pierce Butler.** The island consists largely of a system of dikes and ponds built for the cultivation of rice. The still-visible checkerboard pattern of these old rice fields can be seen in the vertical image in Figure 157. According to House (1954), there were around 5,000 acres of rice farms in the Altamaha River Delta area during the two decades preceding the War Between The States.

The very long and complex story about the historic farming practices on the Altamaha River Delta and the people who did them is told in a remarkable book by Malcolm Ball, Jr. titled *Major Butler's Legacy; Five Generation of a Slaveholding Family.* This fascinating book should be required reading for anyone interested in the history of the south, particularly when it comes to the issue of slavery. Major Pierce Butler (1744-1822) came to America in 1767 as an officer in the British Army, eventually becoming *an aristocratic low-country planter, slave owner, and dealer, who imported slaves from Africa and the West Indies.* Therefore, Major Butler became quite wealthy. Just before the First Civil War, Butler acquired land on St. Simons Island and started Hampton Plantation, primarily for cotton farming, at which time he resigned from the British Army. As his empire expanded, he eventually started rice farming centered on and around Butler Island in the upper delta plain of the Altamaha River Delta. As many as 500 slaves worked on this new Butler Plantation. During his later years, he owned homes at Hampton and in Philadelphia and, strangely, served two terms as South Carolina's first U.S. Senator. After he died in 1822, the Butler Island Plantation was owned by his grandson, Pierce Butler (Mease). This Pierce Butler married a famous British actress named Frances Anne Kemble, who spent a good part of the winter season of 1838-1839 on the Butler Plantation. She wrote a book, *Journal of a Residence on a Georgian Plantation*, based on that experience, which was published in May 1863, 22 years after her visit. Some say that this book was influential in keeping England from entering the War Between The States in support of the Confederacy (Sherpa Guides, 1999). Here is one of her quotes:

> *Scorn, derision, insult, menace – the handcuff, the lash – the tearing away of children from parents, of husbands from wives – the weary trudging in droves along the common highways, the labor of body, the despair of mind, the sickness of heart – these are the realities which belong to the system, and form the rule, rather than the exception, in the slave's experience.*

That ought to get your attention. From reading Ball's book, we also learned that Butler branded his male slaves on the abs with the letters – PB – just like ranchers brand their cattle. To say that this is a sobering book would be an understatement.

Just after you cross the bridge over Butler River, there is a fishing pier that projects out into the river with some picnic tables, a nice spot for looking for birds and a walk out onto the pier.

Next on your right, you will see the tower composed of bricks shown in Figure 162A. This is a 75-foot brick chimney that was a steam powered rice mill built in 1850. Located a ways back of the tower is the two-story home of Col. T. L. Huston, a half-owner of the New York Yankees at the time the residence was built in 1927. For some sort of personal amusement, presumably, he had a Guernsey dairy farm and a successful truck farming operation that shipped iceberg lettuce grown in the restored fields. After the house was built, many baseball players visited

FIGURE 162. Butler Island. (A) A 75-ft brick chimney that was part of a steam powered rice mill built in 1850 located near the southwest bank of the Butler River on the north side of US 17 (see Figure 157). (B) Edge of a refurbished rice pond. The lower bird is an anhinga. Both photographs were taken on 27 March 2010.

Huston, including the great Babe Ruth (HMdb.org, 2008). In fact, the Babe set his single season home run record of 60, which stood for 34 years until Roger Maris hit 61 in 1961, during the very year this house was built, 1927. The Yankee teams of the late 1920s were probably the most powerful and memorable baseball teams ever. Of course, all the records went out the window when the era of steroids was ushered in about the mid 1990s. At the present time, this house and grounds, formerly a holiday spot for those great Yankee teams, is used as a private staff residence for the Altamaha Waterfowl Management Area.

A kiosk near the rice tower states the following:

> *Famous Butler authors. Pierce Butler* (the grandson of Major Butler) *and his daughter Frances, who shared his interest in the south, returned to Butler Island in 1866 and worked to rehabilitate the plantation.* This *Pierce Butler* (whose real name was Pierce Butler Mease, had to drop the Mease so he could inherit this place) *died in 1867, but Frances continued for several years to manage the island. She wrote a book,* ***Ten Years on a Georgia Plantation****, an interesting and valuable account of life in this section during the reconstruction. Owen Wister, famous author of* ***The Virginian*** *and other novels, was the son of Sarah Butler, sister of Frances. He often visited Butler Island Plantation.*

The tone of the writing on this kiosk doesn't exactly match the tone of the writing in the more detailed accounting of the situation as given in *Major Butler's Legacy*, but so it goes.

Continuing on south on U.S. 17 for 0.3 miles beyond the Huston house, turn right onto a dirt road for a drive along side of some of the old rice fields shown in the vertical image in Figure 157. This area is a part of the Altamaha Wildlife Management Area/Altamaha River Waterfowl Area. These Butler Island Refuge impoundments are the result of a cooperative waterfowl and wetland habitat enhancement project between the Georgia DNR and Ducks Unlimited. In this area, some large ponds are maintained for the waterfowl. The photograph in Figure 162B shows the edge of one of the ponds.

We made this short drive of about 1 mile along this dirt road between U.S. 17 and I-95 on 27 March 2010. This was just a quick run through, not necessarily at the right time of the year, but we still saw the following birds: a flock of ibises, turkey vultures, anhingas, cormorants, common egret, coots, red-winged black birds, and a yellow-rumped warbler. After this brief encounter, we concluded that this may be one of the best birding spots in the Georgia Bight! It is almost as good as the Savannah National Wildlife Refuge, offering a great contrast with the rest of the coast, and it is fortuitous that it can be reached so easily.

Next we recommend that you go back out to U.S. 17, take a right, and head south, still on Butler Island. You will note some pullover areas and access to trails in the Butler Refuge on the left. There are some elevated viewing platforms along the trails built on the dikes originally constructed to contain the old rice ponds. There are three places to pull over and at least two viewing platforms. This is another excellent place for birding. On the right, you will pass by about a thousand yards of a fairly mature cypress-tupelo swamp, that can be seen in Figure 163A.

After crossing the bridge over the Champney River, take an immediate left into the **James Allen Williamson Champney River Park**, located on Champney Island. This is a good place to pull over, because you can drive right up to the boat ramp and walk out onto a fishing pier (downstream of the U.S. 17 bridge). The photograph in Figure 163B presents a view looking downstream from the pier, showing a large cypress tree growing along the bank of the Champney River. This area has a definitely different character from anywhere else along the middle Georgia Coast, a classic piedmont (red-water) river delta.

In this same area, on the west side of U.S. 17, is the 34-acre **Ansley-Hodges Memorial Marsh Project** built in 1989 and sponsored by Ducks Unlimited. As noted by Lenz (2002d), this facility has *an excellent observation tower a short walk away from the parking area* (that) *gives bird watchers a perch to view waterfowl and wading birds in the freshwater marsh, including great and snowy egrets, least and American bitterns, and in the spring, black-necked stilts. Common migrating songbirds in the fall include the common yellowthroat, indigo bunting, and swamp sparrow. A 1-mile nature trail, with marked sites and brochures, explains the workings of the managed ecosystem. The road follows the perimeter of the island, looping up to Interstate 95 and back to U.S. 17. Driving slowly on this road, or parking for short excursions, can be rewarding.* This site is also included in the Georgia Wildlife, DNR (2002) discussion of Georgia's Colonial Coast Birding Trail. The specialties they listed include wood stork, swallow-tailed kite, bald eagle, king rail, painted bunting, mottled duck, wood duck, white ibis, and glossy ibis. Other advice they gave includes looking for *common snipe feeding in exposed muddy areas. In summer, look for wood ducks. In fall and winter, black ducks, pintails, green-winged teal, northern shovelers and other waterfowl can be seen in the impoundment. Look for rails darting through the vegetation in the impoundment. Bald eagles can be spotted in winter and spring.*

After leaving this area, if you continue south of U.S. 17, you will be treated to a pleasant drive through a cypress-tupelo swamp about 0.4 miles in length. Next you will go over the bridge that crosses the last of the four main river channels in this part of the upper delta plain of the Altamaha River Delta, the South Altamaha River. Continuing south from the bridge, U.S. 17 crosses 1.3 miles of old rice fields grown up with mostly freshwater marsh grasses. This path of U.S. 17 is plotted on the historic map of the Hofwyl rice fields in Figure 164A.

At the edge of the delta, you will drive up the slight slope to the Pleistocene upland, in this case the Princess Anne Pleistocene barrier island system. Within about half a mile you will come to a road on the left that leads to the **Hofwyl-Broadfield Plantation State Historic Site**. This is an expansive place full of big live oaks covered with Spanish moss. The entry fee is five dollars and it is open 9:00 a.m.-5:00 p.m. Tuesday to Saturday; 2:00 p.m.-5:30 p.m. Sunday. Check for changes in visitation hours (Georgia Wildlife, DNR, 2002). According to a kiosk at the site, rice was planted at Hofwyl Plantation from 1806 to 1913 using fresh water supplied by the Altamaha River. There is quite a bit of information about the African/American heritage in the Visitor Center and elsewhere, as well as in the books sold at the Center.

An historic map of the rice fields at this location is given in Figure 164A, and a vertical image showing what they look like today is given in Figure 164B. A nature trail you would not want to miss runs out along one of

FIGURE 163. Butler Island/Champney River. (A) Oblique aerial view of Butler Island taken on 28 April 2010. See also Figure 157. Arrow A points to Butler River; Arrow B to the birding area on Butler Island discussed in the text; Arrow C to the old rice fields; Arrow D to a fairly mature *cypress-tupelo* swamp; and Arrow E to the boat dock/pier from which the photograph in image B was taken. (B) The Champey River, a fork of the Altamaha River Delta System, looking downstream from near the US 17 bridge on 27 March 2010 (at the James Allen Williamson Champney River Park). A large *bald cypress* tree hugs the river bank.

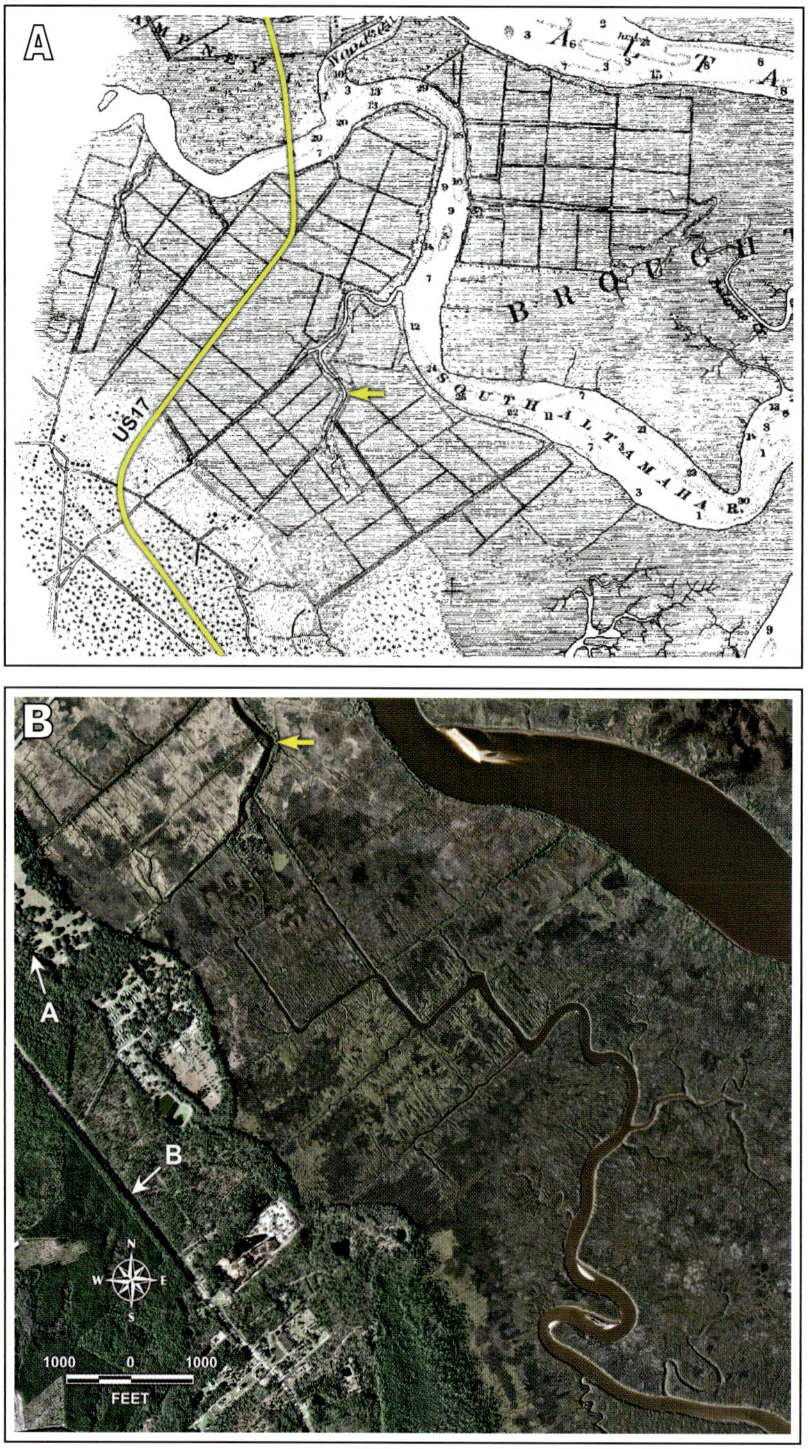

FIGURE 164. Hofwyl-Broadfield Plantation State Historic Site (located on Figure 156). (A) Historic map of the Hofwyl Plantation made circa 1905. The route of the present US 17 is approximately located on the map. (B) Rice fields that were part of the Hofwyl-Broadfield Plantation State Historic Site, located near the landward limit of the lower delta plain of the Altamaha River Delta (see Figure 156). Arrow A points to the plantation building area and arrow B points to US 17. The main channel of the South Altamaha River is visible in the upper right. Yellow arrows point to same location on both images. Digital orthophoto acquired in March 2006, courtesy of USGS.

the old dikes that provide an excellent, elevated view of the former rice fields. The trail is bordered by giant cordgrass, which is very tall at this location. Other grasses and some small bald cypress trees are also present.

An informative kiosk by the trail outlines the history of the plantation:

> *The marsh you see before you was a cypress swamp when William Brailsford began his rice empire here. Hundreds of slaves cleared the land, leveled it to insure proper drainage, and constructed miles of dikes and ditches by hand. Floodgates and trunks were built into the dikes to let water in to flood the rice fields during the growing season.* Flooding took place during a high spring tide, the water was released on a falling tide, and the gates closed at harvest time. *When the rice fields were flooded, foot travel was restricted to pass along the top of the dikes. At the height of production, Hofwyl-Broadfield Plantation encompassed over seven thousand acres worked by 750 slaves. It was said that the plantation produced such a good quality of grain that, according to government records, the superior Broadfield rice on the Charleston market took its name from the Brailsford Plantation. A combination of factors led to the permanent displacement of rice as a cash crop,* namely the War Between The States and *other factors such as crop failures in 1866, 1867, and 1868.* These issues, *combined with lack of capital, were further complicated by a series of disastrous hurricanes, tropical storms, and freshets (river swells caused by heavy rain upstream). And last, the agricultural revolution, which developed in the 1870s and 1880s, led to fresh competition from Texas, Louisiana, and Arkansas rice growers, who used heavier machinery and a superior rice seed imported from Japan.*

These plantation grounds are included in the Georgia Wildlife, DNR (2002) discussion of Georgia's Colonial Coast Birding Trail. Its specialties are wood stork, bald eagle, osprey, glossy ibis, painted bunting, yellow-throated warbler, sharp-tailed sparrow, and northern parula warbler.

There is still even more to see than rice fields and birds on this property. A plantation house (Hofwyl house) sits at an elevation of 15 feet or so. A trail winds around this part of the property, passing by a champion magnolia sweet bay. In addition, there are some huge, moss-draped live oaks, some of the most impressive we have seen anywhere, and some big loblolly pines as well.

LITTLE ST. SIMONS, ST. SIMONS, AND SEA ISLANDS

Introduction

Three sandy islands, Little St. Simons (illustrated in Figures 159B and 165A), Sea Island, and St. Simons Island, border the portion of the delta front of the Altamaha River Delta south of the river mouth (Figure 39B and 156). Little St. Simons Island, Sea Island, and a two-mile-long limb of sand that joins the main body

FIGURE 165. Little St. Simons Island (see vertical image in Figure 159B). (A) Northerly oriented spit feature on the north end of the Island. View looks south. (B) The tidal inlet at the outlet of Hampton River. Arrow A points to vegetated, prograded beach ridges on the south end of Little St. Simons Island. Arrow B points to the remnant of the Silver Bluff Pleistocene barrier island system that composes the bulk of St. Simons Island (Huddlestun, 1986; see Figure 156). Both photographs taken during a low spring tide on 28 April 2010.

of St. Simons Island at an oblique angle, called **East Beach** (the Holocene component of St. Simons Island), compose the outer Holocene fringe of this part of the delta front. Interestingly, East Beach was actually two islands in the early 1900s. An old inlet was located just south of the Coast Guard Station (George Oertel, pers. comm.). The oblique aerial photograph in Figure 166 gives a spectacular view of these three outer islands. These Holocene sand bodies are separated from the main body of St. Simons Island proper by a zone of salt marsh and tidal channels. The northern half of this salt-marsh-dominant zone averages about two miles in width, and contains Hampton River, which feeds a substantial tidal inlet that exits the shore at the north end of Sea Island. Two oblique aerial photographs of the ebb-tidal delta of this inlet are given in Figures 70B and 165B, and it is also shown in the vertical image in Figure 167. The southern part of this marsh/tidal 7-mile-long channel complex tapers gradually in a southern direction to its terminus where the Holocene and Pleistocene limbs of St. Simons Island meet, near the south end of the complex (as shown in Figure 167). An oblique aerial view of this striking salt marsh is given in Figure 95A. The remnant of the Silver Bluff barrier island system, formed about 40,000-50,000 years ago, that composes the main body of St. Simons Island proper, is about 11 miles long and 2 miles wide. This is the largest area developed with human residences, hotels, etc. on any of the barrier islands on the Georgia Coast. The Pleistocene component of Cumberland Island is larger, but it is mostly undeveloped with few man-made structures.

Geomorphology and Beach Erosion

Inspection of the vertical image of **Little St. Simons Island** in Figures 156 and 159B shows that this entire island is composed of a series of near-parallel, prograding beach ridges vegetated with trees. The four Holocene time lines presented on Figure 159B (after DePratter and Thompson, 2013) show that most of the island is <1,400 years old. The consistent relatively recent progradation of Little St. Simons Island is no doubt a reflection of the proximity of the sand source, the mouth of the Altamaha River. There are two, and possibly three, separate cuspate features in the middle of the island, probably representing different depositional episodes within the past thousand years or so. The cusp shape itself is related to the position where the south limb of the ebb-tidal delta located off the mouth of the Altamaha River meets the shore. To the north of the cusp, sand is transported to the north along the beach in the form of spits, such as the one shown in Figure 165A. This transport of sand opposite to the general trend along this coast (i.e., north to south) is a result of wave refraction (bending) around the ebb-tidal delta, as is illustrated in the general model in Figure 36A. The longshore sediment transport direction at the south end of the island is strongly to the south, as the curving, prograding beach ridges in that area clearly show. These curving beach ridges are visible on both the vertical images in Figures 159B and 167, and in the oblique aerial view in Figure 165B. Inasmuch as this island has no human structures along the outer shore, beach erosion is not as serious an issue on Little St. Simons Island as it is on Sea Island and East Beach; therefore, there are no shore protection structures on the island.

The densely populated **Sea Island** is a slender, 5-mile-long prograding barrier island with an average width of about a quarter of a mile (Figures 156 and 167). An unpopulated spit feature 1.7 miles long trails off its southern end. This island has had serious erosion problems in the past. When we visited it well over twenty years ago, a sloping concrete seawall had been built along some of the shore. According to George Oertel, this seawall, which extended from 2nd St. to the Beach Club, was there in the 1970s. At the present time, the sand beach along about a third of the more central part of the island is backed by a riprap revetment that was built in the early 1980s and extends north of 36th St. for 1.3 miles (George Oertel, pers. comm.). As discussed later, the terminal groins near both ends of the populated part of the island have been effective in keeping a sand beach in the central part of the island. The one at the south end is pictured in Figure 87B. The one at the north end is also no doubt aided from time to time by swash bars breaking off from the ebb-tidal delta of Hampton River Inlet and attaching to the beach in that location, as the one pictured in the vertical image in Figure 167 appears to have been on its way to doing (assuming its progress was not hindered by major storm waves).

The beach nourishment projects carried out on Sea Island were summarized for us as follows by George Oertel, who has completed several consulting projects on the island over the years:

1) As noted elsewhere, the Federal Government placed riprap revetments on East Beach to the south that later

FIGURE 166. View looking north of the Holocene barrier islands that make up the outer fringe of the delta front of the Altamaha River Delta (south of the river mouth; Figure 156). In the foreground is East Beach, the Holocene component of St. Simons Island. The island in the middle is Sea Island, and the one in the distance is Little St. Simons Island (see vertical image of this island in Figure 159B). The large inlet in the center is called Gould's Inlet. Photograph taken during a low spring tide on 28 April 2010.

FIGURE 167. Vertical image of Sea Island. Note the ebb-tidal delta of Hampton River Inlet to the northeast and the one that dominates the area on the north end of the Holocene component of St. Simons Island in the southwest corner of the image. The arrow A points to the terminal south groin, which is also pictured in Figure 88B, and arrow B points to a swash bar associated with the ebb-tidal delta of Hampton River Inlet, which appears to be headed toward the beach. Arrow C points to Goulds Inlet. Digital orthophoto acquired in 2009, courtesy of USDA-FSA Aerial Photography Field Office.

became known as "Johnson rocks." However, no sand nourishment was carried out on Sea Island at that time.

2) In 1986, 100,000 cubic yards of sand were added to the outer beach. This sand did not last for long.

3) In 1990, 2,000,000 cubic yards of sand were added to the middle beach. At the same time, a T-head groin was established one half mile south of the north end of the island. As discussed earlier, groins trap sand moving along the coast, and the T-shaped limb keeps the sand from migrating past the end of the groin, trapping it in this "catchment basin."

4) In 1997 the T-head groin and offshore breakwater shown in Figure 87B were established 400 yards south of where the causeway joins the island.

5) It had earlier been established by Oertel and associates that sand on Sea Island migrates away from the middle of the island, moving both north and south. Accordingly, placing the T-head groins at both ends of the island created two "catchment basins" for the sand to accumulate in. This allowed the contractors to create a **sand recycling program** unique to the U.S. east coast. Sand that accumulated at the two catchment basins was periodically moved back toward the center of the island. This recycling was carried out in 1992, 1994, 1995, 1996, 1997, 2001, 2003, and 2007. Approximately 100,000 cubic yards of sand were recycled during each of these projects. Oertel estimates that about 90% of the original volume of nourished sand is still on the beach of Sea Island, demonstrating that beach nourishment can maintain a sand beach if the longshore sediment transport patterns are correctly taken into account and they are creatively dealt with.

East Beach, the Holocene component of St. Simons Island, shown in the foreground of the oblique aerial photograph in Figure 166, could hardly be defined as slender. The huge intertidal bars (swash bars) welding to the beach south of the ebb-tidal delta at the south end of Sea Island (Gould's Inlet) have been key in the growth of this part of the island. The size of these bars is clearly illustrated by the ground photograph in Figure 64C, and the oblique aerial view in Figure 65A. The source of all this sand is an interesting puzzle. It is possible that some sand off the huge ebb-tidal delta of St. Simons Sound to the south is added to this area from time to time. It is also probable that some of the sand used to nourish the beaches at Sea Island has also ended up on this island. In spite of this apparent excess of sand, there are parts of this island that have undergone chronic erosion. The riprap revetments shown in Figure 90 illustrate erosion problems exist along the northern end of East Beach and further south where East Beach joins the main limb of St. Simons Island.

As already mentioned, the heavily forested and populated main body of **St. Simons Island** is composed of a remnant of the Silver Bluff Pleistocene barrier island system (boundaries outlined in Figure 156), in a fashion similar to the origin of the main bodies of Sapelo, St. Catherines, and Ossabaw Islands to the north. This Pleistocene remnant appears to have been larger than the ones on the other islands to the north, probably because of its location so close to the sediment source (Altamaha River mouth). The extreme south end of this Pleistocene remnant of St. Simons Island that faces the open-water of St. Simons Sound has also had chronic beach erosion problems. The entire 1.5 miles of the shore from where the Holocene and Pleistocene components join to the middle of the large golf course at the southwest end of the island are now protected with a riprap revetment.

The marsh and tidal channel system that separates the St. Simons Island proper from the Princess Anne upland area north of the city of Brunswick (Figure 156) averages about 3.5 miles wide. It is composed of the spillover from the lower delta plain of the Altamaha River Delta in the north and the headwaters of St. Simons Sound to the south. It contains the usual complement of salt marshes, with abundant smooth cordgrass and highly meandering tidal channels, in the eastern and southern regions. The northwestern sector, on the other hand, has an abundance of brackish marsh. The width and complexity of this marsh/tidal channel system is clearly illustrated by the vertical image in Figure 168. The tree-like system of tidal channels shown in Figure 108A is located one mile east of U.S. 17 near Brunswick and a little west of the main channel of the MacKay River.

Places to Visit

One way to get to this area is to go south toward Brunswick on U.S. 17. From the entrance to the Wofyl-Broadfield State Historic Site, it is about 3.2 miles to where U.S. 17 (also called Ocean Highway) bends toward the southwest and continues along the edge of the Princess Anne upland (Figures 151 and 156). From that major bend in the road, it is 6 miles to where the Golden Isles Parkway joins U.S. 17 a little ways south of

FIGURE 168. The salt marsh/tidal channel system that separates St. Simons Island from the Princess Anne upland upon which the city of Brunswick is located (Figure 156). Arrow A points to the F.J. Corras Causeway, and arrow B points to Sea Island Road. Digital orthophoto acquired in March 2006, courtesy of USGS.

Cypress Mills. In some spots, there have been houses built out to the edge of the marsh, which is almost entirely black needlerush. In a few places, the marsh occurs on the west side of the road, and in others you drive along with marsh right on side of the road. As we drove along this area on 27 March 2010, we concluded that this was one of the widest zones of black needlerush we had ever seen, an outstanding example. This is presumably the result of the influx of fresh water into the marsh complex that is ultimately derived from the Altamaha River, it being funneled to the south within the lower delta plain (Figure 156).

As shown by the vertical image in Figure 168, the F.J. Corras Causeway, a 4-mile-long, deluxe highway leading to St. Simons Island, has bridges over five major tidal channels: a tributary of Terry Creek, Back River, Little River, Mackay River (also the Intracoastal Waterway), and Frederica River.
(NOTE: One of the many biking trails on the Georgia Coast runs parallel to this highway.)

The F.J. Corras Causeway passes by a very wide and beautiful salt marsh that contains an abundance of smooth cordgrass. Before getting to the high bridge over the second tidal channel, Back River, there is a parking area where you can pull over and check out the marsh, and there is another similar pullover before you get to the next major channel (Little River). These areas are clearly shown on the vertical image in Figure 168. A fishing pier on the Frederica River just north of the road attracts visitors from miles around.

Once off the end of the bridge, you will be driving on the 40,000-50,000 year old Silver Bluff Pleistocene barrier island system, which has maximum elevations of around 4-5 feet (outlined on Figure 156). Therefore, the topography of this island is relatively low in comparison with that of the Princess Anne system, on which the city of Brunswick is located, which is 15-20 feet high.

As noted by Olsen (2006), St. Simons Island is the only one of Georgia's larger barrier islands that has never been completely privately owned by one or a few somehow connected individuals. *It consists of 27,300 total acres including the marsh. It has 12,300 upland acres and three miles of beach. … St. Simons is extremely rich in history, having been inhabited first by the Creek Indians and then by the Spanish, British, and finally by southern plantation owners who grew sea island cotton and live oak timber.* After the War Between The States, St. Simons became a popular resort. *The causeway was built in 1924 and the airport in 1934. The Coast Guard Station, which closed in 1996, was built in 1937. Today most of the island is privately owned residential homes and low-key commercial hotels and condominiums.*

After you exit the bridge, we recommend that you go for about 0.4 miles and turn left onto Sea Island Road at the first major intersection. Then, you will be driving for about a mile through a developed area containing some large mansions. At the end of this drive through high-priced real estate, the road moves out over a broad marsh, with black needlerush along the edge and smooth cordgrass further out. If you want to see a lot of marsh, this is the drive for it!

At the first major intersection, take a left if you are interesting in seeing the middle and northern part of St. Simons Island proper (i.e., the remnant of the Silver Bluff Pleistocene barrier island system). As you do so, you will be driving through an older, well-established development with big live oaks that have been there for a long time, fortunately preserved by the developers of the island. There also is an abundance of Spanish moss all over the place. A little arm of land sticks out in the marsh west of the golf course (Sea Palms Golf), that is crowded with houses, with every one of those located on the water having a dock – 24 of them in a row. One of the docks/boardwalks a little further north is over 1,000 feet long. This is perhaps one of the biggest of these upscale developments on the Georgia Coast. Continuing on this drive, you will eventually come to a roundabout where we recommend you take a left toward Fort Frederica (Figure 156). During our visit there on 27 March 2010, we were marveling about how beautiful the trees were in this area, particularly those surrounding the fort.

Fort Frederica National Monument is located on the west side of the middle of St. Simons Island (the Pleistocene core), on the bank of a medium-sized tidal creek called Frederica Creek. A National Park Service flyer about the fort states the following: Fort Frederica was a strong outpost on the precarious edge of an expanding British empire. *Its very existence was a thorn in the side of a declining Spanish empire with claims to the region.* The fort was established in 1736 to guard Georgia's southern border. At its peak in the 1740s, there were about 70 houses that *lined Frederica's dusty streets. Fort Frederica is best remembered today for repulsing a Spanish invasion of Georgia in 1742.* James Edward Oglethorpe is said to have made the following quote at that time: W*e are resolved to never suffer defeat – we will rather die like Leonidas and his Spartans – if we can but protect Georgia and Carolina and the rest of the Americans from desolation.* The fort was closed after the end of that part of the war between Britain and Spain (in North America), and the town went into oblivion.

[NOTE: **James E. Oglethorpe**: This gentleman, considered to be the **founder of the colony of Georgia**, lived a complicated life, a few comments on which follow (Wikipedia encyclopedia):

1) Born in London in December 1696.

2) Became a British Army officer and fought with distinction in the Austro-Turkish war of 1716-1718.

3) Returned to London and was elected to Parliament in 1722. While there, *he campaigned for the improvement of the circumstances of poor debtors in London prisons. For the purpose of providing a refuge for people who had become insolvent and for oppressed Protestants on the continent, he proposed the settlement of a colony in America between South Carolina and Spanish-held Florida.*

4) He arrived in Charleston, South Carolina late in 1732, and *settled near the present site of Savannah, Georgia*

on 12 February 1733. He negotiated with the Creek tribe for land (stole it?) *and established a series of defensive forts, most notably Fort Frederica, of which substantial remains can still be visited* (and you can visit them yourself, if you so desire). He then returned to England and *arranged to have slavery banned in Georgia.*

5) In 1739, Oglethorpe was responsible for a *number of successful raids on Spanish forts, as well as the unsuccessful Siege of St. Augustine. Although Oglethorpe failed in his attempt to capture St. Augustine, there is a view among some historians that the strike deep into Spanish Florida succeeded in placing the Spanish on the defensive for several years.*

6) Which didn't keep the Spanish from attacking St. Simons Island in 1742. Two very famous battles took place on St. Simons Island on 7 July 1742, the Battle of Gully Hole Creek, won by the British, and the Battle of Bloody Marsh, which the British won by ambushing the Spanish troops while they were preparing dinner.

7) *The Treaty of Aix-la-Chapelle ended the war in 1748 and recognized the status of Georgia as a British colony* (and in 1776, this 28-year-old colony joined 12 others in rebelling against Britain).

8) In 1743, Oglethorpe returned to Britain and went on to bigger and better things. He died in 1785 (just a few years after his colony became part of the United States of America).

9) A book by Blackburn (1994), entitled *James Edward Oglethorpe*, gives many more details on this intriguing historic figure.]

After visiting the fort, we recommend that you go back to the roundabout and continue on toward the north end of the island. A bit further along there is a most interesting kiosk on the side of the road that states the following:

> *From March 9 to May 12, 1736, Charles Wesley, secretary to James Oglethorpe, was Anglican cleric to the founders of the Fort Frederica. His stern discipline earned disfavor among the colonists and Oglethorpe. John Wesley, religious leader of the colony of Georgia, visited Charles in April 1736, preaching in the storehouse. Charles returned to England. John ministered to the Fredericans in four trips before leaving "with utter despair for doing good there" on January 26, 1737.*

On John's return to England in May 1738, the reunited brothers had "heart warming experiences" and later founded the Methodist movement. That is, they couldn't convert the heathen, so, instead, they went home and founded another church that has survived even until this day. According to one poll (*Churches and Church Membership in the United States, 1990*, published by the Glenmary Research Center, Mars Hill, NC) – more than 37 thousand churches and over 11 million members of the United Methodist Church exist in the U.S. We were truly amazed that the guy that founded the Methodist church got his start right there on St. Simons Island. To be a little more specific, the church traces its roots to Reverend John Wesley's evangelistic revival movement in the Anglican Church.

If you continue north from the roundabout for 3 miles you come to major a fork in the road (located by arrow B in Figure 159B). This is where two limbs of the Silver Bluff component of the island join in a subtle Y shape. The limb to the left, which trends north, is considerably more developed than the one to the right, which trends more to the northeast. A salt marsh area about 2.6 miles long occupies the space between these two branches of Pleistocene upland (Figure 159B). A modest sized tidal creek, Butler Creek, is a highly meandering occupant of the marsh. This channel has an abandoned oxbow lake that is filling in with sediment, as is shown by the vertical image in Figure 101A and by the oblique aerial photograph in Figure 101B.

We pondered about why the remnants of the Pleistocene Altamaha Delta are not better preserved here, even though this remnant of the Silver Bluff is rather large. Presumably, much of the original delta was eroded away as the valley expanded during the last lowstand, which peaked about 20,000 years ago, and some of it was probably lost when sea level first reached its present level about 4,500 years ago during the time interval before the river was able to completely fill in the lowstand valley and start building the delta front out into the ocean.

In any event, by taking the left fork at this intersection, you will have a 3.4-mile drive to Butler Point, a mostly up-scale private development (located by arrow A in Figure 159B). It is pretty country down by Butler Point, where a large marina is located on Hampton River (Hampton River Club Marina). In this area, the road winds around the big live oak trees, many of which are heavily draped with Spanish moss. The live oaks at Butler Point are phenomenal, so it is probably worth the drive just to see them. This is not a particularly visitor friendly area, but you can see a tremendous expanse of marsh, as well as numerous birds, such as brown pelicans and cormorants. There appears to be almost no tourist lodgings on this end of St. Simons Island; however, we have stayed in good accommodations on the south end of the island.

As best we can tell from Bell's (1987) map, Major Butler's famous Hampton Plantation, his first purchase on the Altamaha River Delta country, was located close to the present Hampton River Club Marina or more likely where the large golf course is now spread out. In the many years that Major Butler operated this plantation, he had a number of famous visitors, none probably more famous than that of Vice President Aaron Burr on 25 August 1804. This was while Burr was on the run from those who wanted to prosecute him for killing the Secretary of the Treasury Alexander Hamilton in a duel at Weehawken, New Jersey on 11 July 1804. *He stayed five weeks riding out the firestorm of criticism that continued elsewhere, and while there wrote several letters to his daughter, the beautiful Theodosia Burr Alston, who was living on her husband's plantation near Georgetown, South Carolina.* Burr apparently enjoyed his stay, except that he had to endure *the great hurricane of September 1804 that devastated the low country and played particular havoc on the waterfronts of Savannah and Charleston.* Sixteen of Major Butler's slaves died on Butler Island during that storm (Bell, 1987).

How about a visit to Little St. Simons Island, so beautifully illustrated in Figure 159B? This island was originally owned by Major Butler and was inherited by his grandson Pierce Butler (Mease), who took his famous actress wife, Fanny Kemble, to see it and she wrote about it in her book, *Journal of a Residence on a Georgia Plantation.* As noted by Hodnett Cooper (2008), the island is still privately owned, but there are some accommodations available (a small compound of guest cottages and a lodge). Visitors to the still mostly pristine island are limited and reservations are required. *A private boat ferries visitors from Hampton River Club Marina* (located here at Butler Point) *to the island. The cost of accommodations, food, and recreational activities, such as horseback riding, fishing, and kayaking, are all-inclusive.*

On your way to **Sea Island**, you will come down off the Pleistocene "upland" and start out across the mile-wide marsh at that point. This road is very conspicuous on the vertical image of Sea Island in Figure 167. The mature cedars by this road have been there for a while, and along this road is one of the better places on this coast to see certain high-marsh plants (e.g., sea oxeye – *Borichea frutesens*).

According to Bagwell (2008), *the F.J. Torres Causeway, then dubbed the Brunswick-St. Simons Highway, officially opened in July of 1924.... By December of that year, the Long Island causeway* (that led out to what is now called Sea Island), *a simple dirt and mud road through the marsh, was inaugurated. Spectators were soon visiting Long Island by car, spending time on the beach, and viewing the house lots for sale. ... By the 1930s, Howard Coffin and the Sea Island Company would change the coastal island's name from Glynn Isle to Sea Island* (it had been changed from Long Island to Glynn Island in 1926).

Sea Island is private and one must be a guest to get onto the island. Most of the northeast side of the island is "protected" by seawall/revetments with riprap added or used singularly in places (as was discussed earlier). The beach was nourished in 1986 (250,000 cubic yards) and 1990 (2,000,000 cubic yards) using private funds to maintain a dry beach at high tide.

To visit **East Beach**, you should navigate to East Beach Causeway, where you should take a left and drive across the 0.3 miles of salt marsh onto the Holocene limb of St. Simons Island. Once you are off the marsh, go straight ahead for about 400 yards and park in the Coast Guard parking lot, a good spot with clean restrooms, etc., that will provide you access to East Beach.

When we were at this beach during a low spring tide on 27 March 2010, the large intertidal bar (shown in the oblique aerial photograph in Figure 65A and the ground photograph in Figure 64C) was in the process of attaching to the beach. The beach was very wide, the sand surprisingly white, and the sedimentary features all around were fantastic. The top of the bar contained some outstanding antidunes, as well as an array of wind-shadow dunes (see Figures 84A and 84B). The story of the origin of foredunes back of the beach explained in Section I, and illustrated in Figures 83 and 84, was spectacularly displayed on this wide, presently building middle of the Holocene component of St. Simons Island.

From there, we recommend that you go back out of the parking lot and take a right on the first main road you come to. After making that turn, it is 0.8 miles to a parking area on the shore of Gould's Inlet, another spot you shouldn't miss. At this beach, there is a small pier built out to the edge of the inlet channel with a boardwalk that extends across the sand and riprap. When we were last there, there was a pretty wide sand area landward of the riprap. This zone of riprap extends for 340 yards to the south of the pier, as well as to the north end of the island, another 440 yards. The photograph of riprap in Figure 90A was taken from this boardwalk, looking north.

During our visit on 27 March 2010, we walked along the beach to the southeast as far as the outlet to the ponded water in the lee of the large welded intertidal bar shown in Figure 65A. Along the way, we passed by some bubble sand on the landward side of the riprap. Close to the channel outlet, a vertical scarp 8-10 feet high had been eroded into the dunes by waves at high tide (Figure 169A). At the channel outlet, the dunes were being eroded by the currents generated by the falling tide (Figure 169B). In one of the breaks in this dune line is where the zone of heavy minerals pictured in Figure 76C was located. In fact, this is where we collected the white quartzose sand and the dark heavy mineral sand illustrated in Figure 61.

From the pier at Gould's Inlet, we suggest that you take a street across the island to its northwest side and drive south along Ocean Boulevard. The marsh encroaches right up against this road, providing an excellent view of it. Looking across the marsh, you can see the houses on the Pleistocene "upland" across the way. If you want to see the long line of riprap along the south end of East Beach, continue on south past the causeway, and eventually turn left and drive to the beach. There is parking and public beach access at 9th Street, which is where we took the picture of the riprap in Figure 90B. The oblique aerial photograph in Figure 90C also shows the riprap in this area.

An inviting tourist area is located at the very south end of St. Simons Island proper. The St. Simons Island Light Station and the Maritime Museum are areas of considerable interest.

Gould's Inlet and East Beach on the Holocene component of St. Simons Island are cited in the Georgia Wildlife, DNR (2002) Colonial Coast Birding Trail. Specialties in this area include American oystercatcher, black skimmer, painted bunting, bald eagle, least tern, and northern gannet. They suggested that, if you go birding there in the winter, *look for common and red-throated loons, scoters, scaups, and other waterbirds in the ocean*. A special treat in the wintertime is *seeing northern gannets out over the ocean* doing their amazing vertical nosedives.

FIGURE 169. Eroding dunes, Holocene component of St. Simons Island. (A) Wave-cut scarp in the dunes on the north end of the Island by Gould's Inlet (located on Figures 166 and 167). (B) Outlet in runnel area landward of a large welded intertidal bar in the middle of the island that was eroding into the dunes. Photographs taken during a low spring tide on 27 March 2010.

ST. SIMONS SOUND

Introduction

St. Simons Sound is the sixth of the eight major sounds on the Georgia Coast (going south). As can be seen in Figure 170, during ebb tides, the huge outflow at the one-mile-wide entrance to this sound is fed by two primary sources: 1) Waters draining the wide marsh/

FIGURE 170. St. Simons Sound. Arrow A points to the F.J. Torras Causeway that leads to St. Simons Island, arrow B points to the Sidney Lanier Bridge on US 17, and arrow C points to the tree-like tidal channel system pictured in Figure 108B. This sound is located on Figures 151 and 156. The large river that enters the sound from the south is called the Brunswick River. Digital orthophoto acquired in March 2006, courtesy of USGS.

tidal channel complex that separates the Princess Anne upland that Brunswick is located on and St. Simons Island (Figures 168 and 170); and 2) The 20-mile-long drainage network of the Brunswick River that is illustrated in Figure 170. The tidal channels in these systems show a wide range of sinuosity and general connectivity. The spectacular tree-shaped channel network of the headwaters of the South Brunswick River is illustrated in Figure 108B (located by arrow C in Figure 170). As was also true for the five sounds already discussed, because of the large tides and generally depositional nature of this coastline, fairly wide tidal flats and extensive salt and brackish marshes are present.

Places to Visit

If you are travelling from north to south, there are three highways you can take across the lower and middle portions of St. Simons Sound: I-95, U.S. 17, and GA 303 (Figures 151 and 170). Leaving Brunswick and crossing the sound on U.S. 17 is a spectacular mile and a half drive over the phenomenal Sidney Lanier Bridge. One of the reasons this 480-feet-tall bridge was built was because the older, lower bridge was hit by a couple of ships and disabled. *On November 7, 1972 the ship African Neptune struck the bridge, causing parts of the bridge to collapse and causing several cars to fall into the water. Ten deaths were caused by the accident. On May 3, 1987, the bridge was again struck by a ship, this time by the Polish freighter Ziemia Bialostocka* (Wikipedia encyclopedia). The present bridge, which was completed in 2003, affords a fantastic view of the entrance to St. Simons Sound, and the bridge itself is a sight to behold that you can experience from miles away (located by arrow B in Figure 170).

[NOTE: **Sidney Lanier**: A legendary poet of the south who was born in Macon, Georgia in 1842. He was an active soldier and blockade runner for the Confederates during the War Between The States. During the War, he was captured at sea and *incarcerated in a military prison at Point Lookout in Maryland, where he contracted tuberculosis (generally known as "consumption" at the time). He suffered greatly from this affliction for the rest of his life* (Wikipedia encyclopedia). He is most renowned, however, for the poetry he wrote. One of his most noteworthy poems, titled *The Marshes of Glynn* (1878), was about the open coastal marshes in Glynn County, coastal Georgia, the county in which the bridge is located.]

Once you cross the bridge and head west on U.S. 17, you will be driving through 2.4 miles of salt marsh, at the end of which you come up on a low upland called Colonels Island (located in Figure 170). According to Huddlestun (1987), the Princess Anne Pleistocene barrier island system is absent in this area, so we do not know why this upland is at this location. However, once up on it, you will be passing by (on the right) the Colonels Island Terminal operated by the Georgia Ports Authority that, according to Georgia Ports Authority (2009), is *among the largest deepwater agri-bulk operations in the U.S. South Atlantic* that offers *a turnkey service for U.S. Midwest and Southeastern agribusiness.*

Another route would be to get on I-95 at exit 36, near Brunswick, and drive southwest the five miles across the upper portion of St. Simons Sound to exit 29, near Fancy Bluff. At two miles along the way, you will cross the bridge over Turtle River, a tributary of the Brunswick River, which is 0.6 miles wide at this point. There is a salt marsh system over a mile wide before you get to the bridge (Figure 170). Looking to the right as you drive across the bridge, you will see that the head of this estuary is very complex, with a lot of tidal flats upstream (assuming you make the drive at low tide). As soon as you pass the end of this bridge, you will be travelling through a development called Blythe Island, where about 25 docks have been built out into the waters of the Turtle River. In this sound, there is an abundance of docks anywhere there is water deep enough to drive a boat and the land is sufficiently high to build a house on it.

The next major tidal channel you will cross is the South Brunswick River, the main limb of the tree-like complex of tidal channels illustrated in Figure 108B (located by arrow C in Figure 170).

This tidal system called St. Simons Sound is one that Howard and Frey (1980) classified as a **marine system**, casting some doubt as to whether it should be classified as an **estuary**, because of the lack of a significant fresh water stream entering into it. That is, not if one follows Pritchard's (1967) definition of an estuary-drowned river valley with tides and notable fresh water influx.

To see the most landward part of the sound, on 27 March 2010, we drove north on U.S. 82 (located on Figure 170) and made the following observations. For one thing, we noticed that there were a lot of swamps on the north side of the interstate. At about two miles from the intersection where U.S. 17 heads off in a southwest direction, we observed that we were driving up onto a higher upland than we had crossed before on this drive. At that point, we were on the Pamlico Pleistocene barrier island system that formed about 125,000-130,000 years ago and has a present elevation of well over 25 feet.

Assuming you will be making the same upper sound tour that we did, at about 4.5 miles further along, you should turn right on a road that leads to the village of Georgetown, where, after passing through it, at the first intersection you turn right and proceed to the northeast on GA 99. From this intersection, it is 0.34 miles to the bridge crossing Turtle River that runs through a black needlerush marsh about 250 yards wide. There is a small boat ramp on the north side of the road, where you can stop and admire what we called – "a very beautiful marsh up here!" We also concluded that "the headwaters of these sounds are spectacular, really fascinating."

From the Turtle River Bridge, it is about 2.5 miles to Green Creek, where black needlerush appears to extend all the way up into the swamp. From Green Creek it is about three miles to the village of Anguilla, where you encounter the headwaters of the Buffalo River. We thought this was also a nice spot, with a beautiful marsh over a thousand yards wide. Driving on east of the village and looking to the northwest, we could see what appeared to be a massive cypress-tupelo swamp at the southern edge of Paulks Pasture Wildlife Management Area that had been "logged to hell and gone!"

14 COMPARTMENT 4 - JEKYLL ISLAND TO ST. MARYS ENTRANCE

INTRODUCTION

For purposes of discussion, we have subdivided this compartment into four subcompartments (Figure 171):

1) **Jekyll Island** – The smallest of the "combination" barrier islands that is partly developed, but still serves as a state park.

2) **St. Andrews Sound** – One of the larger sounds on the Georgia Coast, with a 2.5-mile-wide entrance that is the outlet for three large tidal river systems.

3) **Cumberland and Little Cumberland Islands** – With an attached completely Holocene barrier island at its northern end that is privately owned (Little Cumberland Island). The rest, the largest of the "combination" barrier islands on the Georgia Coast, is a national seashore.

4) **St. Marys Entrance** – Composed of a combination of the mouth of the St. Marys River and the outlet of the relatively small Cumberland Sound, the main entrance to these two waterbodies, that is jettied and only 0.6 miles wide, marks the southern boundary of the state of Georgia.

JEKYLL ISLAND

Introduction and Geomorphology

Jekyll Island, the origin of which is illustrated in Figure 46, is the smallest of the major barrier islands in Georgia, measuring 7.6 miles in a straight line distance from the northern to southern ends, with a maximum width of about 1.4 miles. However, it has on the order of eight miles of intertidal beaches, with some sand on most of them. It has an area of 5,700 acres, 4,400 of which are uplands (Olsen, 2006). According to Chowns et al. (2008), *prior to about 1480 B.P. Brunswick River entered the Atlantic south of Jekyll Island. Subsequently, rising sea level encouraged the river to follow a more direct route and to empty north of the island.*

As shown on the vertical image and sketch map in Figure 172, like the other major barrier islands on the Georgia Coast, this one has a "core" of sand deposited in the Silver Bluff Pleistocene barrier island system. There are spit forms on both ends of the island of Holocene age (i.e., <4,500 years old), with the one on the north end undergoing some severe erosion in March 2010, and the one at the south end showing signs of continuing accretion (Figure 173B). Around 3.2 miles of the central portion of the eastern shore has historically been eroding into the Silver Bluff sediments (Figure 46; Henry, 2009).

The evidence of the lingering erosion of this east-central part of the island is a continuous line of riprap revetment, pictured in Figures 91 and 174A, that extends for about three miles along the beach. According to *Explorer's Guide To Georgia's Golden Isles* (2010), this revetment was built by the Federal Government after severe erosion caused by a hurricane *Dora* in 1964. These are the famous Johnson "rocks" discussed earlier.

Although there have been discussions about the value of having beach nourishment projects carried out on Jekyll Island (e.g., Kriesel and Turner, 2003), as far as we can tell, no major ones have been executed so far.

Human Habitation

Some of the intriguing human history of this island, as was outlined by Olsen (2006), follows:

1) The Creek Indians, who called the island "Ospo," fished and hunted there.

2) The French Huguenots claimed the island in 1562, changing the name to "Ile de la Somme," and not to be outdone, Spanish Jesuit priests established a mission on the island in 1566.

3) *In 1736, after claiming the island for Britain, General James Oglethorpe established an outpost and renamed it Jekyll. … In 1742, after the defeat of the Spanish, Jekyll was held as a military reservation until it was added to the*

FIGURE 171. Compartment 4, north end of Jekyll Island to the St. Marys River Entrance. Red arrows designate the boundaries of the compartment. Source: ESRI StreetMap USA.

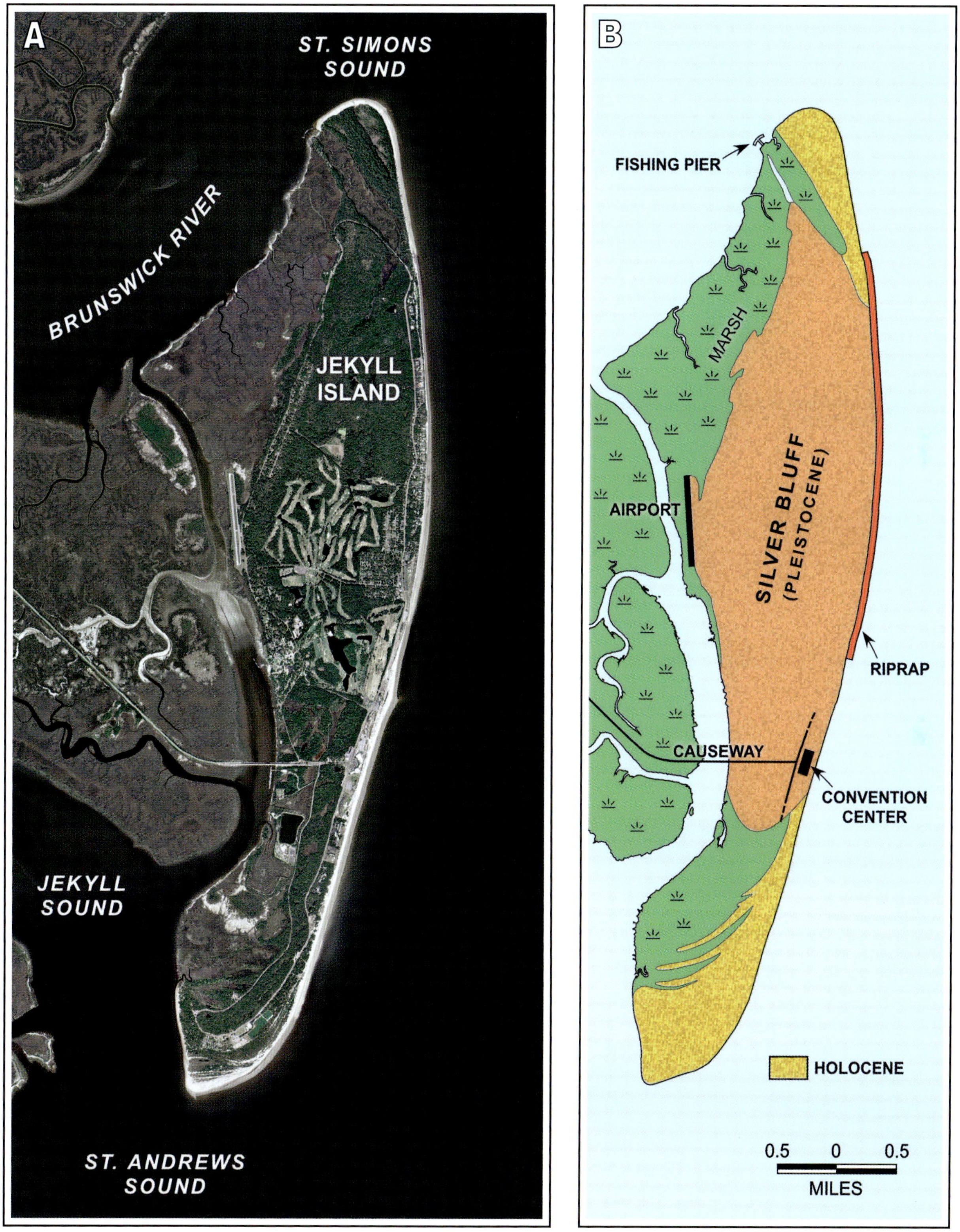

FIGURE 172. Jekyll Island (located on Figure 171). (A) Digital orthophoto acquired in March 2006, courtesy of USGS. (B) Sketch showing some relevant aspects of the Island. Based primarily on Henry (2009).

Parish of St. James by the Georgia legislature in 1765.

4) After that, ownership of the island changed a couple more times, and in the early 1790s, it was purchased by Christopher Poulain duBignon and remained in the duBignon family for nearly a century, during which time sea-island cotton was grown. In his report, Olsen didn't give information on how many slaves were involved in this process.

5) In 1886, the island was purchased by a group of northern millionaires, including the Rockefellers, Morgans, and Vanderbilts, for use as a winter resort.

6) *In 1947, the state of Georgia bought Jekyll for use as a state park. The causeway and bridge were built in 1954, allowing easy access for the general public.*

7) Because Jekyll Island is a state park, Jekyll residents own their homes, but lease the land their homes are built on from the state (Holliday, 2010).

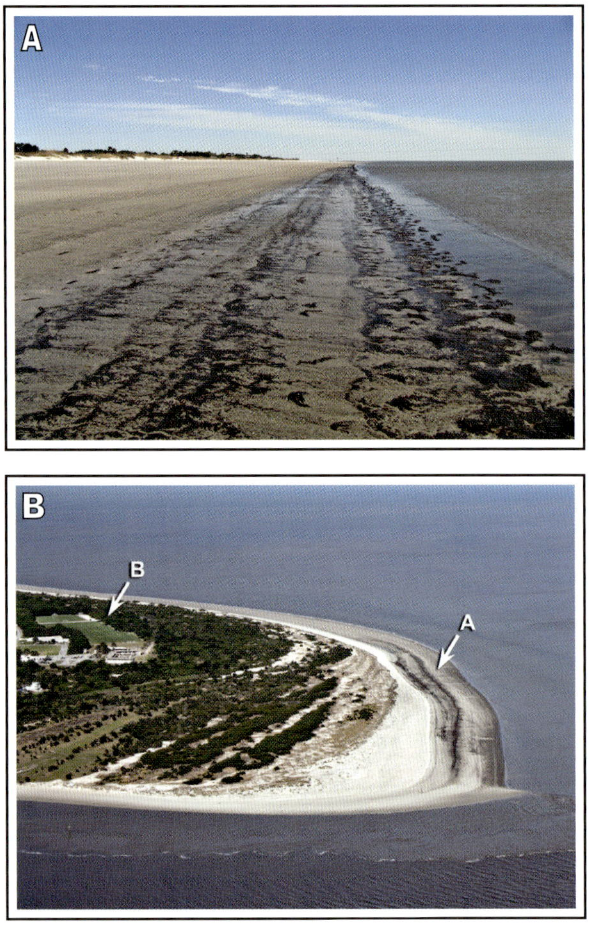

FIGURE 173. Jekyll Island's sand beaches. (A) Ground view of the beach near the Convention Center (Figure 172B) taken during a low spring tide on 26 March 2010. (B) Oblique aerial view of the south end of the Island taken during a low spring tide on 28 April 2010. Photographs of an intertidal bar complex at the location indicated by the arrow A taken during a low spring tide on 26 March 2010 are given in Figures 64B and 72D. Arrow B shows the location of the boardwalk by the soccer fields that is discussed in the text.

Places to Visit

Assuming you are going south on U.S. 17 to get there, leave Brunswick and drive across the amazing Sidney Lanier Bridge. Just past the end of the bridge, turn left onto the Jekyll Island Causeway for a 6-mile drive to the island. Along the way you will have a close-up view of the marshes of Glynn County, made famous by the poet Sidney Lanier (discussed earlier). Just as you start this drive, off to your right will be the headwaters of the Jointer Creek "estuary," actually a classic example of what Howard and Frey (1980) classified as a marine system, rather than conforming to the usual definition of an estuary, because of the lack of fresh water entering the tidal network. The marsh is smooth cordgrass all the way to the upper end of the system.

Looking to the north, in addition to a great view of the Sidney Lanier Bridge, you can see the uplands of Brunswick located on the Princess Anne Pleistocene barrier island system formed about 85,000 years ago. Looking south, it is 3.5 miles to the next Princess Anne upland, Black Hammock on the other side of the Little Satilla River. Therefore, there is a sizeable gap in the Princess Anne barrier island system in this area.

This is an interesting drive, with lots to see, such as the palmetto trees lining the road, numerous "bare spots" at the top of the marsh, an occasional tidal flat, and so on. We concluded that it was quite an engineering feat to make a road like this across the marsh. Looking ahead as you drive along you will be able to see the trees on the Pleistocene center of the island, with its climax maritime forest. You should pull over at the Jekyll Island Visitors Center to get maps, books, etc., as well as to purchase the five-dollar parking pass (for one day, more if you stay longer).

Three Jekyll Island locations are included in the Georgia Wildlife, DNR (2002) Colonial Coast Birding Trail. The first one is the causeway itself, where there are two recommended birding sites located by signs along

FIGURE 174. Erosion zone on Jekyll Island. (A) Riprap revetment near north end of riprap zone shown on the oblique aerial photograph given in Figure 91. This spot is located about 1.5 miles south of the north end of the Island (located by arrow E in Figure 175A). (B) Dead live oak trees littering the high tide line about 200 yards north of the location where the photograph in A was acquired. Both photographs taken at low tide on 26 March 2010.

the road. Specialties for the area include osprey, bald eagle, clapper rail, northern harrier, roseate spoonbill, red knot, black-necked stilt, white ibis, and wood stork. They also gave the following tips: *Shorebirds are best seen at low tide from midsummer through spring. Look for nesting ospreys in spring and summer. Listen for clapper rails and marsh wrens in the salt marshes. Watch for northern harriers flying low over the marsh in winter. Look for roseate spoonbills in summer.* We think spotting those roseate spoonbills is a special treat.

Near the end of the causeway, you will go up onto and over the high bridge across the Intracoastal Waterway. If you go at low tide, you will spot a big, wide mud flat by the waterway.

Obviously, there is a lot to see on Jekyll Island. In this discussion, we describe four of the areas we visited recently and present some ground photographs to illustrate them. One of the more interesting spots is the area just to the north of the Jekyll Island Convention Center, where the beach was definitely depositional in March 2010, with 50-60 yards of well-preserved Holocene dunes. The foredune line was also well developed, with sea oats (*Uniola paniculata*) growing on the top of them (Figure 84D). When we were there, a high berm was present on the landward side of the intertidal zone with seacoast marsh elder (*Iva imbricata*) growing at the backside of the berm (Figure 84C). There was also some wrack (accumulation dead marsh stems, etc.) on the berm top, some buried in the sand as a source bed for the *Iva*. What we saw there was a beautiful example of dune formation and progradation in progress, with all of the stages of dune formation–active berm top, wrack, *Iva*, wind-shadow dunes, and well developed foredunes–being present (see the general model in Figure 83). The intertidal sand beach itself was very wide and flat, as illustrated in Figure 173A.

The next place we visited was a short trail out to the beach a little ways north of the Villas by the Sea Resort, located about a 1.5 miles south of the north end of the island. This spot is close to the north end of the riprap illustrated in Figure 91. A ground photograph of the riprap at this spot is given in Figure 174A (located by arrow E on Figure 175A). This is a highly erosional zone, apparently eroding into the southern edge of the Holocene deposits at the northern end of the island pictured in Figure 175A. There are lots of dead trees exposed both in the intertidal zone (Figure 175B), and at the top of the beach, as is illustrated by the accumulation of dead live oaks shown in Figure 174B.

From there, it is a little over 1 mile to the fishing pier at the north end of the island (shown in the oblique aerial photograph in Figure 176A). You can drive right up to the fishing pier amidst some old live oaks growing on a narrow limb of land. We don't know if this tree-lined projection, that has the highly meandering Clam Creek between it and the Holocene element shown in Figure 175A, is Holocene or Pleistocene.

We recommend that you walk across the bridge over Clam Creek and inspect the end of the Holocene projection with its numerous dead trees exposed in the

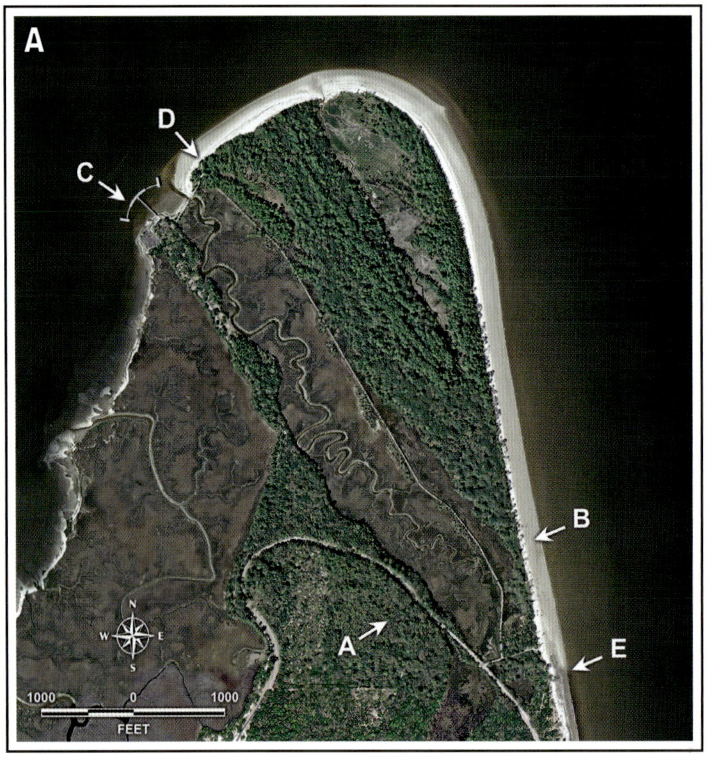

FIGURE 175. North end of Jekyll Island (located on Figure 171). (A) Digital orthophoto acquired in March 2006, courtesy of USGS. Arrow A points to the remnant of the Silver Bluff Pleistocene barrier island system that composes the core of the island (formed 40,000-50,000 years ago; see Figure 172B). Arrow B points to the Holocene component of the northeast end of the island (see Figures 46 and 172B). All of the land northeast of the stream course is Holocene in age, formed less than about 4,500 years ago. Arrow B also points to the approximate location of the area shown in the photograph in B. Arrow C points to the fishing pier (see also the oblique aerial photograph in Figure 176A). Arrow D points to the location where the saw palmetto shown in Figure 176B was exposed in an erosional scarp, and arrow E points to the northern end of the riprap shown in Figures 91 and 174A. (B) Oblique aerial photograph taken during a low spring tide on 28 April 2010 showing the Holocene component of the northeast end of the island that is presently undergoing significant erosion, as evidenced by the trees exposed in the intertidal zone.

FIGURE 176. North end of Jekyll Island. (A) Oblique aerial view (28 April 2010) looking southeast, showing the fishing pier in the foreground. Arrow indicates where the photograph in B was taken. (B) Exposure of horizontally extending roots of saw palmetto (*Serenoa repens*) plants exposed in an eroding scarp on the extreme north end of the island (also located by arrow D in Figure 175A). Scale is ~6 inches.

intertidal zone. There are some very big trees on this Holocene spit, including some giant pine trees, and there is no riprap at the very end in this area. Walk up to the eroding scarp where you can inspect some exposed saw palmetto (*Serenoa repens*) roots. An amazing aspect of these plants is that their large, shallow roots project horizontally, usually a foot or so under the soil, as illustrated in Figure 176B.

Go out on the pier for a great view across St. Simons Sound to St. Simons Island, or maybe to do a little fishing. An eroding marsh with a short, steep sandy tidal flat is located beyond the picnic/parking area to the southwest of the pier.

The north end of the island is the second section of the Jekyll area included in the Georgia Wildlife, DNR (2002) Colonial Coast Birding Trail. They noted that: *Although the beach is more limited than at the south end, the mix of beach, forest and saltwater habitats provides excellent birding opportunities.* Specialties include: least tern, red-throated loon, scoters, American oystercatcher, and black skimmer. They also provided the following tips: *Warblers are best seen during spring and fall migration. Thousands of scoters and scaup raft just offshore in winter. Look for ospreys throughout the year. Look for common and red-throated loons fishing off the beach.*

The south end of the island is perhaps the most interesting area of all, especially with regard to its geomorphology. We suggest that you take River View Drive for a tour around the west side of the island to see some great views across the wide salt marsh. There is some interesting vegetation, myrtle, pines, etc., along this drive, as well as an abundance of Spanish moss. About a tenth of a mile before you reach the causeway, there is a bridge across a very well developed seaside oxeye (*Borrichia frutescens*) marsh.

The next place we recommend you visit is a boardwalk across the dunes located about 0.8 miles from the extreme south end of the island. To get to the boardwalk, go to the soccer fields located along Beach View Drive about two miles south of the end of the causeway. There is a spacious parking area by the soccer fields. The boardwalk crosses almost 200 yards of prograded Holocene dune ridges. When we were there on 26 March 2010, we had a great time walking along this beautiful boardwalk through the low dunes. We noticed that there was no climax maritime forest in the dunes, so they appear to be less than 50 years old. There are, however, sea myrtles on the back lines of dunes so they are at least 15 years old. A little further to the north, these sea myrtles show the excellent salt pruning illustrated in Figure 130. In any event, this is a very highly depositional area, that's clear. Some of these curving beach ridges are shown in Figure 173B (the location of the boardwalk is also shown on this image).

The beach cusps shown in Figure 29C were present at the high-tide line right at the north end of the boardwalk. At that point, the beach was wide and flat, but further south the beautifully developed intertidal bar shown in Figure 64B dominated the intertidal zone. There was a plethora of sedimentary structures associated with the bar, including the rhomboid ripples on top of it pictured in Figure 72D. In fact, if you go out to this spot at low tide, you will be able to see all of the different kinds of structures illustrated in Figures 71 and 72, such as ripples in the intertidal outlet trough and big antidunes all over the place, especially on the top of the intertidal bar. If you walk down to the end of the island, you will be able to look south and see the barren sand dunes on the north end of Little Cumberland Island.

Also, this south end of the island is the third section of the Jekyll area included in the Georgia Wildlife, DNR (2002) Colonial Coast Birding Trail. Specialties include black skimmer, American oystercatcher, marbled godwit, jaegers, south polar skua, northern gannet, piping plover, and glaucous gull. Some tips they gave point out that one should *look for shorebirds around high tide from midsummer through spring. Carefully examine gulls; rare gull species are sometimes found here. Don't disturb resting birds, give them a wide berth. Look for scoters, loons, mergansers, buffleheads and other waterfowl swimming offshore.*

In conclusion, with regard to the geomorphology, Jekyll Island is a somewhat classic story, with: A) The erosion and riprap on the north end, probably because of some major changes in the large ebb-tidal delta shoals off the entrance to St. Simons Sound; B) Some slight accretion in the middle; and C) Conspicuous accretion the further down the island you go.

So, the story on Jekyll Island is pretty straightforward. It's a wonderful, beautiful place to visit, and, as a bonus for the Georgia Coast, easy to get to by driving.

ST. ANDREW SOUND

Introduction

St. Andrew Sound is the seventh of the eight major sounds on the Georgia Coast (going south). As can be seen in Figure 177, during ebb tides, the huge outflow at the entrance to this sound is fed by three primary sources:

FIGURE 177. St. Andrew Sound. The Satilla River, the second largest coastal plain (black-water) river in Georgia, and the Little Satilla River are notable components of this estuarine complex. Arrow A points to where US 17 crosses the White Oak Creek, which is illustrated in Figure 178, and arrow B points to town of Woodbine, which is located on the Satilla River and is illustrated in Figure 179. The locations of the Pleistocene barrier island systems – Silver Bluff (SB), Princess Anne (PA), and Pamlico (P) – are shown (after Huddlestun, 1986). The barrier island sediments (with green color of plants on beach ridges) located seaward of the Silver Bluff (SB) are Holocene, or recent, in age. Landsat image acquired on 23 October 1999, courtesy of Earth Science Data Interface at the Global Land Cover Facility.

1) The combined waters of Jointer Creek and Little Satilla River that drain the wide marsh/tidal channel complex in the lee of Jekyll Island; 2) The watershed of the Satilla River, the second largest coastal plain (black-water) river in Georgia (Table 3); and 3) Cumberland River and its associated marsh/tidal channel complex in the lee of the northern half of Cumberland Island. With a width of 2.5 miles, the entrance to this sound is the second largest of the eight sounds, trailing only the 3.3-miles-wide Wassaw Sound. The average width of the entrances to the eight sounds is 1.7 miles.

Places to Visit

Assuming you are travelling from north to south, there are two possible routes to take to see the area landward of St. Andrew Sound by driving, on either I-95 or U.S. 17. For the I-95 drive, we suggest that you take exit 29 near Fancy Bluff and head south. This is a very pleasant and scenic drive.

Once underway, it is 0.9 miles to the north side of the Little Satilla River valley, where the road runs obliquely across 0.8 miles of marsh associated with this tidal river channel. From the southern edge of the marsh, it is 0.7 miles to exit 28, where you may want to take Dover Bluff Road for about 7.5 miles down to Dover Bluff, which is located on Black Hammock, a remnant of the Princess Anne Pleistocene barrier island system.

At about 2.5 miles along the way, there are a couple of little isolated hummocks with short roads or bridges out to them. This is the kind of place Michel had a fantasy of living on when she was growing up in Mount Pleasant, South Carolina (on the north side of Charleston Harbor).

Back at I-95, get on the highway at exit 28 and head south again. The zone between exits 28 and 14 starts with a very pleasant drive across Hazzards Neck, with the highway running along the landward boundary of some scattered remnants of the Pamlico Pleistocene barrier island system to the southeast, and the wide backbarrier of the Pamlico unit to the northwest. Along the way, you will cross the beautiful White Oak Creek and the Satilla River. Both of these rivers have associated wide marshes with very complex tidal channels that can be clearly viewed from the highway.

To take U.S. 17 south, we recommend that you can also start at exit 29 near Fancy Bluff. It is a little over half a mile going northwest from exit 29 on U.S. 17 until you reach the intersection with GA 303. At that point, stay on U.S. 17 that turns abruptly in a southwesterly direction. From that intersection, it is 2.6 miles to the bridge over the Little Satilla River. There is a boat ramp on the left just after you cross the bridge where you may want to stop to look for birds. This pristine river flows through a beautiful 0.6-mile-wide smooth cordgrass marsh.

If you look to the right as you cross the marsh area, you will be looking toward a massive swamp system at the head of the Little Satilla River that goes by several names–Brailey Swamp, Redcap Swamp, Glencoe Swamp, etc. Some of these swamps have been logged out. They are located on the topographically flat, muddy deposits of the 10 to 12-mile-wide Pamlico backbarrier sequence (Figure 9). This backbarrier system is the widest of the backbarriers in the Pleistocene systems in Georgia. However, once you leave the marsh and drive up the side of the valley at Spring Bluff, you will actually be on a remnant of the Pamlico Pleistocene barrier island system itself, according to Huddlestun (1986). The wide Pamlico backbarrier sequence is located to the northwest of the barrier island proper.

Continuing west on U.S. 17, it is about six miles from Spring Bluff to the town of Waverly, where U.S. 17 makes a right-angle turn to the south. Once you come into the valley of Waverly Creek, you will be crossing through about a half mile of cypress/tupelo swamp. In other areas within the valley, you will see patches of brackish marshes, including stands of black needlerush and giant cordgrass.

From the Waverly Creek Bridge, it is a little less than three miles to the bridge over White Oak Creek (located by arrow A in Figure 177). There is a boat ramp at the north side of the bridge, a great place to see the transition from brackish marsh to swamp. This transition also shows up clearly on the vertical image in Figure 178, which gives a classic example of this change. Continuing south, it is 2.7 miles to Tower Swamp, a chance to see another well developed cypress-tupelo swamp. The slightly upland area in between these two streams is the Wild Neck Peninsula, which is underlain by Pamlico backbarrier sediments. A rails-to-trails path runs alongside this part of U.S. 17.

At another 1.6 miles further, you will drive up across the high bridge over the Satilla River in the town of Woodbine (arrow B in Figure 177). The valley of the Satilla River is fairly narrow at this point. The town of Woodbine, located on the south bank of the Satilla River, is also located on the Pamlico backbarrier complex (according to Huddlestun, 1986). For a better understanding of the Pamlico, see the topographic cross-section from Hoyt and Hails (1974) in Figure 10.

Once in Woodbine, we recommend that you go to Waterfront Park that has a boat ramp, a dock, and a long boardwalk over the marsh along side of this magnificent black-water river, as is illustrated in Figure 179. This is definitely a worthwhile stop, if for no other reason than to see the outstanding giant cordgrass. The boardwalk along the side of the channel, pictured in Figure 179B, intersects with a bike trail along an old railroad trestle, which also has a boardwalk on it, where you can walk out to over the middle of the Satilla River.

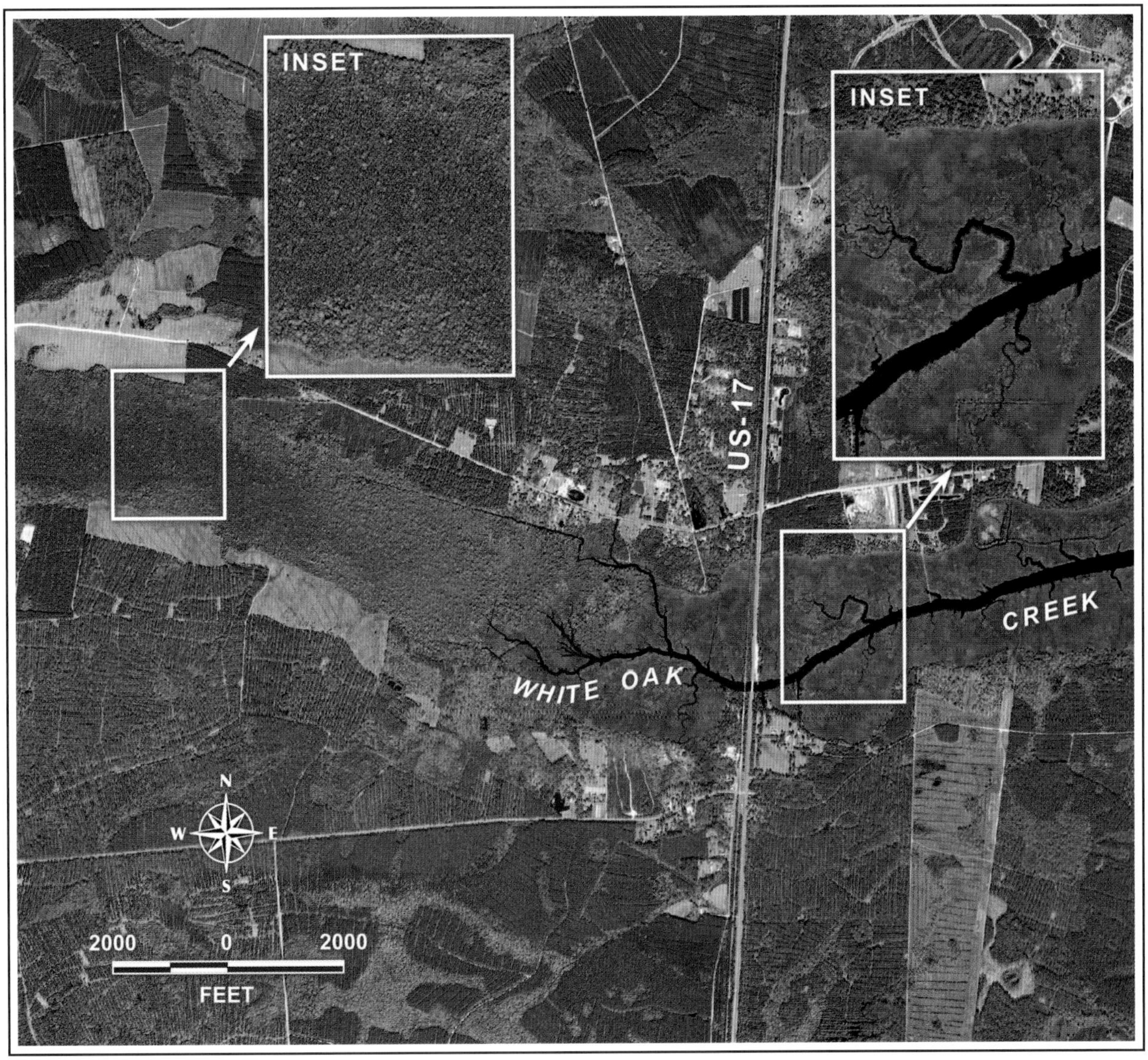

FIGURE 178. Vertical image illustrating the transition of brackish/freshwater marshes (inset on the right) to *cypress-tupelo* swamps (inset on the left) in the headwaters of White Oak Creek, Camden County (located by arrow A in Figure 177). Digital orthophoto acquired in March 2006, courtesy of USGS.

Rice farming started upstream of Woodbine, with very little downstream. House (1954), who gave a short write-up on the subject, noted that there was probably around 5,000 acres of rice farms on this river.

At Woodbine, the channel of the Satilla River is oriented almost directly east/west, with water flowing east, of course. However, as shown in Figure 180A, the river follows a north to south route for about 28 miles where it is located on the landward side of the combined Talbot/Penholoway Pleistocene barrier island systems. This is a striking illustration of the control of the Pleistocene barrier island complexes on the topography and drainage patterns of the coastal plain of Georgia. Throughout this north to south course, the river is highly meandering (Figure 180B). At a straight-line distance of about 15 miles to the southwest of Woodbine, the channel has cut its way to the east through the Pleistocene barrier island systems that range from 45-70 feet in elevation (Huddlestun, 1986), eventually making its way to Woodbine.

FIGURE 179. Waterfront Park on the Satilla River at Woodbine in Camden County (located by arrow B in Figure 177). (A) Vertical image showing the location of the park (arrow A). Arrow B points to the point from which the photograph in C was taken, and arrow C points to a boardwalk along an old railroad trestle that extends out to the middle of the river. Digital orthophoto acquired in June 2008, courtesy of USGS. (B) The boardwalk where it goes under the U.S. 17 Bridge. (C) View looking south of the boardwalk and U.S. 17 Bridge. The grass in the foreground is giant cordgrass (*Spartina cynosuroides*), a plant that thrives in brackish to fresh waters (Figure 93).

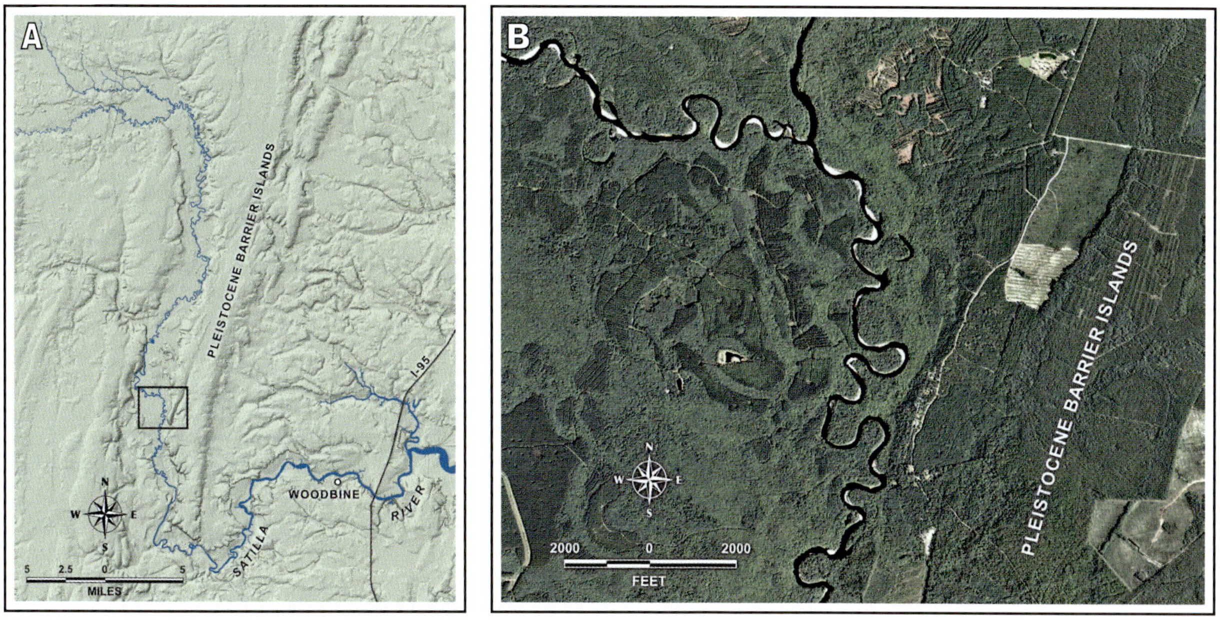

FIGURE 180. The path of the Satilla River, the second largest coastal plain (black-water) river in Georgia. (A) Map showing the river's path to the west of Woodbine, illustrating how the river flows from north to south for 28 miles on the landward side of the combined Talbot/Penholoway Pleistocene barrier island systems. The rectangle gives the location of the image shown in B. (B) The highly meandering channel of the Satilla River west of the Talbot/Penholoway Pleistocene barrier island systems. Note the point bars (white) along the meandering channel. Digital orthophoto acquired in 2009, courtesy of USDA-FSA Aerial Photography Field Office.

CUMBERLAND AND LITTLE CUMBERLAND ISLANDS

Introduction and Geomorphology

In this discussion, we will consider together the two island areas shown in Figure 181. The main body of the system, called Cumberland Island, is a National Seashore and the premier natural gem of the entire Georgia Coast. The privately owned, very northern component of the system is called Little Cumberland Island. Combined, this is the largest of the major barrier islands in Georgia, measuring more than eighteen miles in a straight-line distance from the northern to southern ends, with a maximum width of about three miles. However, it is important to point out that Silver Bluff Pleistocene sediments make up the bulk of the island.

Little Cumberland Island, located at the northern end of the system, is about 3.5 miles long, with a maximum width of 1.3 miles. It has 2.5 miles of beach and covers 2,400 total acres, 1,600 of which are upland (Olsen, 2006). This "island" is entirely Holocene in age.

According to Olsen (2006), Cumberland Island itself has about 17.5 miles of intertidal beaches, which are composed mostly of fine-grained sand. This island is unique to Georgia, in that it has several fields of barren sand dunes along the outer shore. Including the marsh areas, it has an area of 23,000 acres, 15,000 of which are uplands.

As indicated on the vertical image in Figure 181, like the other major barrier islands on the Georgia Coast, this one has a "core" of sand deposited as part of the Silver Bluff Pleistocene barrier island system. Unlike Jekyll Island, however, the Silver Bluff "core" is not exposed to erosion anywhere along the outer shore. Instead, a band composed generally of some kind of dune deposits of Holocene age (i.e., <4,500 years old, but possibly considerably younger than that) borders the seaward side of the island, rarely being more than a few hundred yards wide. As a result of this buffer, beach erosion is not considered to be a problem on this island. As already noted, the fragment called Little Cumberland Island is entirely Holocene, as is the southernmost two miles of the main Island.

FIGURE 181. Vertical image of Cumberland Island. The Silver Bluff "core" of the Island, deposited 40,000-50,000 years ago, is outlined. All of the land seaward of this line is Holocene in age (i.e., <4,500 years old). Arrow A points to Crooked River State Park, arrow B points to the boat dock in St. Marys from which the ferry to Cumberland Island departs, and arrow C points to the Tabby Sugar Works. Remnants of the Pleistocene Princess Anne barrier island system (PA), formed 85,000 years ago, are outlined. Much of the upland north and northwest of St. Marys is a remnant of the Pleistocene Pamlico barrier island system (P), which formed 125,000-130,000 years ago (Huddlestun, 1986). Digital orthophoto acquired in March 2006, courtesy of USGS.

With regard to the overall geomorphology of this island, it is almost a mirror image of Jekyll Island that was discussed in some detail earlier (see especially Figure 172). It is considerably larger than Jekyll, possibly because of an abundance of sediments contributed to this part of the coast by the combined Altamaha, Satilla, and St. Marys Rivers. The backbarrier morphology of the island was discussed in some detail by Jackson, Alexander, and Bush (2007). In that study, they observed that *although the present database is limited to 10 historical shoreline positions, noticeable shoreline change trends emerged and are reasonably linked to various factors such as inlet dynamics, geologic framework, and human activity*. They pointed out that several archaeological sites on the landward side of the island are threatened by erosion.

Human Habitation

Some of the human history of this island, as outlined by Olsen (2006) and Sherpa Guides, follows:

1) The island was first inhabited by a Florida tribe of the Timucuan Indians, who called the island Missoe (meaning sassafras).

2) It was renamed San Pedro by the Spanish, who settled it, set up a mission, and occupied the island for more than a century.

3) It was named Cumberland in 1734 by an Indian chief and, in 1738, the British built Fort William at the southern end of the island and Fort St. Andrews at the northern end. General Oglethorpe built a hunting lodge near Fort William.

4) Between 1765 and 1768, James Cuthbert, Jonathan and Josiah Bryan, John Smith, and James Habersham were given royal grants on the island.

5) When the naturalist William Bartram visited the island in 1774, it was mostly uninhabited.

6) After the First Civil War, several plantations occupied the island, including the Stafford Plantation and Dungeness (home of General Nathaniel Greene).

7) In 1818, General Henry ("Lighthorse Harry") Lee, a friend of Nathaniel Greene, both being heroes of the First Civil War, died on the island and was buried there. His body was moved to Lexington, Virginia in 1913 to "lie beside his famous son."

8) Believe it or not, after the War Between The States, the plantations went out of business, and so did the famous son.

9) In the 1880s, the Pittsburgh millionaire Thomas Carnegie, brother of Andrew, bought part of the island for a winter retreat. His properties were passed down to succeeding generations of Carnegies.

10) About 1900, W. P. Bunkly built the Hotel Cumberland on the northern end of the island called High Point. This property later became the property of the Candler family of Atlanta.

11) Through some rather complicated maneuvers, some by environmentalists, portions of Cumberland Island were donated to the National Park Service in 1970 and, by an Act of Congress in 1972, it became a National Seashore. In 1982, the northern half of the island was added to the National Wilderness Preservation System.

12) Today, the main island is managed by the National Park Service. Little Cumberland Island, with a number of private residences on it, is managed by a property owner's association.

Places to Visit

What better place to visit than Cumberland Island itself? There are no roads to the island, thus visitors may reach it only by boat. The National Parks Service issues permits for day trips and short-term camping. As noted by Olsen (2006), *no more than 300 people are allowed to visit the island per day. There are still several private residences on the island and the Grey Field Inn, a private commercial facility, offers overnight accommodations.*

The most common mode of transportation to the island is by a ferry that leaves from the mainland Visitor Center dock in St. Marys. Some of the information regarding this mode of visiting the island, provided by nps.gov (2010), follows:

1) Visitors can make arrangements to visit the island through contacting the ferry reservation office as much as six months in advance;

2) Ferry fees: $17.00 Adult, $12.00 Child, $15.00 Senior. There are miscellaneous relatively small fees for visiting or camping on the island; and

3) We should point out that the island, which is open 24 hours a day, may also be visited by private boat. *Private boaters may dock at Sea Camp or Plum Orchard. Overnight boaters must anchor offshore* (Georgia Wildlife, DNR, 2002).

We have not spent as much time as we probably should have on these islands. Hayes has overflown them several times doing some preliminary mapping and training; Michel visited Cumberland Island on the ground designing protection strategies for oil spills, with both of our efforts obviously being business oriented. However, our illustrator, Joe Holmes, has visited the area on holiday and gives the following advice.

The Sea Camp Beach campground is a half-mile from the ferry dock, but large carts are provided for campers to transport their gear to the campsites and back. One of the most outstanding features on the island is the tangled canopy of enormous live oak trees. Walking the trails through the forest tunnels can be a very enchanting experience. Closer to the beach, the sand dunes have migrated into the trees burying their trunks and limbs by up to 50 feet. A short trail around the southern end of the island takes you to the Dungeness Ruins, one of the late 19th century mansions built by the Carnegies on Cumberland. It was built on the same site as the Dungeness Plantation originally owned by General Greene. The grounds around the ruins have now become pasture for the wild horses that roam the island. The trail continues past a tidal creek and marsh for some good wildlife viewing, eventually ending at the beach. Bicycles can be rented at the ferry dock if you want to explore more of the island.

As was reported by Clayton et al. (1992), Little Cumberland Island is owned by the Little Cumberland Island Association and is incorporated into the *Coastal Barrier Resources System. Three-quarters of the island is designated wilderness, while the remaining one-quarter is open to limited deed-restrictive development. The island has mostly low-density development with large individual lots. Homes are generally located some distance from the beach and are in little danger from erosion* (in the near future anyway; and, of course, unless a 20-foot hurricane storm surge occurs).

Not surprisingly, Cumberland Island is included in the Georgia Wildlife, DNR (2002) Colonial Coast Birding Trail. They reported that a total of 322 species of birds have been seen on the island. Specialties include: peregrine falcon, painted bunting, red knot, black skimmer, and warblers. Some tips they gave include: *Look for peregrine falcons during fall migration. Painted buntings are common in summer. ... Piping plovers may be spotted on the beach in winter. Bring along food, beverages, sunscreen, rain gear and other necessities because the island has no stores.*

With regard to the area in the lee of Cumberland Island, we recommend that you visit two areas: 1) The tabby ruins of the McIntosh Sugar Works Mill; and 2) Crooked River State Park. Two photographs of the **tabby ruins**, located by arrow C on Figure 181, are given in Figure 182. As described by Livingston (2010), *tabby is a concrete-like substance made of lime, water, powdered oyster shells, and crushed (but chunky) oyster shells.* Obviously, oyster shells were used in the absence of hard rocks or gravel, both of which are rare along the Georgia Coast.

These are the ruins of a sugar works built by John Houston McIntosh at New Canaan Plantation soon after 1825. This was the first horizontal sugar mill ever worked by cattle power (Georgia Historical Commission). He was an interesting guy. Before getting into the sugar cane business, *as a young man, he settled in East Florida and became a leader of the U.S. citizens living there. He was appointed "Governor or Director of the Republic of Florida" in 1812. After a storied career in Florida, he returned to Georgia, and in 1818, served in the Seminole War as General in the Militia* (GeorgiaInfo, 2010a).

After visiting the tabby ruins, we suggest that you continue north on GA 40 Spur for 3.7 miles, where you will reach the shore of the Crooked River, at which you can walk out on a platform by a boat ramp for some views up and down the channel, as well as a get a look at the cutbank shown in Figure 109.

The entrance to the 500-acre **Crooked River State Park**, also located on the bank of the Crooked River (Figure 181), is located near the boat ramp at Elliotts Bluff (at the north end of GA 40 Spur). This is a very nice place, if you are interested in camping and/or birding. A bird blind and a birding platform are available, and you could canoe and kayak to your heart's content in this park, which is located on the Pamlico Pleistocene barrier island system, according to Huddlestun (1986). Eleven cabins are available to rent for overnight stays.

[NOTE: There appears to be some disagreement on the age of the bluff here. According to the map by Hails and Hoyt (1972), the following areas are located on the Princess Anne barrier island system: 1) Part of St. Marys including the St. Marys Airport; 2) Western part of Kings Bay Naval Station; and 3) Elliotts Bluff and at least a good part of Crooked River State Park. However, as already noted, Huddlestun (1986) mapped this area as Pamlico, presumably based on more detailed data gathered later, that is probably the right age.]

Georgia Wildlife, DNR (2002) included this park in the Georgia Colonial Coast Birding Trail, observing

FIGURE 182. The Tabby Sugar Works, located by arrow C in Figure 181, was built by John Houston McIntosh at New Canaan Plantation soon after 1825. (A) Long view through several of the walls. The vertical length (height) of the opening is about 10 feet. (B) Close-up view showing the oyster shells used. Scale is ~6 inches. Both photographs were taken on 26 March 2010.

that it is blanketed with several habitats, including pine flatwoods, salt marsh and maritime forest. Their recommendations included *watch for mergansers and other ducks* on the Crooked River, and *listen and watch for painted buntings in spring and summer*.

By all means, visit the Park Office/Visitor's Center, where an excellent trail map, among other interesting items, is available. While in that area, you will not be able to miss the large pine cones scattered around on the ground under the very common longleaf pines (*Pinus palustris*). A photograph of two of these pine cones is given in Figure 183B.

When we were in the Visitor's Center on 26 March 2010, the following most interesting list of the birds seen in the park during that particular March was available:

1) At the Nature Center: Painted bunting, pine warbler, yellow-throated warbler, gray catbird, mourning dove, black and white warbler, downy woodpecker, and chipping sparrow;

2) At the bird blind: Red-winged black bird, chipping sparrow, white-throated sparrow, cardinal, chickadee, tufted titmouse, eastern towhee, red-bellied woodpecker, song sparrow, brown-headed cowbird, common grackle, and northern parula;

3) At the campground: Pileated woodpecker (?), and yellow-bellied sapsucker; and

4) Along the roadside: Eastern bluebird, redheaded woodpecker, ruby-crowned kinglet, and brown-headed nuthatch.

From a bird list obtained at the Nature Center, we learned that some notable breeding birds (spring/summer) include prothonotary warbler, hooded warbler, summer tanager, painted bunting, orchard oriole, wood thrush, and great-crested flycatcher. Forty-eight species are listed as common. In the park, you can see both birds associated with the salt water (e.g., northern gannet, brown pelican, white ibis, roseate spoonbill, black-necked stilt, great black-backed gull, Caspian tern, and black skimmer), as well as those commonly seen in the woodland setting (already listed). You have probably guessed by now that, as avid birders, we like this place.

There are three main hiking trails in the park, which are described in detail in a trail map and individual trail guides provided by Georgia State Parks & Historic Sites. Most of the following discussion is taken from those documents.

1) **Palmetto Trail (1.25 miles)** – This trail introduces you to the *south Georgia Pine Flatwoods, an ecosystem dominated by longleaf pine, with an understory of saw palmetto*. The longleaf pine, with its heartwood that is immune to termites and resists decay, has needles 10-15 inches long, the longest of any pine tree in the eastern U.S. The reddish-brown bark is fire resistant. *Early settlers distilled pine pitch* (from these trees) *into tar, turpentine and rosin*. Saw palmetto dominates the understory (see photograph in Figure 183A). *Its prostrate trunk lies close to the ground, protecting it from low-intensity fires*, as is illustrated in Figure 176B. *Fire is essential for maintaining this ecosystem, and its plants and animals are well adapted to periodic low-intensity fires. These open woods make it an ideal place to see birds, especially early in the morning or just before sunset*. Birds you may see have already been listed, except for the Georgia State Bird, the brown thrasher. Keep a lookout

FIGURE 183. Crooked River State Park (located by the arrow A in Figure 181). (A) The understory in the pines is saw palmetto (*Serenoa repen*). (B) Cones of the longleaf pine (*Pinus palustris*). Scale is ~6 inches.

for birds soaring overhead, such as the turkey vulture, osprey, and crow.

2) **Sempervirens Trail (1 mile)** – This trail introduces you to a *Coastal Shell Midden Forest. This hardwood forest is unusual because it grows on the decayed remains of oyster shells left by Native Americans hundreds of years ago. The calcium in the rich soil allows several species of plants to grow here that are unusual in South Georgia.* The name sempervirens means 'ever living.' *The huge oak, cherry and hickories in the Midden Forest are the oldest trees in the park.* Hollies, live oaks, and cedars are evergreens, so that even during winter, the forest is green and the canopy is full. Curiously, *little brown bats sometimes roost in the bat box. This bat is about 3 1/2 inches long with a 10-inch wingspan, a glossy brown back, and a grayish-brown belly. It weighs about 1/4 of an ounce. A single bat can eat 500 mosquitoes in an hour.* The soil in this forest was created hundreds of years ago, when the Guale people harvested shellfish from the river and discarded the shells into trash piles, called **middens**. *You can see gentle mounds on the forest floor to this day.*

3) **Bay Boardwalk (1.25 miles)** – We took this trail during our visit and found it to be most interesting. The boardwalk is only a foot or two off the ground in places, and it takes you into a bay forest ecosystem dominated by two species of broad-leafed evergreen trees, the loblolly bay (*Gordonia lasianthus*) and swamp bay (*Persea palustris*). Two things of special interest to us were: 1) Sapsucker holes, horizontal lines of small holes around the tree trunk a distance up off the ground. *These are pecked by a yellow-bellied sapsucker, a small woodpecker. It pecks holes, and eats insects such as ants attracted to the sap. Each sapsucker has a couple of 'favorite trees' within its territory*; and 2) The presence of the red maple (*Acer rubrum*), a common wetland tree with the usual maple leaf shape. Showy clusters of reddish flowers appear in early spring.

Therefore, we conclude that the Crooked River State Park is one special destination that any naturalist travelling in South Georgia would not want to miss.

ST. MARYS ENTRANCE

Introduction

St. Marys Entrance (Figure 184), located at the southern border of the state of Georgia, is the last of the eight major sounds on the Georgia Coast (going south). This is only the second major tidal entrance in Georgia that is "controlled" by jetties, the other one being the entrance to the Savannah River at the northern border of the state (Figure 86). With a width of only 0.6 miles, the entrance to this system is the narrowest of the eight sounds. As can be seen in Figure 184, during ebb tides, the water that flows through the jetties out into the ocean is derived from two primary sources: 1) The St. Marys River itself, the third largest coastal plain (black-water) river in Georgia (Table 3); and 2) The marsh/tidal channel complex in the lee of the southern half of Cumberland Island that is labeled Cumberland Sound on maps and charts (clearly visible on the image in Figure 181).

As shown on the image in Figure 184, the seaward end of the original valley carved by the St. Marys River during the last major lowstand in sea level about 20,000 years ago has been almost completely filled in with the sediment delivered by the modern river, mostly since sea level achieved its present position around 4,500 years ago. However, because these sediments are still confined to within the valley and no lobe of sediment projects out into the open ocean, it does not conform to the normal definition of a river delta building into the world ocean. As already discussed, the Savannah and Altamaha River Deltas have been so defined. Some geologists do refer to these deltas enclosed within the flooded river valleys as "bayhead deltas," but that is a technicality not worthy of discussion here. However, this deposited sediment load within the St. Marys River valley has created a fairly wide zone of spectacular salt and brackish marshes that show up clearly on the color vertical image in Figure 184.

Two kiosks in the town of St. Marys give the following information about manatees and whales in this area: (1) ***Manatees***. *They begin migrating into coastal Georgia in March and early April and remain in our area as late as November. They can be found in tidal creeks, rivers, and sounds along the entire coast of Georgia. They are typically found east of I-95 although they may venture further west in a few rivers and creeks. They feed mostly on smooth cordgrass (Spartina alterniflora), the dominant salt marsh grass, and other vegetation growing on the waters edge. They feed on these plants at high tide and are unable to reach them at low tide. Manatees also feed on the algae that grow on floating objects and in shallow areas;* and (2) **North Atlantic right whales.** *Only 300-400 of them remain, one of the most endangered of all large whales. They migrate between their feeding grounds in the northeastern United States and their only known calving*

FIGURE 184. St. Marys Entrance, showing the entrance jetties. The water that exits the jetties during low tides comes from both Cumberland Sound to the west of Cumberland Island and the St. Marys River. The arrow points to the waterfront of the town of St. Marys. Also shown are the locations of remnants of the Pleistocene Princess Anne barrier island system (PA), formed 85,000 years ago, and the Pleistocene Pamlico barrier island system (P), formed 125,000-130,000 years ago (Huddlestun, 1986). Digital orthophoto acquired March 2006, courtesy of USGS.

area, which is off the southeastern United States. They are usually seen here between December and March, and so they calve off the shore of South Carolina to the southern tip of Florida.

Places to Visit

The upper reaches of the St. Marys River part of the system is a little difficult to get to. However, a great view of the St. Marys River valley, covering about a 1.5 miles of the marshes still within the tidal zone, is provided by a drive on I-95 between exit 1 in Georgia and exit 130 in Florida. The wide channel of the St. Marys River is very beautiful at this spot. Looking on the images on Google Earth, one can see some evidence of rice farming upstream from the I-95 Bridge. One of the plantations, called White Oak, located 4 miles (as the crow flies) west of the I-95 Bridge, had 350 acres of rice paddies around 1860 (White Oak Plantation, 2010), that are still clearly visible on the Google imagery. According to House (1954), there were more than 6,000 acres of "tide-flow rice plantations" along the banks of the St. Marys River during the two decades preceding the War Between The States.

We also recommend that you visit the town of St. Marys itself. On the way, at the bridge over Burrells Creek, you will cross a pretty extensive black needlerush marsh. There is a lot of it! This is probably the case because it is a small tributary of the St. Marys River, so the salinity is not so high there.

The waterfront in St. Marys is definitely another place you wouldn't want to miss. This is where you get on the ferry to go out to the Cumberland Island National Seashore. The Cumberland Island Visitors Center, which has books, maps, etc., is located near the ferry landing.

However, St. Marys is an interesting destination even if you don't go out to the Seashore. This historic section has several architectural attractions, such as First Presbyterian Church founded in 1808. Other tourist attractions, such as a submarine museum, may also be of interest. A waterfront park, St. Marys Waterfront Park (adjacent to Howard Gilman Memorial Park, established in 2001), where you can walk along the edge of the wide smooth cordgrass marsh, as well as out onto a pier, is a great place to spend some time. A photograph of this marsh is shown in Figure 96.

15 THE OKEFENOKEE SWAMP

Being offset from the coast by over 40 miles, this phenomenal giant swamp, shown on the image in Figure 185, will not be treated in as much detail in this book as the other major environmental settings along the Georgia Coast have been. For details on many aspects of this wonderland, including specific places to visit, we recommend that you refer to the brief, but informative, discussion of the swamp in The Longstreet Highroad Guide by Lenz (1999).

The geological origin of the swamp, as illustrated in Figure 185, is that it formed in a low area landward of the oldest, and most landward, Pleistocene barrier island system, called the Trail Ridge in this area (formed during the sea-level highstand called the Wicomico by Georgia geologists). We also quoted Johnson et al. (1974) earlier as follows: *A great barrier island, now called Trail Ridge, formed in front of the area now know as the Okefenokee Swamp, which was then a very large salt marsh.* This process started (i.e., the barrier island was deposited) sometime between 1.0 and 1.7 million years ago according to Winker and Howard (1977), and that barrier island was formed when sea level was as much as 95-100 feet higher than it is today (Olsen, 2006). The Trail Ridge was said by Burdette et al. (2012) to be 2.2 million years old based on the ESROD technique, a sand-grain dating method.

The following items of interest about the Okefenokee were provided by discussions of the Okefenokee National Wildlife Refuge, that was created in 1937, by the U.S. Fish and Wildlife Service (2010b), and by the Georgia Wildlife, DNR (2002) treatment of Georgia's Colonial Coast Birding Trails:

1) The Indian word Okefenokee means the "**Land of the Trembling Earth**." This land trembles because it is underlain by a vast deposit of **peat**, a sediment composed of >70% organic matter (such as leaves, stems, etc.). After a period of weathering, this organic material becomes a waxy brown mass with the consistency of peanut butter. With little mineral-containing sediments coming into this low-lying area, and a period of perhaps more than a million years for the organic matter to accumulate, the peat deposits are widespread. A former colleague of ours at the University of South Carolina Geology Department, Dr. Art Cohen and his students, spent years coring and sampling these peat deposits. One reason for the scientific interest in this peat is that its origin is analogous in many ways to that of some of the widespread coal deposits found around the world.

2) This vast peat bog with its 70 islands is one of the oldest and best-preserved freshwater systems in America. Comprised of 354,000 acres of land, this is the largest national wildlife refuge in the eastern United States.

3) The longleaf pine, cypress, carnivorous sundews, and other plants make up different habitats from dry upland forests to open wetlands.

4) The swamps, forests, "prairie wetlands" and waterways, which are filled with alligators, provide habitat for more than 234 species of birds.

5) And, finally, there are many ways to explore the Okefenokee. *Guided boat tours take visitors through cypress forests, historic canals and open prairies. Water trails and platforms allow people to canoe for the day or stay overnight* deep within the huge wilderness. *Winding boardwalks and trails lead through unique habitats to observation towers and viewing platforms. Opportunities for nature photography, hunting and fishing are readily available.*

We have been birding in the Okefenokee Swamp, and this is an outstanding place to do so. Georgia Wildlife, DNR (2002) listed the following specialties: Sandhill crane, red-cockaded woodpecker, prothonotary warbler, northern parula, and Bachman's sparrow. They concluded further that *this is a great spot to view wading birds, waterfowl and alligators. Winter is the best time to view sandhill cranes, and spring is an ideal time to spot endangered red-cockaded woodpeckers along Swamp Island Drive.*

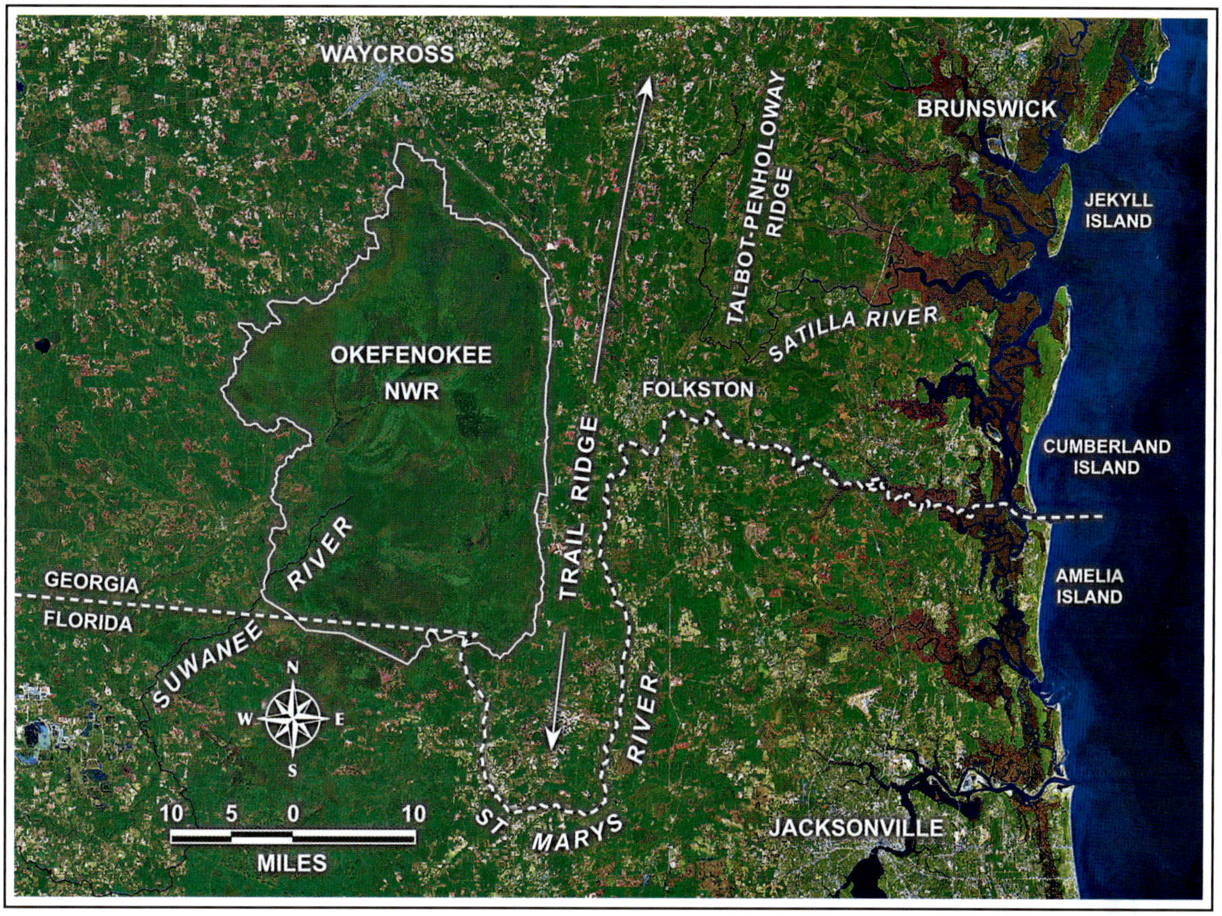

FIGURE 185. The Okefenokee Swamp and surroundings. This image illustrates the geological explanation of the origin of the swamp, which formed in the lee of the oldest, and most landward, of the Pleistocene barrier island systems on the Georgia Coast, the Trail Ridge. This barrier island system was proposed to have been deposited around 1.0-1.7 million years ago (by Winker and Howard, 1977). Our interpretation of the origin of the swamp is based on a sketch map by Pirkle, Pirkle, and Reynolds (1993). Two well known rivers have headwaters in the swamp: 1) The St. Marys, which flows south for 16-17 miles (straight-line distance) before turning east and flowing along a valley through the Trail Ridge system then turning back to the north; and 2) The Suwanee River, which flows south to the west Florida shore. Landsat image mosaic acquired 1999-2001, courtesy of Earth Science Data Interface at the Global Land Cover Facility.

High priority habitats associated with this ecosystem complex include *freshwater "prairies," pine flatwoods, pondcypress savannah, wet pine savannahs, titi swamp, herb and shrub bogs.* Examples of rare species known from the Okefenokee Swamp include *sandhill crane, Sherman's fox squirrel, flatwoods salamander, Florida water rat, striped newt, wood stork, Florida black bear, Rafinesque's big-eared bat, Florida orange grass, and Okefenokee giant pitcher plant* (Southern Coastal Plain, 2010).

This is a special place, and we hope you will be able to find the time to visit there soon.

16 THE FUTURE OF THE GEORGIA COAST

A little over fifteen years ago, Hayes (1994) published a paper on the barrier islands of the Georgia Bight. That paper had a closing statement entitled "Future of the Barrier Islands." As the future of the barrier islands goes, so goes the shoreline of the Georgia Coast. The following discussion is a reproduction of the somewhat dire predictions made in that paper. Nothing much has happened in the past fifteen years that would change these deductions.

As evidence continues to accumulate in favor of the global warming hypothesis, a concept no longer in doubt, by the way, the prognosis for the future of the barrier islands of the Georgia Bight is not good. Based on a mapping study to determine the potential impact of three levels of projected sea-level rise in the Charleston Harbor area, Kana et al. (1984) (a paper co-authored with the writers of this book) determined that Sullivans Island, a mixed-energy, prograding barrier island located just north of the harbor, would be converted to a narrow, landward-migrating barrier island around the year 2075 under the maximum sea-level rise scenario considered (7 feet in 80 years, a number considerably higher than the predictions we commonly make now – 2.5 to 3.5 feet in 100 years). This heavily populated island is now about 2,000 feet wide. It, and all the other prograding barrier islands in the state of South Carolina, would be eroded away under such conditions of rapid sea-level rise. The Pleistocene (Silver Bluff) heart of the bigger islands to the south in Georgia would no doubt convert to transgressive barrier islands more slowly because of the added volume of sediment to remove and the elevation of the sand. The new, landward-migrating (transgressive) barrier islands would be lodged against the interfluves (high areas between original river channels) between the expanded estuaries, and the waves would once again carve serious scarps into the Pleistocene cores of Hilton Head Island, St. Catherines, St. Simons, Sapelo, Jekyll, and Cumberland Islands, just as they did around 5,000 years ago.

Furthermore, sand supply, the other half of the erosion/deposition equation, has been significantly impacted already in some localities by power dams and jetty systems. Also, under the rapid sea-level rise scenario, many of the important offshore sand sources, such as submerged delta lobes and major ebb-tidal deltas, could no longer contribute to the coastal sand budget as much as they formerly did because of the increasing water depths.

All but three of the barrier islands described in this book (Tybee, St. Simons, and Sea Island) are at least partially protected as national seashores, state parks, state and national wildlife refuges, and other environmentally oriented holdings. With regard to private property and other human development, these areas will not be a cause of serious concern, even if sea level does rise significantly and all the prograding islands are converted to landward-migrating washover terraces, eroding Pleistocene scarps, etc., except for the loss of wetlands and overwintering habitat for waterfowl. However wetland loss might be compensated for by the possible growth of wetlands in the expanded sounds and estuaries. On the other hand, significantly eroding beaches would also result in turtle nest failures, loss of habitat for shorebirds, eroding marshes, etc.

The developed part of the coast is another story, and that is unfortunate, because there has been considerable progress made in the past 20 years of dealing with beach erosion problems on the developed shores of the Georgia Bight, principally in two areas:

- Fairly stringent beach-protection and conservation laws have been passed. However, some of these types of legislation are not being seriously implemented, as yet, and some have been challenged legally, even to the level of the U.S. Supreme Court. Classic setbacks

have been known to happen, such as the state of South Carolina's decision to basically abandon all regulation and encourage the construction of an "emergency dune" 9 feet high along 63 miles of the developed coast of the state a few days after hurricane *Hugo* (1989). Most of the sand was scraped from the lowest intertidal zone. As we write this on 29 May 2010, the governor of the state of Louisiana has been advocating, and even starting, a "sand berm" along the Louisiana Coast to "protect the marshes" from the oil being leaked by the *Deepwater Horizon* blowout. No serious, informed scientific planning that considers how barrier islands actually function was needed, it would seem. Typical. Incidentally, co-author Michel had to spend so much time in Louisiana planning cleanup strategies for that spill that it slowed down the completion of this book.
[NOTE: Now, one year and four months after the spill (August, 2011) it has been clearly shown that this "sand berm" was an abominable waste of time and money. No science needed, indeed.]

- There has been a move toward "soft engineering" solutions to erosion problems in developed areas, rather than building of hard structures, such as groins, seawalls, and rock revetments. An example of a soft solution is a project managed by our group at Captain Sams Inlet, South Carolina, where a huge volume of sand was made available to a severely eroding beach by relocating a small tidal inlet (Kana and McKee, 2003; Hayes and Michel, 2008).

And, finally we add the following comment by Clayton et al. (1992): *The future of the Georgia barrier islands, especially the developed islands of Tybee, Sea, St. Simons, and Jekyll, depends on the ability of the state, counties, and local communities to evaluate the needs of and potential challenges to the islands. Differences in local laws and their implementation strongly affect the quality of life on the individual barrier islands.* This is basically an appeal for stronger management strategies to cope with *beach erosion, coastal storms, construction and post-storm reconstruction, beach access, and sea-level rise*, some of which is underway at the present.

Under rapidly rising sea level, however, it seems likely that planners will take a more defensive posture and that the recent advances mentioned earlier will be lost. Major defensive engineering structures will be built and many of the features we now know as barrier islands will become a series of isolated, heavily armored headlands. That is, unless the world's population gets serious about the problem of global warming, which doesn't seem very likely at this point.

RECOMMENDED ADDITIONAL READING

Bell, M., Jr., 1987. Major Butler's Legacy: Five Generations of a Slaveholding Family. Univ. of Ga. Press, Athens, GA, 673 pp.

An amazing account of life on the Georgia Coast, starting around 1770 and continuing through and for some time after the War Between The States. Excellent detailed account of the question of slavery, the people who caused it, and those that endured it. If you ever travel along the coast in the vicinity of the Altamaha River Delta, you will find this book to be an excellent companion.

Berendt, J., 1997, Midnight in the Garden of Good and Evil. Vintage Books, 400 pp.

Good book to read to get you in the right frame of mind for visiting the historic districts of Charleston and Savannah (the setting for the book). The makers of a movie by the same name went out of their way to avoid doing justice to this book.

Clayton, T.D., Taylor, L.A., Jr., Cleary, W.J., Hosier, P., Graber, P.H.F., and Pilkey, O.H., 1992. Living With the Georgia Shore. Duke Univ. Press, 296 pp.

One of the earlier books in the Pilkey series on "Living With the Coast." Detailed discussions on the origin of the islands, specifics on erosional history of different areas, and so on. This is definitely a book anyone owning property on this coast should have.

Davis, R.A. Jr., and FitzGerald, D.M., 2004. Beaches and Coasts. Blackwell Science Limited, Oxford, U.K., 419 pp.

The best modern textbook for giving the novice an introduction to the physical processes of the coastal zone.

Gore, A., 2006. An Inconvenient Truth. Rodale Pub., Emmaus, PA., 325 pp.

Well-illustrated discussion of most of the issues related to global warming, such as melting of glaciers and ice fields and increased CO_2 in the atmosphere.

Hatcher, R.D., Jr., 2005. Southern and Central Appalachians. In: Selley, R. C., Cocks, R., and Plimer, I. (eds.), Encyclopedia of Geology. London, Elsevier Ltd., pp. 72-81.

Beautifully illustrated and thorough treatment of the origin of the Appalachian Mountains.

Hobbs, C.H., 2012. The Beach Book, Science of the Shore: Columbia Univ. Press, 195 pp.

Recently published thorough coverage of many aspects of beaches from a global perspective, including issues related to beach erosion.

Kana, **T.W., 2011. Coastal Erosion and Solutions: A Primer. 2nd Edition. Coastal Science and Engineering, Inc., Columbia, SC, 37 pp.**

Concise introduction to the problem, with emphasis on erosion on the South Carolina Coast. Many well-illustrated examples of the processes and products of coastal erosion in the state.

Lenz, R.J., 1999. Longstreet Highroad Guide to the Georgia Coast & Okefenokee. Longstreet, Atlanta, GA, 336 pp.

Excellent detailed discussions of all sections of the Georgia Coast and the Okefenokee Swamp. Numerous drawings of the different biota one might see during a visit to the coast. Great source of information on a variety of topics.

Pilkey, O.H., and Fraser, M.E., 2003. A Celebration of the World's Barrier Islands. Columbia Univ. Press, New York, 309 pp.

Beautifully illustrated coverage of barrier islands in general, with several references to those in the Georgia Bight. Excellent introduction for one just getting interested in barrier islands.

Sibley, D.A., 2003. The Sibley Field Guide to Birds of Eastern North America. A.F. Knopf, New York, 347, pp.

This is the field guide we use. Outstanding coverage with beautiful artwork.

Tiner, R.W. and Rorer, A., 1993. Field Guide to Coastal Wetland Plants of the Southeastern United States. Univ. of Massachusetts Press, 336 pp.

Covers the wide variety of plants present in the coastal zone of Georgia.

GLOSSARY

Amphidromic system – Separate, more-or-less circular, gyrating tidal systems (standing waves) with a nodal point in the middle. They are a large rotary tidal system created by the behavior of tides as giant low waves that are affected by the shape of the world's oceans and the rotation of the Earth. The height of the tide depends on its location within the system. Tides at the axis of the system (nodal point) are nil, but they increase outward with distance from the axis. In the Northern Hemisphere, the tides rotate counterclockwise about the axis; in the Southern Hemisphere, the rotation is clockwise. Both diurnal (one complete rotation of the tides daily) and semi-diurnal (two complete rotations of the tides daily) can coexist in the same ocean basin.

Antidunes – Linear and parallel, trochoid-shaped features typically composed of sand generated by rapid flow of water in the upper flow regime. In the process, a relatively thin layer of rapidly moving water flows as a standing wave across the top of the antidune surface. The standing waves in the water are in phase with the sediment surface. The form itself moves up current, but the individual sand grains move down current in a series of stops and starts. They have a typical spacing of about 15-20 inches on beaches, where they are commonly associated with flat beds. See Figures 73 and 74.

Asthenosphere – Part of the Earth's mantle below the lithosphere that behaves in a plastic manner. The rigid and more brittle lithosphere rides over it (Hyndman and Hyndman, 2006).

Astronomical tides – Periodic changes in water levels caused primarily by the gravitational attraction of the Moon and Sun on the world ocean. See Figures 20, 21, and 22.

Backwash of a wave – That portion of the water that rushes up the beachface under the impetus of a breaking wave, doesn't percolate into the beachface, and returns back down the slope of the beachface under the influence of gravity.

Barrier islands – Elongate, shore-parallel accumulations of unconsolidated sediment (usually sand), some parts of which are situated above the high-tide line (supratidal) most of the time, except during major storms. They are separated from the mainland by bays, lagoons, estuaries, or wetland complexes and are typically intersected by deep tidal channels called tidal inlets. They do not occur on coasts with a tidal range greater than 12 feet (macrotidal shores). See Figures 35 and 36.

Barycenter – The center of mass of the combined Earth-Moon system. The Earth and the Moon revolve together around the barycenter, which is displaced a distance from the center of the Earth toward the Moon, and is always located on the side of the Earth turned momentarily toward the Moon, along a line connecting the individual centers of mass of the Earth and Moon (Wood, 1982). See Figure 21.

Base level – The lowest level to which a river flows.

Beach – An accumulation of unconsolidated sediment transported and molded into characteristic forms by *wave-generated* water motion. The landward limit of the beach is the highest level reached by average storm waves, exclusive of catastrophic storm surges, and the seaward limit is the lowest level of the tide. See Figures 62, 63, and 66.

Beach cusps –The scalloped seaward margin of a beach berm consisting of a regularly spaced sequence of indented bays and protruding horns of a relatively small scale (rarely more than a few tens of feet between horns). They are thought to be primarily depositional features (i.e., on the seaward flank of a prograding depositional berm) that form as a result of the interaction of shore-normal standing waves (edge waves) with shore-parallel, reflective waves, the same general process that produces rip currents (but at a smaller scale) (Guza and Inman, 1975). See Figure 29.

Beach drift – The transport of sediment along the beachface by the combined processes of uprush and backwash. The uprush from a wave that approaches the beachface obliquely will move sand grains (or pebbles) at an angle across the beachface. Once the approaching uprush loses its momentum, the sand grains (or pebbles) are transported in a perpendicular direction back down the beachface by the backwash under the influence of gravity. The ultimate result is that the sediment particles move along the beach face in a sawtooth pattern away from the direction of wave approach. See Figure 28 (lower left).

Beach nourishment – Artificially adding sand to a beach to counter beach erosion. The sand can come from a variety of sources, with offshore sand bars being a common one.

Beach ridges – Linear lines of ridges of sand that typically occur on prograding barrier islands and wave-dominated deltas. They started as abandoned foredunes that were originally located immediately landward of the high tide line at the beach, hence the term. See Figures 34C, 83, and 84.

Beachface – The zone of wave uprush and backwash during mid to high tide. It usually slopes seaward at an angle of 5-10 degrees and is commonly the seaward face of the berm. See Figures 63 and 67.

Beachrock – Solid rock usually formed recently by rapid solidification of beach sand through the precipitation of calcium carbonate from sea water into the pore spaces between the sand grains, a process more likely to occur where the water is warm.

Bedform – General term for features formed on a sand bed by moving water, such as ripples and sand waves. See Figure 68.

Berm on a sand beach – Wedge-shaped sediment mass built up along the beach by depositional wave action in between storms. It typically has a relatively steep seaward face (5-10 degrees; the beachface) and a gently-sloping landward surface (2-3 degrees; the berm top). See Figures 63 and 67.

Boulder – Sediment particle that is greater than 256 mm (~10 inches) in size.

Brackish marsh – Low-lying wetland containing grassy vegetation that is regularly flooded by brackish waters during mid to high tides. Salinities average less than 15 parts of "salts" per thousand parts of water (ppt).

Breakwaters – See offshore breakwaters in this glossary.

Bubble sand – Sand with a high degree of open pore spaces suggesting a sponge-like texture. It is best developed on upper intertidal sand flats and beachfaces where air is trapped between the water table in the sand and ocean water coming over the surface as a result of the rising tide. The water table in the beach and tidal flat sediments is elevated more slowly than the ocean surface is during the rising tide, which creates the trapping mechanism (air between two distinct layers of water saturated sand; Reel, 1982). See Figure 81.

Cenozoic era – Period of geological time that extended from 65.5 million years ago to the present. It is subdivided into three periods, Paleogene, Neogene, and Quaternary (Table 1).

Closure point – Term defined by coastal engineers as the most seaward distance for sand to move into the nearshore zone (off the intertidal beach in an offshore direction) during storms. In effect, it is the seaward limit for sand to move during the normal beach cycle that takes place in response to the passage of storms and intervening periods of relative quiescence. Many coastal geologists and some engineers dispute the validity of this concept. See Figure 62.

Coastal geomorphology – Scientific discipline focused on the understanding of how the different coastlines of the world originated, as well as their resulting forms and organization.

Coastal Plain Physiographic Province in Georgia – The portion of the state seaward of the fall line. It has a maximum width of about 200 miles and covers over 60 percent of the state's area. It has some relief in its upper portions, with hills that reach a maximum elevation of around 750 feet. However, the lowest portions of the Coastal Plain, with their surface sediments of Pleistocene and Holocene age, are considerably flatter. See Figure 6.

Coastal plain river – In the Georgia Bight, it is defined as a river that has most of its watershed contained within the Coastal Plain Physiographic Province. Sometimes called black-water rivers, because much of its waters are derived from *cypress/tupelo* swamps, which contain tannic acid and suspended organic matter, which gives it a black or dark brown color. See Figures 7 and 110B.

Cobble – Sediment particle that is between 64-256 mm (2.5-10 inches) in size.

Collision coasts – Coasts defined by Inman and Nordstrom (1971) as those that occur along the leading edges of continental plates, such as the ones along much of the west coasts of North and South America. They usually are mountainous and commonly contain rocky shores with short, steep rivers emptying into the sea. See Figure 17.

Competency of flow – The ability of the moving water to transport sediment.

Conglomerates – Sedimentary rocks made up mostly of gravel-sized particles or clasts.

Continental drift – Theory proposed by Alfred Wegener (1912) that states that the present continents on the Earth's surface were joined together and later moved apart, based to some extent on the almost perfect fit of the east coast of South America with the west coast of Africa, plus other comprehensive evidence. Lacking in his theory was a provable mechanism to move the continents apart. See Figure 3B.

Continental shelf – The gently sloping seabed between the beach and the shelf break, where there is an abrupt increase in slope, at depths measured in hundreds of feet. See Figures 116 and 117.

Continental slope – Steeply descending planar surface that adjoins the outer margin of the continental shelf. See Figure 116.

Core of the Earth – The central, interior portion of the Earth that is composed mostly of iron and makes up approximately 30% of the Earth's volume. It consists of two layers, an outer extremely hot zone, which is molten, and in inner layer, also very hot, that is solid because of the intense pressure at those depths within the Earth. See Figure 16.

Coriolis effect – An effect whereby a body of air, water, etc. moving in a rotating frame of reference experiences a force that acts perpendicular to the direction of motion and to the axis of rotation. On the rotating Earth, this effect deflects moving bodies to the right in the northern hemisphere and to the left in the southern hemisphere. It was named after a French engineer G. G. Coriolis (1792-1843).

Crust of the Earth – The solid upper layer of the Earth composed of a variety of igneous, metamorphic, and sedimentary rocks. The crust under the oceans, which averages about 6 miles thick, is composed mostly of rocks that contain an abundance of iron and magnesium-bearing minerals (e.g., basalt), whereas the crust on the continents, which averages 20-30 miles thick, is composed of rocks made up of lighter minerals (e.g., granite) (Wikipedia encyclopedia). See Figure 16.

Cut bank – The side of a meandering river channel on the outer flank of a meander bend where the current undermines the bank and causes it to erode. Typically, the deepest part of the channel is located near the cut bank, where the maximum amount of erosion is taking place. See Figure 105A.

Cypress-tupelo swamp – Zones II and III of the bottomland hardwoods that occur on flood plains of rivers in the coastal plain of the southeastern United States. These two zones make up the most frequently flooded areas of the flood plains (more than 50% of the years of record and typically >25% duration during the growing season). These swamps are dominated by two readily recognizable tree species, bald cypress (*Taxodium distichum*) and water tupelo (*Nyssa aquatica*). See Figures 114 and 115B.

Delta – An irregular progradation (usually out into the open ocean in the case of the Georgia Bight) of the shore at the mouth of a sediment-laden stream. See Figures 60 and 39B.

Delta front – The seaward margin of the major piedmont river deltas in the Georgia Bight. It consists of sand-dominant bulge in the shore, composed of barrier islands and tidal inlets.

Depositional coasts – Coastal type defined by Hayes (1965) as one characterized by: 1) Coastal zone made up chiefly of broad coastal and deltaic plains; 2) Bedrock composed generally of Tertiary and Quaternary sediments (young geologically speaking); 3) Tectonically subsiding areas; and 4) Many large and long rivers emptying into the sea. They most commonly occur on the trailing edges of continental plates or along the coasts of epicontinental seas, such as the Black and Caspian Seas.

Diurnal inequality of the tides – Condition where, although the tides are roughly semi-diurnal, four significantly different levels (rather than two) occur during one tidal day (referred to as higher-high, lower-high, higher-low and lower-low tides). One possible cause of this is the tilt of the Earth's axis, because the depth of the water under the tidal bulge will be different on opposite sides of the Earth when the Moon is not directly over the equator (Figure 21C). Another probable cause is a combination of tides from separate diurnal and semidiurnal amphidromic systems, which appears to the case on the Central California coast (Ed Clifton, pers. comm.).

Downdrift – The direction in which the individual sediment grains are moving along the shore within the longshore sediment transport system. See Figures 30 and 31.

Ebb-tidal currents – Currents generated during the falling tide.

Ebb-tidal delta – Lobe of sediment, usually composed of sand, much of which has been transported from the scoured-out throat of a tidal inlet by ebb-tidal currents. The sand is deposited on the seaward side of the inlet throat as the ebb currents slow down upon entering the open ocean. The morphological components of ebb-tidal deltas include a main ebb channel (with slightly stronger ebb currents than flood currents) flanked by linear sand bars on both sides and a terminal lobe at the seaward end. The main ebb channel is bordered by a platform of sand (swash platform) dominated by swash bars (intertidal bars built up by wave action). The swash platform is separated from adjacent beaches by marginal flood channels (which are dominated by flood currents). See Figure 52.

***El Niño* event** – A globally coupled, poorly understood ocean-atmosphere phenomena that results in: 1) Relatively low pressure over the Pacific Ocean; 2) A weakening of the trade winds (during normal conditions, these strong winds generate a very strong westerly flow of the surface waters to the west in the equatorial region) that contributes to an increase in water level on the California Coast (as much as 8-12 inches); and 3) The thermal expansion of the warmer water because of a decrease in upwelling along the coast. During these events, unusually heavy rains typically accompany an increased number of winter storms. More extratropical storms and rainstorms occur along the Georgia Coast during *El Ninos*.

Ensoulment – Concept that suggests that humans may participate with the Earth as if it were a living being (Cajete, 1995).

Entrainment – The selective transport of sediment particles by water or air.

Environmental Sensitivity Index (ESI) – System for mapping coastal environments, the first version of which was proposed for the coast of Lower Cook Inlet, Alaska by Hayes, Michel, and Brown in 1977. It ranks coastal habitats on a scale of 1-10 based upon their sensitivity to oil-spill impacts. Exposed rocky shores are ranked 1 (least sensitive) and sheltered salt marshes are ranked 10 (most sensitive). ESI maps, used for oil-spill response and other types of environmental assessments on a worldwide basis, also include detailed biological and human-use information.

Epiphyte – A plant, such as a tropical orchid or a staghorn fern, that grows on another plant upon which it depends for mechanical support, but not for nutrients, such as Spanish moss (Free Online Dictionary).

Estuary – Coastal water body that, according to Pritchard (1967), has three defining qualities: 1) A flooded river valley that was formed during the lowstand of sea level that terminated about 12-16 thousand years ago; 2) A water body with a substantial freshwater influx; and 3) A water body subject to tidal fluctuations. See Figure 57.

Evaporates – Class of sedimentary rocks usually, but not always, formed as the result of evaporation of seawater, typically under high surface temperatures. Evaporite deposits are most commonly composed of halite (NaCl) and gypsum [$CaSO_4 \cdot (2H_2O)$].

Fall line – A line of contact, usually between the Coastal Plain and Piedmont Physiographic Provinces, where stream courses descend abruptly as rapids cutting into the older bedrock of the Piedmont Province and flow on into the lowlands of the Coastal Plain Province, to never again erode into material older than Cretaceous sedimentary rocks. The gradients of the streams in the coastal plain are also considerably less than those in the piedmont. See Figures 1 and 7.

Fetch – The length of water surface over which the wind blows to form waves. See Figure 24B.

Flat bed – The surface of a sand bed created when sand grains follow long, linear paths typically under strong currents in shallow water, such as typically occurs during wave uprush and backwash along the surface of the beachface (Froude numbers of ≥1.0). This results in stacked horizontal layers of sand and smooth, relatively flat surfaces on the beach.

Flocculation – Process that takes place in estuaries when colliding clay particles tend to adhere together under the influence of the increased concentration of cations, such as Na+, Ca++ and Mg++, derived from the sea water. As a result, the clay particles, which may have diameters as little as one micron, cluster into groups (**flocs**) that may have diameters measured in hundreds of microns. Many of the flocs commonly sink to the bottom at slack tide. See Figure 57B.

Flood plain – Many river channels run through valleys of different widths that were carved during the lowstand of sea level during the last major glaciation that terminated about 16,000 years ago. During floods, sediment rich (muddy) water spills out of the channel and builds up a flat plain between the channel and the sides of the valley. These flat-lying deposits occupy a zone known as the flood plain. See Figures 111 and 112.

Flood-tidal currents – Currents generated by the rising tide.

Flood-tidal delta – Lobe of sediment, usually composed of sand, most of which has been transported from the scoured-out throat of a tidal inlet by flood-tidal currents. The sand is deposited on the landward side of the inlet throat as the flood currents slow down upon entering the more open bay on the landward side of the inlet. See Figure 50.

Flow regime – A concept worked out by hydraulic engineers to distinguish among features that develop in sediment (most appropriate for sand) along the bed of a stream (or beach, etc.) under the influence of flowing water. When the water is flowing slowly (or is relatively deep), the bed changes from small ripples to larger megaripples and sand waves (asymmetric features that move in the same direction as the current is flowing) under increasing flow strength. This first part of the sequence, during which the value of the Froude number (the ratio of water velocity to the square root of the gravitational constant times water depth) is <1.0, is called the **lower flow regime**. As the Froude number increases to values >1.0, by either increasing the velocity of flow or decreasing the water depth, the sand bed goes into the **upper flow regime**, where either flat beds or antidunes are formed.

Foraminifera – Small (usually less than 1 mm in size), single-celled marine organisms that float freely in the ocean in large numbers. They secrete a calcareous shell, or test.

Foredunes – Wind-blown sand dunes that form immediately adjacent to the backbeach. They are most commonly vegetated. See Figures 83 and 84.

Gaia hypothesis – An ecological theory that proposes that the planet Earth functions as a single living organism (Lovelock, 1972).

Gastropod – A univalve; a class of mollusks typically having a one-piece coiled shell and flattened muscular foot with a head bearing stalked eyes (WordNet Search).

Geomorphology – Scientific discipline devoted to understanding the origin and three-dimensional shape of the landforms of the Earth.

Georgia Bight – Major, landward-bending arc in the coast of the southeastern U.S. that extends for over 740 miles between Cape Hatteras, North Carolina and Cape Canaveral, Florida. Also called the South Atlantic Bight. See Figure 4.

Ghost crab (*Ocypode quadrata*) – Fast moving crabs that dig J-shaped burrows at 45 degree angles to depths of three feet or so just above the high tide line on the sand beaches of barrier islands. *Despite this semi-terrestrial existence, the ghost crab's link to the sea is a critical one that provides not only moisture for its gills, but also the necessary means for development of larvae that are hatched from eggs that the females release at the water's edge* (Knott, 2010). See Figure 77.

Ghost shrimp (*Callianassa major*) – Burrowing shrimp about four inches long that creates a catacomb-like system of burrows at depth below the surface of the mid to lower intertidal zone (and the nearshore zone) on the sand beaches of barrier islands. See Figures 78 and 79.

Glacial rebound coasts – Coasts that continue to rise even to this day as a result of the removal of the vast ice sheets associated with the last Pleistocene glaciation.

Global warming – The concept that at the present time, and in the relatively recent past, a global warming trend is in progress. Evidence for this includes: a) Rapid melting of glaciers around the world as well as melting of ice in the Arctic Ocean; b) Actual temperature measurements; and c) The steady increase in the concentration of CO_2 in the atmosphere. Some prefer the term climate change. See Figure 15.

Gondwana – Super continent that existed between 300-400 million years ago. It consisted of the present areas of South America, Africa, India, Australia, and Antarctica. It was separated from another supercontinent, Laurasia, by the Rheic Ocean. See Figure 3A.

Grain flow – The movement of sediment under gravity where the sediment motion is caused by direct grain-to-grain contact (encyclopedia.com). This differs from suspension transport, where the sediment moves under the influence of gravity while being carried in suspension by turbulent flow in a body of water.

Granite – Light-colored, coarsely crystalline igneous rock that is composed of 20-60% quartz, with most of the remaining component being between 10 and

65% of a mixture of alkali feldspar and plagioclase. It may also contain a few dark-colored minerals, such as hornblende and biotite.

Granule – Sediment particle that is 2-4 mm in size.

Gravel – Sediment with a median diameter greater than 2 mm. Gravel is defined according to the Wentworth (1922) scale, which includes four classes – *granules* (median diameter = 2-4 mm), *pebbles* (median diameter = 4-64 mm), *cobbles* (median diameter = 64-256 mm), and *boulders* (median diameter >256 mm).

Groin – Linear, relatively short, shore-protection structure built perpendicular to the beach for the purpose of widening the beach by trapping a portion of the sand in the littoral drift. See Figure 87.

Half-life – The amount of time it takes half of a radioactive parent element to decay to a daughter element. For example, it takes 1.25 billion years for half of Potassium 40 (the radioactive parent) to decay to Argon 40 (its daughter).

Halophytes – Plants adapted to grow in salty soil (Basan and Frey, 1980).

Heavy minerals – Minerals with specific gravities significantly greater than quartz, which has a specific gravity of 2.65 (the specific gravity of water is 1.00; i.e., a cubic centimeter of water has a mass of 1 gram). Examples include magnetite (specific gravity of 5.1), garnet, and hornblende. See Figure 76.

High marsh – Zone between neap high tide and spring high tide populated by a variety of plants, including glasswort (*Salicornia virginica*), sea oxeye (*Borrichia frutescens*), and salt grass (*Distichlis spicata*) in the southeast U.S. See Figures 58 and 92.

Highstand – The period when sea level was high for a significant amount of time, possibly measured in thousands of years, as during intervals between the four major glaciations of the Pleistocene epoch, called interglacials.

Holocene epoch – The segment of geological time that has occurred since the end of the last glaciation, starting about 12,000 years ago (Table 1).

Hornblende – Dark-colored silicate mineral of the ferromagnesium class that is a common constituent of both igneous and metamorphic rocks with a range in compositions.

Hurricane – Storm of tropical origin with a cyclonic wind circulation (counterclockwise in the Northern Hemisphere) of 74 miles per hour (mph) or higher (≥12 on Beaufort scale) (Dunn and Miller, 1960, p. 9). The North Atlantic member of the tropical cyclone family that includes the "typhoon" of the western North Pacific, the "cyclone" of the Indian Ocean, and the "willy-willy" of Australia (Tannehill, 1956). Tropical cyclones are the most powerful and destructive of all storms.

Hydraulic equivalence – Size–density relationship that governs the deposition of mineral particles from flowing (or standing) water. Two particles of different sizes and densities are said to be hydraulically equivalent if they are deposited at the same time under a given set of conditions (Encyclopedia Britannica; e.g., if dropped into a cylinder of water, the different particles would settle to the bottom at the same speed despite their differences in size).

Hydrodynamic regime (*hydro* = water; *dynamics* = kinetic energy of the water) – For purposes of interpreting the morphology of coastal features, the hydrodynamic regime is commonly expressed as the ratio of wave energy to tidal energy (i.e., how big the waves are versus how large the tidal range is). It exerts a major control on the geomorphological makeup of depositional coasts.

Hydrograph – Graphical representation of stage, flow, velocity, or other characteristics of water at a given point along a river as a function of time (Answers.com).

Igneous rocks – Rocks that crystallized from an extremely hot molten mass of material (magma). One class of igneous rocks, called granitic rocks (e.g., granite and diorite), develops by slow cooling of the molten mass at great depths within the Earth's crust. As a result of the slow cooling, large crystals of individual minerals form that can be as much as an inch or so in diameter. A second class of igneous rocks, called volcanic rocks (e.g., basalt and rhyolite), crystallize rapidly as a result of molten lava being extruded suddenly onto the Earth's surface. The crystals of these minerals that form so rapidly are, for the most part, too small to be seen by the naked eye.

Ilmenite – Black heavy mineral that is rich in titanium (Fe_2TiO_3). An important economic mineral that occurs in some abundance in Pleistocene barrier island systems, such as the Trail Ridge deposits along the Georgia/Florida border.

Intertidal bar – Asymmetrical, linear sand bar that occurs in the intertidal zone of sand beaches. They are initiated in the shallow nearshore zone by breaking waves and migrate across the intertidal zone until they finally weld against the beachface of a high tide berm. Earlier terminology referred to these features as ridges. See Figures 63B and 64.

Intertidal trough – Topographically low area on the landward side of an intertidal bar. Earlier terminology referred to these features as runnels, as in ridge-and-runnel systems. See Figures 63B and 66.

Invertebrates – Animals that do not have an internal skeleton made of bone, such as anemones, sea stars, jellyfish, clams, snails, crabs, shrimp, and worms.

Jetties – Long parallel structures built at a river mouth, a tidal inlet, or even an artificially dug harbor, to stabilize the channel, prevent its shoaling by sediments, and protect its entrance from waves. They are commonly composed of large pieces of rock of boulder size called riprap. See Figure 86 and 184.

Laurasia – Supercontinent that existed between 300-400 million years ago. It consisted of the present areas of North America, Greenland, Europe, and part of Asia. It was separated from another supercontinent, Gondwana, by the Rheic Ocean. See Figure 3A.

Law of Superposition – Sedimentary layers are deposited in a time sequence, with the oldest at the bottom and the youngest at the top.

Limestones – Sedimentary rocks composed of calcium carbonate that usually develop in marine waters by a combination of chemical precipitation and accumulation of the hard parts of marine organisms, such as seashells.

Lithosphere – The rigid outer rind of the Earth approximately 40-60 miles thick. It composes the tectonic plates (Hyndman and Hyndman, 2006).

Littoral drift – Volume of sand moving along the shore in the longshore sediment transport system.

Longshore current – Current formed when an obliquely approaching wave piles up water in the surf zone that flows away from the direction of the approaching wave. The resulting current runs parallel to the shore. See Figure 28.

Longshore sediment transport – The movement of sediment along the beachface and within the nearshore zone when waves approach the shore at an oblique angle. This movement of sediment is usually produced by the combined processes of beach drift and longshore currents. See Figure 28.

Low marsh – The zone of the marsh between mid tide and neap high tide, mostly populated by smooth cordgrass (*Spartina alterniflora*) in Georgia. See Figures 58 and 92.

Lower delta plain – The lower portion, typically about half, of the major piedmont river deltas in the Georgia Bight. It is composed of a wide triangular belt of mud-dominant habitats. This muddy zone separates the more sand-rich upper delta plain from the sand-dominant delta front.

Lowstand – Term that refers to the period when sea level was low for a significant amount of time, possibly measured in thousands of years, as during the four major glaciations of the Pleistocene epoch. For example, the lowstand during the last glaciation, the Wisconsin, that occurred around 20,000 years ago, was about 350-400 feet lower than it is today.

Lowstand valleys – Valleys carved by rivers flowing across the coastal plain and continental shelf during the low sea levels that occurred during the glaciations of the Pleistocene epoch. See Figure 7.

Low-tide terrace – A relatively flat surface that is located seaward of the toe of the beachface and slopes offshore at a small angle (2-3 degrees). The landward margin is usually near mean sea level and the seaward margin near mean low water. See Figure 63B.

Macrotidal – Term originally used by Hayes (1976) to define types of depositional coasts (Figure 18). Most macrotidal depositional coasts have large estuarine systems with wide tidal flats and no barrier islands. Davies (1973) defined macrotidal coasts as those with a tidal range (vertical distance between high and low tide) of >4 meters (approximately 12 feet), but he did not make generalizations that related coastal geomorphology to tidal range.

Mantle – The thick layer of material below the relatively thin Earth's crust. It contains more of the heavier elements, such as iron and magnesium, than the crust, is about 1,800 miles thick, and makes up approximately 70% of the Earth's total volume. It is composed mostly of the igneous rock peridotite. See Figure 16.

Marsh – Wetland populated with herbaceous (grasses, sedges, and rushes) plants. See Figures 58 and 92.

Meandering channel – Channel with a highly winding path, like the loops one throws into a rope. Such channels tend to have: 1) Relatively flat slopes; 2) A high ratio of suspended to bedload sediments; and 3) A steady discharge. Meandering tidal channels are by far the most common channel type in the coastal water bodies of the Georgia Coast. See Figures 103 and 104.

Megaripples in sand – Asymmetric, linear, or cuspate features moved along a sandy sediment surface by flowing water with spacing between their crests of 2-18 feet. On the shore, they are seen more commonly in tidal channels, where the water is deeper and the currents are stronger. See Figures 68 and 69.

Mesotidal – Term originally used by Hayes (1976) to define types of depositional coasts (Figure 18). Most mesotidal depositional coasts are characterized by short, drumstick-shaped barrier islands with complex tidal flats and coastal wetlands on their landward flanks. Davies (1973) defined mesotidal coasts as those with a tidal range (vertical distance between high and low tide) of 2-4 meters (approximately 6-12 feet), but he did not make generalizations that related coastal geomorphology to tidal range.

Mesozoic era – Period of geological time that extended from 251 to 65.5 million years ago. It is subdivided into three periods, Triassic, Jurassic, and Cretaceous (Table 1).

Metamorphic rocks – Rocks that result from dramatic changes in igneous and sedimentary rocks affected by heat, pressure, and water that usually creates a more compact and more crystalline condition. These changes usually take place at significant depths below the Earth's surface.

Microtidal – Term originally used by Hayes (1976) to define types of depositional coasts (Figure 18). Most microtidal depositional coasts have long barrier islands backed by lagoons as well as some wave-dominated deltas. Davies (1973) defined microtidal coasts as those with a tidal range (vertical distance between high and low tide) of <2 meters (approximately 6 feet), but he did not make generalizations that related coastal geomorphology to tidal range.

Mid-ocean ridge – Long ridge with a deep rift along its crest that usually runs the length of, as well as along the middle of, one of the major oceans (with the Pacific Ocean being an exception; Alt and Hyndman, 2000). These ridges mark the lines where tectonic plates pull away from each other. As new lava flows spill out onto the ocean floor along fissures in these ridges, new oceanic crust is created.

Milankovitch cycle – Theory of climate change proposed by Milankovitch (1941) that related such change to astronomical (orbital forcing) transformation. *This theory postulates that as the Earth travels through space around the Sun cyclical variations in three elements of Earth-Sun geometry combine to produce variations in the amount of solar energy that reaches Earth* (Kaufman, 2002). These three elements that have cyclic variations are eccentricity (the shape of the Earth's orbit around the Sun), obliquity (the tilt of the Earth's axis away from the orbital plane), and precession (change in orientation of the Earth's rotational axis – like the wobble of a spinning top) (Howard, 2002).

Mixed-energy coasts – Term used by Hayes (1976) to define types of depositional coasts (Figure 18). Most mixed-energy depositional coasts are characterized by short, drumstick-shaped barrier islands with complex tidal flats and coastal wetlands on their landward flanks. They usually have tides in the mesotidal range (6-12 feet).

Mixed-energy delta – River delta in which marine processes mold the outer margins of the delta, with both waves and tides playing an important role. The two significant river deltas on the Georgia Coast, the Savannah River Delta and the Altamaha River Delta, are of this type.

Mollusc – Shellfish; invertebrate having a soft unsegmented body usually enclosed in a shell (WordNet Search).

Mud – Sediment with a median diameter finer than 0.0625 millimeters (mm). It includes two classes – silt (median diameter 0.00391-0.0625 mm) and clay (median diameter <0.00391 mm). Silt rubbed between your teeth feels gritty, whereas clay feels smooth, a practice recommended only for pristine areas.

Mudflats – Tidal flats composed primarily of muddy sediments. They occur mostly in the more sheltered (from wave action) parts of coastal water bodies.

Neap tides – The minimum tides of the month that exist when the gravitational attraction of the Moon and Sun on the water bodies of the world ocean are working in opposition, during the first and third quarter of the Moon (condition called quadrative). See Figure 20.

Neogene period – Period of geological time during the Cenozoic era that extended from 23 to 2.6 million years ago. See Table 1.

Offshore breakwaters – Offshore segments of riprap or heavy concrete blocks of miscellaneous shapes several tens of feet long that are detached from and usually oriented parallel to the beach (but not always). They reduce wave energy on their landward sides, which causes sand moving along shore in the longshore sediment transport system to accumulate in their lee, commonly forming tombolos.

Orthoclase feldspar – Silicate mineral with an abundance of aluminum and potassium (e.g., $KAlSi_3O_8$).

Oxbow lake – Lake that has filled in an abandoned meander bend of a meandering stream. A common component of the flood plains of rivers in the Georgia Coastal Plain. See Figures 101A, 101B, and 112.

Paleogene period – Period of geological time during the Cenozoic era that extended from 65.5 – 23 million years ago. See Table 1.

Paleozoic era – Period of geological time that extended from 542 to 251 million years ago. It is subdivided into six periods, Cambrian, Ordovician, Silurian, Devonian, Carboniferous (Mississippian/Pennsylvanian), and Permian (Table 1).

Pamlico Pleistocene barrier island system – Highstand barrier island complex that formed around 125,000-130,000 years ago at an elevation of 20-25 feet above the present sea level. See Figures 9 and 10.

Pangaea – Single, large mass consisting of all the land area on the Earth's surface that formed about 300 million years ago, during the Carboniferous period of the Paleozoic era. It was created when two supercontinents, Gondwana and Laurasia, collided and welded together. In the process, the Appalachian Mountains were formed. See Figure 3A.

Pantheism – View that the universe (nature) and God are identical (Picton, 1905).

Peat – Sediment composed of >70% organic matter (such as leaves, stems, etc.). After a period of weathering, this organic material becomes a waxy brown mass with the consistency of peanut butter.

Pebble – Sediment particle that is 4-64 mm in size.

Penholoway Pleistocene barrier island system – Highstand barrier island complex that formed around 780,000 years ago at an elevation of 70-75 feet above the present sea level. See Figures 9 and 10.

Peridotite – Coarsely crystalline, dark-colored igneous rock that consists mainly of olivine and other ferromagnesian minerals. It is the principal rock type in the mantle.

Period of a wave – The amount of time it takes two succeeding wave crests to pass a single point.

Permafrost – Permanently frozen subsoil.

Permeability of sediments – Measure of the ease with which a fluid flows through open pore spaces (e.g., between sand grains or gravel clasts).

Piedmont river – In the Georgia Bight, it is defined as a river that has a significant amount of its watershed contained within the Piedmont and/or other upland Physiographic Provinces. Sometimes called red-water rivers, because much of its waters have drained lateritic soils, giving it a reddish color. See Figure 110A.

Plagioclase feldspar – White to light gray silicate mineral ranging in composition from $NaAlSi_3O_8$ to $CaAl_2Si_2O_8$. Common constituent of coarsely crystalline igneous rocks such as diorite, tonalite, and granodiorite.

Plate tectonics – The theory named by J. Tuzo Wilson that states that individual blocks of the lithosphere, called plates, compose the outer portion of the Earth's surface. The individual plates either move apart (as away from mid-ocean ridges), into each other (as in subduction zones), or parallel to each other in a few places (as along the transform faults). The plates move slowly (2-4 inches per year) above (riding on) the portion of the Earth's mantle called the asthenosphere, which behaves in a plastic manner. See Figures 3B and 17.

Plateau-shield coast – Coastline defined by Hayes (1965) as one characterized by plateaus and relatively narrow continental shelves. The bedrock along the coast is typically composed of ancient basement complexes of granite and gneiss (in shield areas) and Paleozoic and Mesozoic sedimentary rocks (in plateau areas). They are tectonically stable and some moderately long rivers may be present, depending upon the climate.

Pleistocene epoch – Period of geological time that was initiated approximately 2.6 million years ago and terminated about 12,000 years ago (Table 1). During that time, at least four major glacial episodes occurred, during which ice covered much of the northern half of the continent of North America. The last one, the Wisconsin Glaciation, peaked about 20,000 years ago.

Plunging waves – Waves that take on a cylindrical shape (Hawaii Five-O type) when they break, falling abruptly down (collapsing) with considerable force. See Figure 26A.

Point bar – An arcuate-shaped bar, commonly composed of sand, that occurs on the opposite side of the channel from the cut bank in a meandering stream. See Figure 105A.

Porosity of sediments – The ratio of the open pore spaces between the individual sediment grains or gravel clasts to the space taken up by the sand and gravel clasts themselves (when they are stacked together).

Princess Anne Pleistocene barrier island system – Highstand barrier island complex that formed around 85,000 years ago at an elevation of 10-15 feet above the present sea level. See Figures 9 and 10.

Precambrian time – Period of geological time that extended from 4.6 billion years ago to 542 million years ago. It is subdivided into two eras, Proterozoic and Archaeozoic. (Table 1).

Prograding barrier island – Seaward-building barrier islands typically composed of multiple, relatively parallel linear ridges of sand topped by vegetated sand dunes that originally formed as front-line dunes on the backbeach (called **foredunes**). As the island progrades, a thick wedge of sand results. See Figures 34C, 35-lower half, and 36.

Quadrative – The times when the gravitational forces of the Moon and Sun work in opposition to each other with regard to generating tides in the world ocean (first and third quarter of the Moon). This condition produces the smallest tides of the month (neap tides). See Figure 20.

Quaternary period – That part of geological time that extends from 2.6 million years ago to the present (Table 1). It is composed of two epochs: 1) Pleistocene epoch – 26 million years ago to 12,000 years ago; and 2) Holocene epoch – 12,000 years ago to the present.

Radioactive decay – Process whereby naturally occurring radioactive materials break down into other radioactive materials at known rates, or half lives.

Relative sea level change – The overall, long-term change in sea level at a location regardless of whether it is caused by global water-level changes or the rising/sinking of the land.

Resurrection fern (*Polypodium polypodioides*) – A species of creeping, coarse-textured fern native to the Americas and Africa. The evergreen fronds of this fern are 10 inches high by 2 inches wide. It is an epiphyte like Spanish moss (Wikipedia encyclopedia).

Revetment – A facing that supports an embankment (die.net). It can be composed of a variety of materials including concrete, riprap, wood, and sand bags.

Rhythmic topography – Bulges of sand on the beach that move down the shore like the sinusoidal loops one might throw along a rope. They are created by waves that approach the beach at an oblique angle and are typically spaced 10s to 100s of feet apart. See Figure 28 (upper right).

Rip currents – Wave-generated currents that most commonly flow straight off the beach and tend to be very regularly spaced. They form most commonly during conditions when the waves come straight on shore (not at an angle usually, but there are some exceptions to this general rule). Also, there is typically some degree of reflection of the waves off the beachface when they are forming. Most rip currents owe their formation to the occurrence of edge waves – standing waves with their crests normal to the shore, according to Bowen and Inman (1969) and Komar (1976). See Figure 32.

Ripples in sand – Asymmetric, linear or cuspate features mostly composed of sand that are moved along the sediment surface by flowing water (or wind) with a spacing between their crests of less than 2 feet. On the beach, they usually occur in the intertidal troughs on the landward side of intertidal bars, where the water depths are greater than elsewhere on the beach. See Figures 68A, 71, and 72.

Riprap – Large pieces of rock of boulder size used in shore-protection structures, such as groins, revetments, and jetties. See Figures 87C and 90.

River delta – Protruding, delta-shaped (triangular) mass of sediment that builds into a standing body of water off a river mouth. See Figure 60.

River-dominated delta – River delta in which a strong river supply of sediment overwhelms the dynamic processes at the shore and a lobe of sediment, such as the birdfoot of the Mississippi River Delta, protrudes out into the sea. See Figure 60A.

Salt marsh – Low-lying wetland containing salt-tolerant vegetation that is regularly flooded by marine waters during mid to high tides. Salinities range from 15-36 parts of "salts" per thousand parts of water (ppt). See Figures 58 and 92.

Salt pruning – An occurrence along many coasts where the trees or shrubs back of the shore bend away from a persistent prevailing wind, with growth only occurring on the side of the tree or shrub that is protected from the salt spray. Thus, the limbs extend downwind like flags blowing in the wind. This presumably has something to do with the dehydrating effects of the salt spray, but there may be other explanations. See Figure 130.

Saltation – Mode of sand transport by the wind, whereby

the sand grains bounce off the sand bed and "fly" for some distance at heights usually not more than a couple of feet.

Sand – Sediment with a median diameter between 0.0625 and 2 mm. Individual sediment grains smaller than this are difficult to see with the naked eye.

Sand bypassing system – Mechanical systems, such as land-based dredging plants and pumping systems, that move the sand from the updrift to the downdrift sides of jetties, or other features, such as man-made harbors, that block the natural longshore transport of sand along the shore.

Sand waves – Asymmetric, linear features (ripple shaped) moved along a sandy sediment surface by flowing water with a spacing between their crests of >18 feet. On the outer shore, they are mostly found in deep tidal channels or tidal inlets. See Figure 70.

Sandstones – Sedimentary rocks made up mostly of sand-sized particles of minerals, most notably quartz and rock fragments.

Seafloor spreading – Process first described by Hess (1962) after he observed that the floor of the Pacific Ocean was in motion and "expanding." This happens in response to volcanic flows being added to the sea floor along mid-ocean ridges at the same time as similar volcanic materials are carried to great depths in subduction zones along the outer margin of the oceans.

Seas – Waves in the area where the wind that forms them is blowing. Under sea conditions, the water surface is choppy and the waves are asymmetrical. Whitecaps are common. See Figure 24B.

Seawalls – Vertical, hard structures very commonly built along the backbeach to protect man-made structures. See Figure 88.

Sedimentary rocks – Rocks most commonly formed by weathering and erosion of pre-existing rocks on the Earth's surface. This process creates sediment that can be transported by water, or wind in some cases, to be deposited in masses of significant size. Once deposited, these sediments may become buried where they are consolidated by chemical cementation and other processes to form solid rocks.

Seismograph – An instrument used to measure seismic waves that runs continuously, recording all incoming waves. Because of their different speeds, the seismic P waves from a specific earthquake will arrive ahead of the seismic S waves.

Setback lines – A "soft" solution for dealing with potential beach erosion in which a line is established behind the shoreline, seaward of which building is prohibited. They work best and are easiest to implement in areas that have not been developed as yet. The criteria used to establish such lines include historical analysis of changing positions of the high-tide line, preservation of the zone of foredunes, and defining areas of flooding and storm wave uprush.

Shale – Sedimentary rock made up mostly of silt and/or clay that is fissile (i.e., composed of very fine, multiple layers).

Shelf break – Point where the outer edge of the continental shelf joins a steeply descending planar surface called the continental slope. See Figures 116 and 117.

Sherd – Broken pottery found in archaeological sites.

Shoreface - Gently sloping offshore submerged face of the barrier island. On the Georgia Coast, most storm waves, with the exception to those generated by major hurricanes, start to "feel the bottom" in maximum water depths of around 25-30 feet. This depth typically defines the base of an offshore environment called the shoreface, which, in this area, is considered to be the absolute base (outer limit) of the nearshore portion of the barrier islands. Therefore, this is the zone between the low tide line and the point (water depth) where normal storm waves cease to agitate the bottom sediment, enough to keep silt and clay in suspension.

Significant wave height – Measure that portrays the average height of the waves that comprise the highest 33% of waves in a given sampling period.

Silica tetrahedron – The basic building block of all silicate minerals, which consists of a single silicon cation surrounded by four oxygen anions. Silicon cations, which have a charge of +4, are less than 1/3 the size of the oxygen anions, which have a charge of –2. As a result of these size differences, one silicon cation fits "snugly" between 4 oxygen anions, forming the tetrahedron.

Silicates – Group of minerals that are the fundamental building block of the Earth's crust, with the atoms of silicon and oxygen combined making up about 75% of it. The basic structure of the silicate minerals is the silica tetrahedron.

Silver Bluff Pleistocene barrier island system – Highstand barrier island complex that formed around 40,000-50,000 years ago (Muhs et al., 2004) at an elevation of 4.5 feet above the present sea level (Hails and Hoyt, 1968; 1969). See Figures 9 and 10.

Slip face – Steeply sloping face of a migrating wedge of sand, such as a bedform created by flowing water (e.g., ripples and sand waves or on the downwind side of a sand dune created by blowing wind). The angle of slope of the slip face, which is essentially the angle of repose of the sand (dip angle achieved when a pile of sand is created by simply dumping the sand out of a container, such as a truck bed), averages around 22 degrees in water and a little over 30 degrees in air. See Figures 64C, 67A, and 68C.

Spanish moss – Fibrous gray-green epiphyte (*Tillandsia usneoides*) found in southern North America, and elsewhere, that often hangs in large, beardlike, silvery-gray masses from trees and other plants. However, it is not parasitic or structurally intertwined with its host.

Spilling waves – Waves that start disintegrating into foaming lines fairly far offshore and continue to foam their way to shore, gradually decreasing in height as they go. See Figure 26B.

Spit – Linear projection of sediment across an embayment, stream outlet, and so on that builds in the direction that the longshore transport system is moving the sediment. See Figures 30 and 54.

Spit-elongation hypothesis – Mode of formation of barrier islands by means of the elongation of sand spits away from a headland with permanent segmentation of the spits occurring as they grow as a result of the formation of permanent tidal inlets through them during storms, creating individual islands (Gilbert, 1989; Fisher, 1967; see Figure 40).

Spring tides – The maximum tides of the month that exist when the gravitational attraction of the Moon and Sun on the water bodies of the world ocean are working together, during full and new moons (condition called syzygy). See Figure 20.

Stillstand – Period of time when sea level remained near some level, as, for example, the present stillstand which started about 4,500 years ago.

Storm surge – Elevation of water level associated with passage of a storm, with hurricane storm surges being the largest experienced on the Georgia Coast. During hurricanes, the elevation of the water is caused primarily by wind set-up, but the decrease in atmospheric pressure that accompanies the storm is also a factor in some cases. Storm surges of over 20 feet have occurred along the shore of the Georgia Bight.

Storm profile – Term suggested by Paul Komar, who was working on the Oregon coast, to indicate a flat beach profile eroded by storm waves. Hayes and Boothroyd (1969) used the term post-storm beach. See Figure 63B.

Surf zone – The nearshore zone off the beachface where the waves usually break. See Figure 25 (Lower).

Surging waves – Waves that approach shore and wash directly up the beachface without breaking.

Swamp – Wetland populated with woody plants.

Swash bars – Landward oriented, asymmetric, generally intertidal sand bars created on the landward side of breaking waves. Very commonly associated with the margins, or on the surface, of ebb-tidal deltas. Figures 36A and 36C.

Swash zone – Area where the water released by a broken wave washes across the portion of the beach known as the beachface. The released water at first rushes up the beachface until it loses its momentum, at which point it stops momentarily before returning back down the beachface under the influence of gravity. Usually, some of this water is lost through percolation into the beachface sediments.

Swell – Waves that have passed beyond the area where they were formed. Swell waves are typically longer on the average than those in the area of formation. The swell surface is regular and more-or-less symmetrical (whitecaps are rare). See Figure 24B.

Swell profile – Term suggested by Paul Komar, who was working on the Oregon coast, to indicate a beach with a well-developed depositional berm (formed between storms). Hayes and Boothroyd (1969) used the terms late accretional or mature profile. See Figure 63B.

Syzygy – The times when the gravitational forces of the Moon and Sun work together with regard to generating tides in the world ocean (full and new Moon). The largest tides of the month (spring tides) occur during syzygy. See Figure 20.

Tabby – Concrete-like substance *made of lime, water, powdered oyster shells, and crushed (but chunky) oyster shells* (Livingston, 2010). This material was one of the primary substances used to build dwellings in the early days of European settlement of what is now the southeastern U.S. See Figure 182.

Talbot Pleistocene barrier island system – Highstand barrier island complex that formed around 200-240,000 years ago at an elevation of 40-45 feet above the present sea level. See Figures 9 and 10.

Tectonic (lithospheric) plates – A dozen or so segments of the lithosphere that covers the Earth's outer part (Hyndman and Hyndman, 2006). The plates that comprise the continents (e.g., the North American Plate) are made up of relatively light material composed of silica, oxygen, and so on. The plates that constitute the ocean floors are composed of relatively heavy material that contains abundant iron, magnesium, and so on. See Figure 3B.

Tectonism – The forces that produce deformation of the Earth's crust, such as folding or faulting of the rocks, with the motion of tectonic plates commonly being the driving mechanism for this deformation.

Tertiary period – Period of geological time during the Cenozoic era formerly used to cover the time interval from 65.5 to 2.6 million years ago. This period is now subdivided into the Neogene and Paleogene periods (Table 1).

Tidal flat – Flat intertidal surface covered and uncovered as the tides rise and fall. It is usually a depositional surface with the sediments supplied by tidal currents. In more exposed areas, where the waves are somewhat active and the tidal currents are strong, the flats are composed of sand. In more sheltered areas, they are composed mostly of mud.

Tidal inlet – Major tidal channels that intersect barrier islands, or separate two barrier islands, usually to depths of 10s of feet. See Figures 50 and 51.

Tidal prism – Volume of water that flows in and out of a tidal inlet during a single tidal cycle.

Tidal range – The vertical distance between high and low tide. The spring tidal range in Georgia varies from about 8.2 feet at the South Carolina border to about 7.0 feet at the Florida border

Tidal river point bar – An arcuate-shaped bar, commonly composed of sand, that occurs on the opposite side of the channel from the cut bank in a meandering tidal river. However, in contrast to point bars in freshwater streams, there is usually a high-level, relatively shallow channel dominated by flood currents that separates the bar from the channel margin. See Figure 107.

Tide-dominated coasts – Term used by Hayes (1976) to define types of depositional coasts. They are coasts characterized by such features as open mouthed, multi-lobate river deltas, abundant large estuarine complexes, and extensive tidal flats and salt marshes. Most tide-dominated coasts are macrotidal (tidal range >12 feet). They usually occur on coasts with a combination of relatively small waves and large tides.

Tide-dominated delta – River delta that consists of a series of funnel-shaped water bodies at multiple river mouths with a number of shore-perpendicular, tidal sand ridges that extend offshore of the river mouths. See Figure 60C.

Tombolo – Triangle-shaped body of sediment (usually sand) that forms in the lee of an offshore structure, such as a small island or shipwreck. Waves approaching shore bend around this offshore structure, with the result that two opposing longshore sediment transport directions are created that converge in the shelter of the offshore structure, resulting in the formation of the tombolo. See Figure 27.

Trailing edge coasts – Coasts defined by Inman and Nordstrom (1971) as those that occur on the trailing edges of continental plates. They are typically characterized by wide coastal plains, numerous river deltas, and barrier islands. Rocky coasts are extremely rare.

Trail Ridge/Wicomico Pleistocene barrier island system – Highstand barrier island complex that formed around 1.0-1.7 million years ago (Winker and Howard, 1977) at an elevation of 95-100 feet above the present sea level (Hails and Hoyt, 1968; 1969). Burdette et al. (2012) gave an age of 2.2 million years based on the ESROD technique. See Figures 9 and 10.

Transgressive barrier island – Landward-migrating barrier island composed of coalescing washover fans, or a washover terrace, that is overtopped at high tides, usually several times a year. In the process of migration, the entire washover terrace complex moves landward, leaving an eroded nearshore zone in its wake. See Figures 34A, 34B, and 35 (upper row).

Transgressive-regressive interfluve hypothesis – Process by which prograding, Holocene barrier islands were formed that involved several steps: 1) Landward migrating primary barrier island (presumably of the transgressive barrier island type) moved "quickly" across the continental shelf as sea level rises; 2) These migrating barrier islands were typically situated on interfluves between lowstand river valleys; and 3) When sea level stopped rising so fast 4,500 years ago, the barrier islands began to prograde seaward. See Figure 43.

Updrift – The direction from which the individual sand grains moving along the shore within the longshore transport system have come. See Figure 28.

Upper bottomland hardwood forests – Zones IV and V of the bottomland hardwoods that occur on flood plains of rivers in the coastal plain of the southeastern U.S. These two zones are less frequently flooded than the adjacent cypress-tupelo swamps (<50% of the time), and have a much more diverse population of plant species. See Figures 114 and 115A.

Upper delta plain – The upper portion, typically about half, of the major piedmont river deltas in the Georgia Bight, the landward limit of which coincides roughly with the "head of the tides." Typically contains highly sinuous (meandering) channels as well as a number of abandoned channels. Point bars composed of coarse-grained sand may be present along the main channel, as well as oxbow lakes in abandoned channels with minor connections to it.

Washover fan – Feature found commonly on transgressive barrier islands where waves wash all the way across the top of the island during storms or during very high spring tides. This process creates a flattened-off surface along which sand is transported across the island's surface and into the standing water behind it. The resulting sand deposit commonly has a fan-like shape. See Figures 34A and 34B.

Wave-dominated coasts – Term used by Hayes (1976) to define types of depositional coasts. They are characterized by such features as deltas with smooth outer margins and abundant long barrier islands. Most wave-dominated coasts are microtidal or low-mesotidal (tidal range less than 6-8 feet). They are usually found on coasts with a combination of relatively large waves and small tides.

Wave-dominated delta – River delta on a wave-dominated coast, where the rivers have enough sediment to build a bulge out into the ocean and the waves rework the sand into a smooth outer margin composed of prograding beach ridges. See Figure 60B.

Wave diffraction – The phenomenon by which energy is transmitted laterally along a wave crest. When part of a wave is interrupted by a barrier, the effect of diffraction is manifested by propagation of waves into the sheltered region within the barrier's geometric shadow. A good example would be waves passing between two offshore breakwaters and into the sheltered areas landward of the breakwaters, with the diffracted waves bending into a bow-like shape.

Wave height – Vertical distance between the peak of the wave crest and the bottom of the wave trough. See Figure 24A.

Wave length – Horizontal distance between wave crests. See Figure 24A.

Wave refraction – Any change in the direction of a wave resulting from the bottom contours. For example, a submerged hump would cause the two ends of the wave crest located away from the hump to bend toward each other. Such differences in the offshore topography can cause the energy of the waves to be focused in some areas along the shore and less energetic in others. See Figure 27.

Wave uprush – The water that moves up the beachface under the impetus of a breaking wave.

Wetland – *Land where an excess of water is the dominant factor determining the nature of soil development and the types of animals and plant communities living at the soil surface. It spans a continuum of environments where terrestrial and aquatic systems intergrade* (Cowardin et al., 1979). Examples include marshes and swamps.

Wicomico/Trail Ridge Pleistocene barrier island system – Highstand barrier island complex that formed around 1.0-1.7 million years ago (Winker and Howard, 1977) at an elevation of 95-100 feet above the present sea level. Burdette et al. (2012) gave an age of 2.2 million years based on the ESROD technique. In Georgia, much of this barrier island system has been eroded away, notably toward the northeast. See Figures 9 and 10.

Wind-shadow dune – Wind-blown accumulation of sand grains in the lee of some obstacle, such as a pile of wrack or a clump of beach grasses. See Figures 83, 84A, and 84B.

Wrack – An accumulation of plant stems and roots usually found near high-tide lines on the top of beach berms and elevated intertidal bars, or, in some cases, along the landward margin of marshes. Stems of the smooth cordgrass (*Spartina alterniflora*) is a common constituent of wrack on the Georgia Coast. See Figure 84A.

Young mountain range coasts – Defined by Hayes (1965) as coastlines that abut high mountains created by relatively recent orogenic (mountain building) activity. In most places, the land is continuing to rise. The bedrock is usually dominated by relatively young sedimentary and volcanic rocks. Mostly short, steep rivers empty into the sea along such coasts. They are most common on the leading edges of continental plates.

Zone of dynamic change (ZODC) – Zone along the shore within which the sand moves back and forth. The landward limit of this is the point at which the foredunes are eroded during "normal" storms (not hurricanes or major northeasters). Accordingly, during these "normal" storms, the sand formerly deposited on the intertidal beach, as well as up to several feet of the foredunes, is usually transported into the subtidal area, which, in some areas, may contain an offshore bar. The seaward limit of the zone is what engineers call the closure point. See Figure 62.

Zone of the turbidity maximum – Zone in the middle reaches of an estuary that contains an abundance of fine-grained sediment in suspension, because of the mixing of salt water with fresh water containing a profusion of suspended clay particles, which tend to flocculate. See Figure 57B.

REFERENCES CITED

Adherants.com., 2005. Largest religious groups in the United States. Available at: http://www.adherents.com/rel_USA.html.

Alexander, C.R., 2008. (special issue editor). Coastal and Marine Geology of the Southeastern United States. Southeastern Geology, 45(3).

Allen, J.R.L., 1968. Current ripples – their patterns in relation to water and sediment motion. North-Holland Publ. Co., Amsterdam, Holland, 433 pp.

Allen, J., and Komar, P., 2000. Are ocean wave heights increasing in the eastern North Pacific? EOS, 81:561-567.

apt, 2012. Extinction of the dinosaurs? Available at: http://www.pbs.org/wgbh/evolution/extinction/dinosaurs/asteroid.html.

Ask GeoMan, 2012. Densities of Earth's layers. Available at: http://jersey.uoregon.edu/~mstrick/AskGeoMan/geoQuerry57.html.

Bagwell, 2008. Jekyll Island History. Available at: http://www.jekyllislandhistory.com/earlyseaisland.shtml.

Baldwin, W.E., Morton, R.A., Schwab, W.C., Gayes, P.T., and Driscoll, N.W., 2004. Maps showing the stratigraphic framework of South Carolina's Long Bay from Little River to Winyah Bay. U.S. Geological Survey, Open-File Report: OF 2004-1013.

Baldwin, W.E., Morton, R.A., Putney, T.R., Katuna, M.P., Harris, M.S., Gayes, P.T., Driscoll, N.W., Denny, J.F., and Schwab, W.C., 2006. Migration of the Pee Dee River System inferred from ancestral paleochannels underlying the South Carolina Grand Strand and Long Bay inner shelf. GSA Bulletin, 118(5/6):533-549.

Barry, J.M., 1980. Natural Vegetation of South Carolina. Univ. of South Carolina Press, Columbia, SC, 214 pp.

Basan, P.B., and Frey, R.W., 1977. Actual-palaeontology and neoichnology of salt marshes near Sapelo Island, Georgia. In: Crimes T.P, and Harber, J.C. (eds.), Trace fossils 2. Geog. Jour., Spec. Issue 9:41-70.

Bascom, W.H., 1951. The relation between sand size and beachface slope. Amer. Geophys. Union, Trans. 32:866-874.

Bascom, W.H., 1954. Characteristics of natural beaches. Proc. 4th Conf. on Coastal Eng., pp. 163-180.

Beck. T.M., and Legault, K., 2012. St. Augustine Inlet, Florida: Application of the coastal modeling system. Report 2, Coastal Inlets Research Program, U.S. Army Corps of Engineers, ERD/CHLTR-12-14, pp. 1-84.

Bell, M., Jr., 1987. Major Butler's Legacy: Five Generations of a Slaveholding Family. Univ. of Georgia Press, Athens, 673 pp.

Berendt, J., 1997. Midnight in the Garden of Good and Evil. Vintage Books, 400 pp.

Berger, A., Imbrie, J., Hays, J., Kukla, G., and Saltzman, B., 1984. (eds.), Milankovitch and climate: Understanding the response to astronomical forcing. D. Reidel Pub. Co., Dordrecht, 895 pp..

Beyene, A., and Wilson, J.H., 2006. Comparison of wave energy flux for northern, central, and southern coasts of California based on long-term statistical wave data. Energy, 31:1856-1869.

Biggs, R.B., 1978. Coastal bays. In: Davis, R.A., Jr. (ed.), Coastal Sedimentary Environments. Springer Verlag, New York, pp. 69-99.

Bindoff, N.L., and eleven others, 2007. Observations: Oceanic climate change and sea level. In: Climate change 2007: The physical science basis. Contribution of Working Group 1 to the Fourth Assessment Report of the Intergovernmental Panel on Climate Change. Cambridge University Press, Cambridge, United Kingdom and New York, pp. 387-482.

Bird, E.C.F., 1996. Beach Management. Coastal Morphology and Research. John Wiley, Chichester, 292 pp.

Bishop, G.A., Rollins, H.B., Thomas, D.H., Beratan, K.K., Booth, R.K., Camann, E.J., and Vento, F.J., 2011a.

Geoarchaeology of St. Catherines Island, Georgia: Proceedings of the Fourth Caldwell Conference, St. Catherines Island, Georgia, March 27-29, 2009. (Anthropological papers of the American Museum of Natural History, no. 94).

Bishop, G.A., Meyer, B.K., Vance, R.K., and Rich, F.J., 2011b. Geoarchaeological research at St. Catherines Island: Defining the geological foundation. In: Bishop, et al. (eds.), Geoarchaeology of St. Catherines Island, Georgia: Proceedings of the Fourth Caldwell Conference, St. Catherines Island, Georgia, March 27-29, 2009. (Anthropological papers of the American Museum of Natural History, no. 94), pp. 79-99.

Blackwelder, B., Pilkey, O.H., and Howard, J.D., 1979. Late Wisconsin sea levels on the southeast U.S. Atlantic shelf based on in-place shoreline indicators. Science, 204:618-620.

Blackburn, J., 1994. James Edward Oglethorpe. R. Bemis Publishing, Ltd., Marietta, Georgia, 184 pp.

Blake, E.S., and Gibney, E.J., 2011. The deadliest, costliest, and most intense United States tropical cyclones from 1851 to 2010 (and other frequently requested hurricane facts). NOAA Technical Memorandum NWS NHC-6. Available at: http://www.nhc.noaa.gov/pdf/nws-nhc-6.pdf.

Booth, R.K., and Rich, F.J., 1999. Identification and paleoecological implications of a late Pleistocene pteridophyte-dominated assemblage preserved in brown peat from St. Catherines Island, Georgia. Castanea, 64(2):120–129.

Boothroyd, J.C., 1969. Hydraulic conditions controlling the formation of estuarine bedforms. In: Hayes, M.O. (ed.), Coastal Environments – NE Massachusetts and New Hampshire: Field trip guidebook for Eastern section of SEPM, May 9-11, 1969, pp. 417-427.

Bowen, A.J., and Inman, D.L., 1966. Budget of littoral sands in the vicinity of Point Arguello, California. Coastal Eng. Res. Center, Tech. Memo. No. 19, 56 pp.

Bowen, A.J., and Inman, D.L., 1969. Rip currents: Laboratory and field observations. Journal of Geophysical Research, 74:5479-5490.

Broecker, W.S., 2002. The Glacial World According to Wally. Eldigo Press, Lamont-Doherty Earth Observatory of Columbia University, Palisades, New York, 346 pp.

Broecker, W.S., Thurber, D.L., Goddard, J., Ku, T.L., Matthews, R.K., and Mesolella, K.J., 1986. Milankovitch hypothesis supported by precise dating of coral reefs and deep-sea sediments. Science, 159:297-300.

Brown, D., 1970. Bury My Heart at Wounded Knee: An Indian History of the American West. H. Holt and Co., New York, 487 pp.

Browns Guide to Georgia, 2009a. Fort Morris Historic Site. Available at: http://brownsguides.com/kids/fort-morris-historic-site/.

Browns Guide to Georgia, 2009b. The Liberty Trail. Available at: http://brownsguides.com/diytours/the-liberty-trail/.

Bruun, P., 1962. Sea level rise as a cause of shore erosion. Am. Soc. Civil Eng. Proc., Jour. Waterways and Harbors Div., 88:117-130.

Bues, S.S., and Morales, M., 1990. Grand Canyon Geology. Oxford Univ. Press, New York, 9 pp.

Burdette, K.E., Rink, W.J., Mallinson, D.J., Means, G.H., and Parham, P.R., 2013. Electron spin resonance optical dating of marine, estuarine, and aeolian sediments in Florida. Quaternary Research, 79(1):66-74.

Cajete, G.A., 1995, Ensoulment of nature. In: Hirschfelder, A. B. (ed.), Native Heritage: Personal Accounts by American Indians, 1790 to the Present. Macmillan Pub. Co., 298 pp.

Carter, R.W.G., and Woodroffe, C.D. (eds.), 1994. Coastal Evolution. Cambridge Univ. Press, Cambridge, 517 pp.

Chowns, T.M., Schultz B.S., Griffin, J.R., and Crook, M.R., Jr., 2008. Relocation of Brunswick River and other estuaries on the Georgia, USA Coast as a consequence of Holocene transgression. In: Alexander, C.R. (ed.), Coastal and Marine Geology of the Southeastern United States. Southeastern Geology, 45:3:143-159.

Clark, P.U., and Mix, A.C., 2002. Ice sheets and sea level of the Last Glacial Maximum. Quat. Science Rev., 21:1-7.

Clark, J.R., and Benforado, J. (eds.), 1981. Wetlands of bottomland hardwood forests – Proceedings of a workshop on bottomland hardwood forest wetlands of the southeastern United States. Developments in Agricultural and Managed-Forest Ecology, 11:335-357.

Clayton, T.D., Taylor, L.A., Jr., Cleary, W.J., Hosier, P., Graber, P.H.F., and Pilkey, O.H., 1992. Living with the Georgia Shore. Duke Univ. Press, Durham, North Carolina, 296 pp.

Coastal Georgia Experience, 2007. Available at: http://www.coastalgeorgiaexperience.com/site/502999/page/92872.

Coleman, J.M., and Wright, L.D., 1975. Modern river deltas: Variability of processes and sand bodies. In: Broussard, M.L (ed.), Deltas, Houston Geol. Soc., Houston, Texas. 2nd ed., pp. 99-150.

Colquhoun, D.J., 1969. Geomorphology of the lower coastal plain of South Carolina. South Carolina Div. Geology, Pub. MS-15, 36 pp.

Colquhoun, D.J., 1974. Cyclic surficial stratigraphic units of the middle and lower coastal plains, central South Carolina. In: Oaks, R.Q., and DuBar, J.R. (eds.), Post-Miocene stratigraphy, central and southern Atlantic coastal plain. Utah State Univ. Press., Logan, pp. 179-190.

Colquhoun, D.J., and Brooks, M.J., 1986. New evidence from the southeastern U.S. for eustatic components in the late Holocene sea levels. Geoarchaeology, 1:275-291.

Colquhoun, D.J. et al., 1983. Surface and subsurface stratigraphy, structure and aquifers of the South Carolina Coastal Plain. Report for S.C. Dept. of Health and Environmental Control, Dept. Geology, Univ. of South Carolina, Columbia, 78 pp.

Comet, P.A., 1996. Geological reasoning: Geology as an interpretive and historical science. Discussion, Geol. Soc. Amer. Bull., 108(11):1508-1510.

Concerned Coastal Geologists, 1981. Saving the American beach. Skidaway Inst. Ocean. Conf., Savannah, Georgia, March 1981. 12 pp.

Cooke, C.W., 1931. Seven coastal terraces in the southeastern states. Washington Acad. Sci. Jour., 21:503-513.

Cooke, C.W., 1932. Tentative correlations of American glacial chronology with the marine time scale. Washington Acad. Sci. Jour., 22:310-313.

Cornelius, D., 2009. Small earthquake shakes Georgia neighbors. Available at: http://www.associatedcontent.com/article/2465642/small_earthquake_shakes_georgia_neighbors.

Cote, R.N., 2006. City of Heroes. Corinthian Books, Mt. Pleasant, South Carolina, 542 pp.

Cowardin, L.M., Carter, V., Golet, F.C., and La Roe, E.T., 1979. Classification of Wetlands and Deepwater Habitats in the United States. U.S. Dept. Interior, Fish & Wildlife Service, FWS/OBS-79/31, 131 pp.

Curray, J.R., 1964. Transgressions and regressions. In: Miller, R.L. (ed.), Papers in Marine Geology - Shepard Commemorative Volume. Macmillan and Co., New York, pp. 175-203.

Dahlen, M.K., Bell, R., Richardson, J.I., and Richardson, T.H., 2000. Beyond D-0004: Thirty-four years of loggerhead (*Caretta caretta*) research on Little Cumberland Island, Georgia, 1964-1997. In: Abreu-Grobois, F.A., Briseno-Duenas, R., Marquez, R., and Sarti, L. (compilers), Proceedings of the Eighteenth International Sea Turtle Symposium. NOAA Technical Memorandum NMFS-SEFSC-436, pp. 60-62.

Darwin, C., 1859. On the Origin of Species by Means of Natural Selection, or the Preservation of Favoured Races in the Struggle for Life. John Murray, London, 1st edition, 502 pp.

Davies, J.L., 1973. Geographical Variation in Coastal Development. Hafner Pub. Co., New York, 204 pp.

Davis, R.A., Jr., and FitzGerald, D.M., 2004. Beaches and Coasts. Blackwell Publishing, Malden, MA, 419 pp.

Davis, R.A., Jr., and Hayes, M.O., 1984. What is a wave-dominated coast? Mar. Geol., 60:313-329.

Dean, R.G., and Walton, T.C., 1975. Sediment transport processes in the vicinity of inlets with special reference to sand trapping. In: Cronin, L.E. (ed.), Estuarine Research, Geol. and Eng., Academic Press, New York, 2:129-150.

de Beaumont, E., 1845. Lecons de geologie practique. In: Bertrand, P. (ed.), Septieme Lecon, Paris, France, pp. 223-252.

DePratter, C.B., 1977. Environmental changes on the Georgia Coast during the prehistoric period. Early Georgia, 5(1-2):1-14.

DePratter, C.B., and Howard, J.D., 1980. Indian occupation and geologic history of the Georgia Coast: A 5,000 year summary. In: Howard, J.D., DePratter, C.B., and Frey, R.W. (eds.), Excursions in Southeastern Geology: Archaeology of the Georgia Coast. Georgia Dept. Natural Resources, Atlanta, pp. 1-65.

DePratter, C.B., and Howard, J.D., 1981. Evidence for a sea level lowstand between 4500 and 2400 years B.P. on the southeast coast of the United States. Jour. Sed. Petrology, 51:1287-1295.

DePratter, C.B., and Thompson, V.D., 2013. Chapter 6. Past shorelines of the Georgia Coast. In: Thompson, V.D., and Thomas, D.H. (eds.), Life Among The Tides, Recent Archaeology on the Georgia Bight, Anthropology Papers of The American Museum of Natural History, 98:145-167.

Dietz, R.S., and Holden, J.C., 1970. Reconstruction of Pangaea: Breakup and dispersion of continents, Permian to Present. Jour. Geophys. Res., 75:4939-4956.

Dole, R.B., and Stabler, H., 1909. Denudation. U.S. Geological Survey Water Supply Paper 234, pp. 78-93.

Dörjes, J., 1972. Georgia coastal region, Sapelo Island, USA. Sedimentology and biology. VII. Distribution and zonation of macrobenthic animals. Senckenbergiana Maritima, 4:183-216.

Dörjes, J., Frey, R.W., and Howard, J.D., 1986. Origins of, and mechanisms for, mollusk shell accumulation on Georgia beaches. Senckenbergiana Maritima, 18(1-2):1-43.

Dunn, G.E., and Miller, B.I., 1960. Atlantic Hurricanes: Louisiana State Univ. Press, Baton Rouge, 326 pp.

Dyer, K.R., 1973. Estuaries; A Physical Classification. John Wiley, New York, 140 pp.

Eckert, A.W., 1993. A Sorrow in Our Hearts, The Life of Tecumseh. Bantam Books, New York, 1,088 pp.

Edling, R., 2010. Darien, Georgia. Available at: http://www.civilwaralbum.com/misc5/darien1.htm.

Edwards, A.L., 1997. Georgia seashells. Available at: http://amylyne.myweb.uga.edu/GSC/seashellGA.html.

Edwards, J.M., and Frey, R.W., 1977. Substrate characteristics within a Holocene marsh, Sapelo Island, Georgia. Senckenbergiana Maritima, 9:215-259.

Edwards, R.L., and Merrill, A.S., 1977. A reconstruction of the continental areas of eastern North America for the times 9,500 B.P. and 12,500 B.P. Archaeology of Eastern North America, 5:1-43.

Eggar, A.E., 2006. Minerals III: The Silicates. Available at: http://www.visionlearning.com/library/module_viewer.php?mid=140.

Einstein, A., 1954. The cause of the formation of meanders in the courses of rivers and of the so-called Baer's Law. In: Ideas and Opinions. Bonanza Books, New York, pp. 249-253

Emiliani, C., 1955. Pleistocene temperatures. J. Geol. 63:538-78.

Enotes, 2009. Sea-floor spreading. Available at: http://enotes.com/earth-science/sea-floorspreading.

Explore Tybee.com, 2009. Welcome to Georgia's Golden Isles. Available at: http://www.exploretybee.com/Fort_Screven.shtml.

Explorer's Guide To Georgia's Golden Isles, 2010. Jekyll Island, Georgia. Available at: http://www.explorestsimonsisland.com/Jekyll_Island.html.

FDEP/RPI. 1996. Environmental Sensitivity Index (ESI) Maps of the Shoreline of the Coast of East Florida. By Research Planning, Inc. under contract to the Florida Department of Environmental Protection, 59 maps.

Federov, A.V., and Philander, S.G., 2000. Is *El Niño* changing? Science, 288:1997-2002.

Finkl, C.W., and Pilkey, O.H. (eds.), 1991. Impacts of hurricane *Hugo*: September 10-22, 1989. Jour. of Coastal Research, Special Issue 8, 356 pp.

Fisher, J.J., 1962. Geomorphic expression of former inlets along the Outer Banks of North Carolina. M.S. thesis, Univ. of North Carolina, Chapel Hill, 120 pp.

Fisher, J.J., 1967. Origin of barrier island chain shoreline, Middle Atlantic States. Geol. Soc. Am. Spec. Paper 115:66-67.

FitzGerald, D.M., 1977. Hydraulics, morphology, and sediment transport at Price Inlet, South Carolina. Ph.D. Thesis, Dept. Geol., University of South Carolina, Columbia, 84 pp.

FitzGerald, D.M., and Hayes, M.O., 1980. Tidal inlet effects on barrier island management. Coastal Zone '80, Amer. Soc. Civil Eng, New York, pp. 2355-2379.

FitzGerald, D.M., Nummedal, D., and Kana, T.W., 1976. Sand circulation patterns of Price Inlet, South Carolina. In: Proc. 15th Coastal Engineering Conf., American Society Civil Eng., New York, pp. 1868-1880.

Fleming, K., Johnston, P., Zwartz, D., Yokoyama, Y., Lambeck, K., and Chappell, J., 1998. Refining the sea-level curve since the Last Glacial Maximum using far- and intermediate-field sites. Earth and Planetary Science Letters, 163:327-342.

Frazer, N.B., 1983. Survivorship of adult female loggerhead sea turtles, *Caretta caretta*, nesting on Little Cumberland Island, Georgia, USA. Herpetologica, 39(4):436-447.

Frazer, N.B. 1985. Survival from egg to adulthood in a declining population of loggerhead turtles, *Caretta caretta*. Herpetologica 42(1): 47-55.

Frey, R.W., and Basan, P.B. (1978, 1985) Coastal salt marshes: In: Davis, R.A. (ed.), Coastal Sedimentary Environments:

Springer-Verlag, New York, pp. 225-301.
Frey, R.W., and Basan, P.B., 1981. Taphonomy of relict Holocene salt marsh deposits, Cabretta Island, Georgia. Senckenbergiana Maritima, 13: 111–155.
Frey, R.W., and Howard, J.D., 1969. A profile of biogenic sedimentary structures in a Holocene barrier island-salt marsh complex. Trans. Gulf Coast Assoc. Geol. Soc., 19:427-444.
Frey, R.W., and Howard, J.D., 1988. Beach and beach-related facies, Holocene barrier islands of Georgia. Geology, 125, pp. 621-640.
Frey, R.W., and Mayou, T.V., 1971. Decapod burrows in Holocene barrier island beaches and washover fans, Georgia. Senckenberg. Marit. 3:53-77.
Frey, R.W., Howard, J.D., and Pryor, W.A., 1978. Ophiomorpha: its morphologic, taxonomic and environmental significance. Paleogeogr. Paleoclimatol. Paleoecol., 23:199-229.
Gallop, C.D., Cheng, H., Taylor, F.W., and Edwards, R.L., 2002. Direct determination of the timing of sea level change during Termination II. Science, 295:310-313.
Galloway, W.E., 1975. Process framework for describing the morphology and stratigraphic evolution of deltaic depositional systems. In: Broussard, M.L. (ed.), Deltas, Houston Geol. Soc., Houston, Texas. 2nd Edition, pp. 87-98.
Gayes, P.T., 1991. Post-Hurricane *Hugo* nearshore side-scan sonar survey: Myrtle Beach to Folly Beach, South Carolina. In: Finkl, C.W. and Pilkey, O.H. (eds.), Impacts of Hurricane *Hugo*: September 10-22, 1989. Jour. of Coastal Research, Special Issue 8:95-111.
Gayes, P.T., Scott, D.B., Collins, E.S., and Nelson, D.D., 1992 , A Late Holocene sea-Level fluctuation in South Carolina. In: Fletcher, C.H., III and John F. Wehmiller, J.F. (eds.), Quaternary Coasts of the United States: Marine and Lacustrine Systems, Society for Sedimentary Geology Special Publication No. 48:155-160.
Gayes, P.T., Schwab, W.C., Driscoll, N.W., Morton, R.A., Baldwin, W.E., Denny, J.F., Wright, E.E., Harris, M.S., Katuna, M.P., Putney, T.R., and Johnstone, E., 2003: Sediment dispersal pathways and conceptual sediment budget for a sediment starved embayment; Long Bay, South Carolina. In: Coastal Sediments '03, 5th Annual Symposium on Coastal Engineering and Science of Coastal Sediment Processes, Clearwater Beach, FL, May 18-23, 2003. East Meets West Productions, Corpus Christi, TX, 14 pp.
GeorgiaInfo, 2010a. John Houston McIntosh. Available at: http://georgiainfo.galileo.usg.edu/gahistmarkers/johnhmcintoshhistmarker.htm.
GeorgiaInfo, 2010b. Barrier island sequence district. Available at: http://georgiainfo.galileo.usg.edu/physiographic/BIS.htm.
Georgia Ports Authority, 2009. Colonels Island Terminal. Available at: http://www.gaports.com/Default.aspx?tabid=356.
Georgia Southern University, 2010. Applied Coastal Research Laboratory. Available at: http://cost.georgiasouthern.edu/geo/acrl.html.
Georgia State Parks & Historic Sites, 2006. Wormsloe State Historic Site, Ga. Dept. Natural Resources.
Georgia Travel, 2010. St. Catherines Island visitors information. Available at: http://www.geckosgeorgia.com/st-catherines-island.php.
Georgia Wildlife, DNR, 2002. Georgia's colonial coast birding trail. Available at: http://www.georgiawildlife.com/ColonialCoastBirdingTrail.
Gilbert, G.K., 1889. The topographic features of lake shores. U.S. Geological Survey 5th Annual Report, Washington, D.C., pp. 69-123.
Giles, R.T., and Pilkey, O.H., 1965. Atlantic beach and dune sediments of the southeastern United States. Jour. of Sed. Petrology, 35(4):900-910.
Glaeser, J.D., 1978. Global distribution of barrier islands in terms of tectonic setting. Jour. Geol., 86:283-298.
Glen, W., 1975. Continental Drift and Plate Tectonics: C.E. Merrill Pub. Co., Columbus, Ohio, 188 pp.
Gore, A., 2006. An Inconvenient Truth. Rodale Pub., Emmaus, Pennsylvania, 325 pp.
Gore, P.J.W., 1999. Radiometric dating. Available at: http://facstaff.gpc.edu/~pgore/geology/geo102/radio.htm.
Great Southern Publishers, Inc., 2010. Map and Walking Tour of Darien, Georgia, 4 pp.

Griffin, M.M., 1981. Feldspar distribution in river, delta, and barrier sands of central South Carolina. M. S. Thesis, Dept. of Geology, University of South Carolina, Columbia, 122 pp.

Griffin, M.M., 1986. Coastal processes. Plate No. 6. In: Kellam, M.F. (ed.), Environmental Geologic Atlas of Coastal Georgia. Dept. of Nat. Resources, Envt. Protection Div., Georgia Geologic Survey, Atlanta, GA.

Griffin, M.M., and Henry, V.J., 1984. Historical changes in mean high water shoreline of Georgia, 1857-1982. Georgia Department of Natural Resources Bulletin 98, 96 pp.

Griggs, G., 1986. Littoral cells and harbor dredging along the California Coast. Environmental Geology, 10:7-20.

Griggs, G.B., Patsch, K., and Savoy, L., 2005. Living With the Changing California Coast. Univ. of California Press, Berkeley, CA, 540 pp.

Gross, T.F., and Werner, F.E., 1997. Tidal circulation modeling of a Georgia coastal sound. Proc. 1997 Georgia Water Resources Conf., Athens, GA, U.S. Geological Survey, University of Georgia, pp. 383-388.

Guza, R.T., and Inman, D.L., 1975. Edge waves and beach cusps. Journal of Geophysical Research, 80:1285-1291.

Hails, J.R., and Hoyt, J.H., 1969. An appraisal of the evolution of the Lower Coastal Plain of Georgia. U.S.A. Trans. Inst. British. Geog., 44:133-139.

Hails, J.R., and Hoyt, J.H., 1972. The nature and occurrence of heavy minerals in Pleistocene and Holocene sediments of the Lower Georgia Coastal Plain. Jour. Sed. Petrol., 42:3:646-666.

Hall, L., 1992. Trochoids, roses, and thorns--beyond the spirograph. College Math. J., 23:20-35.

Hand, B.M., Wessel, J.M., and Hayes, M.O., 1969. Antidunes in the Mt. Toby Conglomerate (Triassic), Massachusetts. Jour. of Sediment. Research, 39:4:1310-1316.

Hands, E.B., 1977. Implications of submergence for coastal engineers. Coastal Sediments '77, Proc., Amer. Soc. Civil Eng., New York, pp. 149-166.

Hatcher, R.D., Jr., 2005. Southern and central Appalachians. In: Selley, R.C., Cocks, R., and Plimer, I. (eds.), Encyclopedia of Geology. Elsevier Ltd., London, pp. 72-81.

Hayes, M.O., 1964. Lognormal distribution of inner continental shelf widths and slopes. Deep-Sea Research, 11:53-78.

Hayes, M.O., 1965. Sedimentation on a semiarid, wave-dominated coast (south Texas); with emphasis on hurricane effects. Ph.D. Thesis, Dept. of Geol., University of Texas, Austin, 350 pp.

Hayes, M.O., 1967. Relationship between coastal climate and bottom sediment type on the inner continental shelf. Marine Geology, 5:111-132.

Hayes, M.O., (ed.), 1969. Coastal Environments – NE, Massachusetts and New Hampshire: Field trip guidebook for Eastern section of SEPM, May 9-11, 462 pp.

Hayes, M.O., 1976. Lecture Notes. In: Hayes, M.O. and Kana, T. W. (eds.), Terrigenous Clastic Depositional Environments. Technical Report 11-CRD, Geol. Dept., Univ. of South Carolina, Columbia, pp. I-1 – I-131.

Hayes, M.O., 1977. Development of Kiawah Island, South Carolina. Coastal Sediments '77, Am. Soc. Civil Eng., New York, pp. 828-847.

Hayes, M.O., 1979. Barrier island morphology as a function of tidal and wave regime. In: Leatherman, S. (ed.), Barrier islands, from the Gulf of St. Lawrence to the Gulf of Mexico. Academic Press, New York, pp. 1-27.

Hayes, M.O., 1980. General morphology and sediment patterns in tidal inlets. Sediment. Geol., 26:139-156.

Hayes, M.O., 1985a. Beach erosion. In: Clark, J.R. (ed.), Coastal Resources Management: Development Case Studies. National Park Service and U.S. Agency for International Dev., pp. 67-201.

Hayes, M.O., 1985b. Atlantic USA – South. In: Bird, E.C.F. and Schwartz, M.L. (eds.) The World's Coastline. Van Nostrand Reinhold, pp. 207-211.

Hayes, M.O., 1994. Georgia Bight. Chapter 7. In: Davis, R.A., Jr. (ed.), Geology of the Holocene Barrier Island System, Springer-Verlag, Berlin, pp. 233-304.

Hayes, M.O., 1999. Black Tides. Univ. of Texas Press, Austin, Texas, 287 pp.

Hayes, M.O., 2009. Barrier islands. In: Gillespie, R., and Clague, D., (eds.), Encyclopedia of Islands, Univ. Cal. Press, Berkeley, pp. 82-88.

Hayes, M.O., 2011. Coastal Heroes. Pandion Books, Columbia, SC, 323 pp.

Hayes, M.O., and Boothroyd, J.C., 1969. Storms as modifying agents in the coastal environment. In: Hayes, M.O.

(ed.), Coastal Environments – NE Massachusetts and New Hampshire: Field trip guidebook for Eastern section of SEPM, May 9-11, pp. 290-315.

Hayes, M.O., and Kana, T.W. (eds.), 1976. Terrigenous Clastic Depositional Environments. Technical Report 11-CRD, Geol. Dept., Univ. of South Carolina, Columbia, 315 pp.

Hayes, M.O., and Sexton, W.J., 1989. Fieldtrip Guidebook T371: Modern clastic depositional environments, South Carolina. 29th Int. Geol. Cong., American Geophysical Union, Washington, D.C., 85 pp.

Hayes, M.O., and Michel, J., 2008. A Coast for All Seasons: A Naturalist's Guide to the Coast of South Carolina. Pandion Books, Columbia, SC, 285 pp.

Hayes, M.O., and Michel, J., 2010. A Coast to Explore: The Coastal Geology and Ecology of Central California. Pandion Books, Columbia, SC, 338 pp.

Hayes, M.O., Michel, J., and Brown, F.J., 1977. Vulnerability of coastal environments of Lower Cook Inlet, Alaska to oil spill impact. In: Proc. 4th International Conference on Port and Ocean Engineering Under Arctic Conditions, 26-30 September 1977, Memorial Univ. of Newfoundland, 12 pp.

Hayes, M.O., Kana, T.W., and Barwis, J.H., 1980. Soft designs for coastal protection at Seabrook Island, SC. Proc. 17th Conf. Coastal Engineering, Amer. Soc. Civil Eng., New York, Chapter 56, pp. 897-912.

Hayes, M.O., Michel, J., and Montello, T.M., 1993. ROPME Sea oil spill nearshore geochemical processes study. Volume 1. Distribution and weathering of oil in intertidal and subtidal sediments for Year 1 (1992). Prepared by Research Planning, Inc. for Marine Spill Response Corporation, Washington, D.C., MSRC Technical Report Series 93-002.1, 230 pp.

Hayes, M.O., Sexton, W.J., and Sipple, K.N., 1994. Fluid-bearing capacity of strandline deposits-implication for hydrocarbon exploration (Abstract). Am. Assoc. Pet. Geol. Bull., 68:485.

Hayes, M.O., Wilson, S.J., FitzGerald, D.M., Hulmes, L.J., and Hubbard, D.K., 1975. Coastal processes and geomorphology. Environmental Inventory of Kiawah Island, Environmental Research Center, Inc., Columbia, SC, 165 pp.

Hayes, M.O., FitzGerald, D.M., Hulmes, L.J., and Wilson, S.J., 1976. Geomorphology of Kiawah Isand, South Carolina. In: Hayes, M.O., and Kana, T.W. (eds.), Terrigenous Clastic Depositional Environments. Technical Report 11-CRD, Geol. Dept., Univ. of South Carolina, Columbia, pp. II-80 – II-100.

Hayes, M.O., Henry, V.J., Moslow, T.F., Wojtal, A.M., Zarillo, G.A., and Lens, L.F., 1980. Coastal environments of Georgia and South Carolina. In: R.W. Frey (ed.), Excursions in Southeastern Geology. Geological Society of America Guidebook V.II, American Geological Institute, Arlington, Virginia, pp. 281-310.

Hayes, M.O., Sexton, W.J., Colquhoun, D.J., and Eckard, T.L., 1993. Evolution of the Santee/Pee Dee Delta Complex, South Carolina, USA. In: Kay, R. (ed.), Deltas of the World, Am. Soc. Civil Eng., New York, pp. 54-64.

Hebert, P., 1980. Atlantic hurricane season of 1979. Available at: http://www.aoml.noaa.gov/general/lib/lib1/nhclib/mwreviews/1979.pdf.

Henry, V.J., 2009. Geology of the Georgia Coast – Coastal Processes. The New Georgia Encyclopedia. Available at: http://www.georgiaencyclopedia.org/nge/Article.jsp?id=h-2777.

Henry, V.J., and Hoyt, J.H., 1968. Quaternary paralic and shelf sediments of Georgia, Southeastern Geology, 9:195-214.

Henry, V.J., and Idris, F.M., 1992. Offshore minerals assessment studies on the Georgia continental shelf – Phase 2: Seismic stratigraphy of the TACTS area and evaluation of selected sites for economic hard minerals potential: Georgia Dept. of Natural Resources Proj. Rept. 18, Atlanta, GA, 143 pp.

Henry, V.J., and Kellam, J.A., 1988. Seismic investigation of the phosphate-bearing, Miocene-age strata of the continental shelf of Georgia. Dept. of Nat. Resources., Georgia Geol. Survey, Bull. 109, 43 pp.

Henry, V.J., McCreery, C.J., Foley, F.D., and Kendall, D.R., 1981. Ocean bottom survey of the Georgia Bight. In: Popenoe, P. (ed.), Environmental Geologic Studies on the Southeastern Atlantic Outer Continental Shelf, 1977-1978, U.S. Geological Survey Open-File Report 81-582-A, U.S. Geological Survey, Woods Hole, Massachusetts, pp. 6-1 - 6-85.

Hess, H.H., 1962. History of ocean basins. In: Engel., A.E.J., James, H.L., and Leonard, B.F. (eds.), Petrologic Studies: A Volume in Honor of A.F. Buddington, Geology Society of America, pp. 599-620.

Hine, A.C., and Snyder, S.W., 1985. Coastal lithosome preservation: evidence from the shoreface and inner continental shelf off Bogue Banks, North Carolina. Mar. Geol., 63:307-330.

Hine, A.C., Riggs, S.R., and Ellington, M.D., 1986. Geologic framework, Cenozoic history, and modern processes of sedimentation on the North Carolina continental margin. In: Textoris, D.A. (ed.), SEPM Field Guidebook, Southeastern U.S. 3rd Annual Midyear Meeting, SEPM, Raleigh, NC, pp. 132-194.

HMdb.org, 2008. Butler Island Plantation. Available at: http://www.hmdb.org/marker.asp?marker=10802.

Hodge, J.A., 1981. Erosion of the North Santee River Delta and development of a flood-tidal delta complex. M.S. thesis, Dept. of Geology, Univ. of South Carolina, Columbia, 151 pp.

Hodler, T.W., and Schretter, H.A., 1986. The Atlas of Georgia. University of Georgia Press, Athens.

Hodnett Cooper, 2008. Coastal Georgia Area Information: Available at: http://www.hodnettcooper.com/st-simons-island-area-info.htm.

Holliday, P.M., 2010. Blowing up the bridge vs. rolling out the welcome mat: The battle between development and conserving natural resources on St. Simons Island. Sherpa Guides, The Natural Georgia Series: Barrier Islands. Available at: http://www.sherpaguides.com/georgia/barrier_islands/blowing_up_the_bridge/

Hoppe, L.M., and Smith, D.R., 2009. Development of nor'easters during *El Niño* years. Available at: http://ams.confex.com/ams/pdfpapers/103397.pdf.

Hosier, P.E., 1975. Dunes and marsh vegetation: In: Environmental Inventory of Kiawah Island, Environmental Research Center, Columbia, SC, pp. D-1 – D-96.

House, A.V., 1954. Labor management problems on Georgia rice plantations, 1840-1860. Agricultural History, 28:4:149-55.

Howard, J.D., 1969. Primary physical and biogenic structures of estuarine sediments: Doboy and Sapelo Sounds, Georgia. (Abstract): Southeastern Sectional Mtg., Geol. Soc. America, Columbia, SC, pp. 37-38.

Howard, J.D., and Reineck, H.E., 1972. Georgia coastal region, Sapelo Island, USA, sedimentology and biology. IV Physical and biogenic sedimentary structures of the nearshore shelf. Senckenbergiana Maritima, 4:81-123.

Howard, J.D., and Dorjes, J., 1972. Animal-sediment relationships in two beach-related tidal flats, Sapelo Island, Georgia. J. Sediment. Petrol., 42:608-623.

Howard, J.D., and Frey, R.W., 1975. Estuaries of the Georgia Coast, USA. I Introduction. Senckenbergiana Maritima, 7:1-31.

Howard, J.D., and Frey, R.W., 1980. Holocene depositional environments of the Georgia Coast and continental shelf. In: Howard, J.D., DePratter, C.B., and Frey, R.W. (eds.), Guidebook 20: Excursions in Southeastern Geology: the Archaeology-Geology of the Georgia Coast. Annual Meeting, Geol. Soc. Amer., Atlanta, pp. 66-134.

Howard, J.D., and DePratter, C.D., 1980. Field trip guide to the archaeology-geology of the Georgia Coast. In: Howard, J.D., DePratter, C.B., and Frey, R.W. (eds.), Guidebook 20: Excursions in Southeastern Geology: the Archaeology-Geology of the Georgia Coast. Annual Meeting, Geol. Soc. Amer., Atlanta, pp. 234-253.

Howard, J.D., Frey, R.W., and Reineck, H.E., 1973. Holocene sediments of the Georgia coastal area. In: Frey, R.W. (ed.), The Neogene of the Georgia Coast. Guidebook, 8th Annual Field Trip, Georgia Geol. Soc., Athens, pp 1-58.

Howard, J.D., Mayou, T.V., and Heard, R.W., 1977. Biogenic sedimentary structures formed by rays. Jour. of Sed. Petrology, 47:1:339-346.

Howard, J.D., DePratter, C.B., and Frey, R.W., (eds.), 1980. Guidebook 20: Excursions in Southeastern Geology: the Archaeology-Geology of the Georgia Coast. Annual Meeting, Geol. Soc. Amer., Atlanta, 254 pp.

Howard, K., 2002. The Milankovitch theory. Available at: www.emporia.edu/earthsci/student/howard2/milan.htm.

Hoyt, J.H., 1962, High-angle beach stratification, Sapelo Island, Georgia. Jour. Sed. Petrol., 32:309-311.

Hoyt, J.H., 1967a. Occurrence of high angle stratification in littoral and shallow neritic environments, Central Georgia Coast. Sedimentology, 8:229-238.

Hoyt, J.H., 1967b. Barrier island formation. Geol. Soc. Amer. Bull., 78:1427-1432.

Hoyt, J.H., 1968. Geology of the Golden Isles and Lower Georgia Coastal Plain, In: Maney, S.D. (ed.), The Future of the Marshlands and Sea Islands of Georgia. Georgia Natural Areas Council and Coastal Area Planning and Development Commission, pp. 18-32.

Hoyt, J.H., and Henry, V.J., 1963. Rhomboid rill marks, an indication of current direction and environment. Jour. of Sed. Petrol., 33:604-608.

Hoyt, J.H., and Weimer, R.J., 1963. Comparison of modern and ancient beaches, central Georgia Coast. Am. Assoc. Petrol. Geol. Bull., 47:529-5531.

Hoyt, J.H., and Henry, V.J., 1964, Development and geologic significance of soft beach sand: Sedimentology, 3:44-51.

Hoyt, J.H., and Henry, V.J., 1965, Significance of inlet sedimentation in the recognition of ancient barrier islands. 19th Field Conference, Wyoming Geol. Assoc. Guidebook, Wyoming Geol. Assoc., Laramie, pp. 190-194.

Hoyt, J.H., and Hails, J.R., 1967. Deposition and modification of Pleistocene shoreline sediments in coastal Georgia: deposition and modification. Science, 155:3769:1541-1543.

Hoyt, J.H. and Hails, J.R., 1974. Pleistocene stratigraphy of Southeastern Georgia. In: Oaks, R.Q., and DuBar, J.R. (eds.), Post-Miocene Stratigraphy, Central and Southern Atlantic Coastal Plain. Utah State Univ. Press., Logan, pp. 191-205.

Hoyt, J.H., Weimer R.J., and Henry V.J., 1964. Late Pleistocene and Recent sedimentation, central Georgia Coast, USA. Developments in Sedimentology, Deltaic and Shallow Marine Deposits, 1:170-176.

Hoyt, J.H., Henry, V.J., and Howard, J.D., 1966. Pleistocene and Holocene sediments, Sapelo Island, Georgia and vicinity. Southeastern Section, Geol. Soc. America, Mtg. at Sapelo Island, Georgia, Guidebook for Field Trip No. 1, pp. 1-27.

Hubbard, D.K., 1977. Variations in tidal inlet processes and morphology in the Georgia Embayment. Tech Rept. 14-CRD, Dept. Geol., Univ. of South Carolina, Columbia, 79 pp.

Hubbard, D.K., Oertel, G., and Nummedal, D., 1979. The role of wave and tidal currents in the development of tidal-inlet sedimentary structures and sand body geometry: examples from North Carolina, South Carolina, and Georgia. Jour. of Sed. Petrol., 49:1073-1092.

Huddlestun, P.F., 1986. Geomorphology. Plate No. 2. In: Kellam, M.F., (ed.) 1986, Environmental Geologic Atlas of Coastal Georgia. Dept. of Nat. Resources, Envt. Protection Div., Georgia Geologic Survey, Atlanta, GA

Imbrie, J., Hays, J.D., Martinson, D.G., McIntyre, A., Mix, A.C., Morley, J.J., Pisias, N.G., Prell, W.L., and Shackleton, N.J., 1984. The orbital theory of Pleistocene climate: Support from a revised chronology of the marine ∂^{18}O record. In: Berger, A., Imbrie, J., Hays, J., Kukla, G., and Saltzman, B. (eds.), Milankovitch and Climate: Understanding the Response to Astronomical Forcing. D. Reidel Pub. Co., Dordrecht, pp. 269-305.

Inman, D.L., and Nordstrom, C.E., 1971. On the tectonic and morphologic classification of coasts. Jour. Geol., 79:1-21.

Inman, D.L., and Jenkins, S.A., 2005. Climate patterns in the coastal zone. In: Schwartz, M.L., (ed.), Encyclopedia of Coastal Science. Springer, The Netherlands, pp. 243-246.

International Whaling Commission, 2001. Report of the workshop on the comprehensive assessment of right whales: A worldwide comparison. Journal of Cetacean Research and Management, Special Issue 2:1-60.

IPCC, 2007. Fourth Assessment Report of the Intergovernmental Panel on Climate Change. Cambridge University Press, Cambridge and New York, 1,075 pp.

Jackson, C.W., Alexander, C.R., and Bush, D.M., 2007. Back-barrier shoreline change history: Cumberland Island, GA 1957-2002. Southeastern Geology, 45(2):73-85.

Jackson, C.W., Alexander, C.R., and Bush, D.M., 2012. Application of the AMBUR R Package for Shoreline Change Analysis: Jekyll Island, Georgia. Computers and Geoscience, 10.1016/j.cageo.2011.08.009 (published online – published in print 3/2012).

Jenkins, M., 2010. Mason Inlet Relocation Project, Wrightsville Beach, North Carolina. Applied Technology & Management (ATM). Available at: http://www.appliedtm.com/projects/coastal/mason-inlet.htm>

Johnson, S.S., Hillestad, H.O., Shanholtzer, S.F., and Shanholtzer, G.F., 1974. Cumberland Island: An Ecological Survey of the Coastal Region of Georgia. National Park Service Scientific Monograph No. 3.

Kahn, R., 1972. The Boys of Summer. Harper/Collins Publishers, New York, 442 pp.

Kana, T.W., 1976. Coastal processes and sediment transport at Price and North Inlets, South Carolina. M. S. Thesis, Dept. of Geology, Univ. of South Carolina, Columbia, 144 pp.

Kana, T.W., 1977. Suspended sediment transport of Price Inlet, South Carolina. Coastal Sediments '77, American Society of Civil Engineers, pp. 366-382.

Kana, T.W., 1988. Beach erosion in South Carolina. Report for South Carolina Sea Grant Consortium, Charleston, 55 pp.

Kana, T.W., 1989. Erosion and beach restoration at Seabrook Island, South Carolina. Shore and Beach, 57(3):3-18.

Kana, T.W., 2006. Myrtle Beach restoration: A success story? South Carolina Business & Economic Review, Moore School of Business, USC, 52(4):19-22.

Kana, T.W., and Mason, J.E., 1988. Evolution of an ebb-tidal delta after inlet relocation. In: Aubrey, D.G., and Weishar, L. (eds.), Hydrodynamics and Sediment Dynamics of Tidal Inlets. Springer-Verlag, New York, pp. 382-411.

Kana, T.W., and Gaudiano, D.J., 2001. Regional beach volume changes for the central South Carolina Coast. Technical Report, Dept. of Geol. Sciences, Univ. of South Carolina, Columbia, SC, 124 pp.

Kana, T.W., and McKee, P., 2003. Relocation of Captain Sam's Inlet – 20 years later: Coastal Sediments '03, Amer. Soc. Civil Eng., 12 pp.

Kana, T.W., Hayter, E.J., and Work, P.A., 1999. Mesoscale sediment transport at southeastern U.S. tidal inlets: Conceptual model applicable to mixed energy settings. Jour. of Coastal Research, 15:303-313.

Kana, T.W., White, T.E., and McKee, P., 2004. Management and engineering guidelines for groin rehabilitation: Jour. of Coastal Research, v. 33, p. 57-82.

Kana, T.W., Michel, J., Hayes, M.O., and Jensen, J.R., 1984. The physical impact of sea-level rise in the area of Charleston, SC. In: Barth, M.C., and Titus, J.G. (eds.), Greenhouse Effect and Sea Level Rise, Van Nostrand Reinhold Co., New York, pp. 105-150.

Kaufman, Y., 2002. On the shoulders of giants - Milutin Milankovitch (Earth Observatory, NASA). Available at: http://earthobdservatory.nasagov.Library/Grants/Milankovitch.

Kaufman, B., and Magill, K.E., 1989. Hydrological changes in Charleston Harbor. Coastal Zone '89, Proc. 6th Symposium of Coastal and Ocean Management, Amer. Soc. Civil Eng., New York, pp. 2,640-2,649.

Kellam, M.F., 1986. (ed.) Environmental Geologic Atlas of Coastal Georgia. Dept. of Nat. Resources, Envt. Protection Div., Georgia Geologic Survey, Atlanta, GA, 10 Plates.

Knott, D., 2010. Atlantic Ghost Crab. Available at: http://www.dnr.sc.gov/cwcs/pdf/Ghostcrab.pdf.

Knowlton, A.R., Kraus, S.D., and Kenney, R.D., 1994. Reproduction in North Atlantic right whales (*Eubalaena glacialis*). Canadian Journal of Zoology, 72:1,297-1,305.

Komar, P.D., 1976. Beach Processes and Sedimentation. Prentice-Hall, Inc., Englewood Cliffs, New Jersey, 429 pp.

Komar, P.D., and Wang, C., 1984. Processes of selective grain transport and the formation of placers on beaches. Jour. of Geol., 92:637-655.

Komar, P.D., and Li, Z., 1988. Pivoting analyses of the selective entrainment of sediments by shape and size with application to gravel threshold. Sedimentology, 33:425-436.

Kraus, S.D., Kenney, R.D., Knowlton, A.R., and Ciano, J.N., 1993. Endangered right whales of the southwestern North Atlantic. U.S. Department of the Interior, Minerals Management Service, Herndon, VA, MMS 93-0024, 77 pp.

Kriesel, W., and Turner, S., 2003. A community economics perspective on coastal erosion management. American Agricultural Economics Association Annual Meeting, Montreal, Canada, July 27-30, 2003, 28 pp.

Kukla, G.J., 2000. Paleoclimate: The last interglacial. Science, 287(5455):987-988.

Kumar, N., and Sanders, J.E., 1974. Inlet sequences – a vertical succession of sedimentary structures and textures created by the lateral migration of tidal inlets. Sedimentology, 21:291-323.

Land, L.S., 1964. Aeolian cross-bedding in the beach dune environment, Sapelo Island, Georgia: Jour. Sed. Petrol., v. 34, p. 389-394.

Land, L.S., and Hoyt, J.H., 1966. Sedimentation in a meandering estuary. Sedimentology, 6:191-207.

Langley, S.K., Alexander, C.R., Bush, D.M., and Jackson, C.W., 2003. Modernizing shoreline change analysis in Georgia using topographic survey sheets in a GIS environment. Jour. of Coastal Research, 38:168-177.

Larson, E., 1999. Issac's Storm: A Man, a Time, and the Deadliest Huricane in History. Crown Publishing Group, 336 pp.

LeConte-Woodmanston, 2010. LeConte-Woodmanston Plantation and Botanical Gardens: Available at: http://www.leconte-woodmanston.org/index-aboutus.html.

LeGrand, H.E., 1961. Summary of geology of Atlantic Coastal Plain. Am. Assoc. Pet. Geol. Bull., 45:1,557-1,571.

Lenz, R.J., 1999. Longstreet Highroad Guide to the Georgia Coast & Okefenokee. Longstreet, Atlanta, GA, 336 pp.

Lenz, R.J., 2002a. Longstreet Highroad Guide to the Georgia Coast & Okefenokee. Available at: http://www.sherpaguides.

com/georgia/coast/natural_history/natural_processes.html.

Lenz, R.J., 2002b. Sherpa Guides – Ossabaw Island. Available at: http://www.sherpaguides.com/georgia/coast/northern_coast/ossabaw_island.html.

Lenz, R.J., 2002c. Sherpa Guides – Shellman Bluff. Available at: http://www.sherpaguides.com/georgia/coast/central_coast/shellman_bluff.html.

Lenz, R.J., 2002d. Sherpa Guides – Altamaha River: Available at: http://www.sherpaguides.com/georgia/coast/central_coast/altamaha_river_bioreserve.html.

Lin, S.H., 2012. Many horseshoe crabs and birds on Delaware Bay. Available at: http://www.shltrip.com/Horseshoe_Crab.html.

Linsley, D.M., Bishop, G.A., and Rollins, H.B., 2008. Stratigraphy and geologic evolution of St. Catherines Island. In: Thomas, D.H. (ed.), Native American Landscapes of St. Catherines Island. Theoretical framework. Anthropological Papers of the American Museum of Natural History 88(1):26–41.

Liritzis, I., 2011. Surface dating by luminescence: an overview. Geochronometria, 38(3):292-302.

Livingston, J., 2010. Tabby sugar works ruins. SmugMug, Inc. Available at: http://actionshots4u.smugmug.com/History/Tabby-Sugar-Works-Ruins/998682_LipgZ#46187695_CjKNF.

Lovelock, J.E., 1972. Gaia as seen through the atmosphere. Atmospheric Environment, 6:579-580.

Luminescence Laboratory, 2012. Optical stimulated luminescence. Available at: http://www.uic.edu/labs/ldrl/osl.html.

Lyell, C., 1845. Travels in North America, Vol. 1. Wiley & Putnam. New York, 231 pp.

Matz, D.B., and Hayes, M.O., 1966. Sedimentary petrology of Beacon sediments. Antarctic Jour., United States, 1:134-135.

McCarney-Castle, K., Voulgaris, G., and Kettner, A.J., 2010. Analysis of fluvial suspended sediment load contribution through anthropocene history to the South Atlantic Bight coastal zone, U.S.A. Journal of Geology 118:399-416.

Meade, R.H., 1969. Errors in using modern stream-load data to estimate natural rates of denudation. Geological Society of America Bulletin, 80:1,265-1,274.

Meisburger, E.P., and Field, M.E., 1974. Geomorphology, shallow structure, and sediments of Florida inner continental shelf, Cape Canaveral to Georgia. Tech, Memo 54, U.S. Army Corps of Engineers, Coastal Engineering Research Center, Fort Belvoir, Virginia, 119 pp.

Meyer, B.K., Bishop, G.A., and Vance, R.K., 2011. An evaluation of shoreline dynamics at St. Catherines Island, Georgia (1959-2009) utilizing the digital shoreline analysis system (USGS). Geological Society of America Abstracts with Programs, 43(2):68.

Michel, J. (ed.), 2013. South Atlantic Information Resources: Data Search and Literature Synthesis. U.S. Department of the Interior, Bureau of Ocean Energy Management, Regulation, and Enforcement, Gulf of Mexico OCS Region, New Orleans, LA, OCS Study BOEM 2013-01157.

Milankovitch, M., 1941. Kanon der Erdbestrahlung und seine Andwendung auf das Eiszeiten-problem. Royal Serbian Academy, Belgrade. (English translation: 1998, Canon of Insolation and the Ice Age Problem. With introduction and biographical essay by Nikola Pantic. Alven Global, 636 pp.

Mitsch, W.J., and Gosselink, J.G., 2000. Wetlands, 3rd edition. John Wiley and Sons, New York, 920 pp.

Monroe, J.S., and Wicander, R., 1998, Physical Geology, 3rd edition. Wadsworth Publishing. Company, Belmont, California, 646 pp.

Moslow, T.F., 1980. Stratigraphy of mesotidal barrier islands. Ph.D. Thesis, Dept. Geol., Univ. of South Carolina, Columbia, 186 pp.

Moslow, T.F., and Heron, D.S., 1978. Relict inlets: Preservation and occurrence in the Holocene stratigraphy of southern Core Banks, NC. J. Sediment. Petrol., 48:1275-1286.

Moslow, T.F., and Tye, R.S., 1985. Recognition and characterization of Holocene tidal inlet sequences. In: Oertel, G.F., and Leatherman, S.P. (eds.), Barrier Islands. Mar. Geol. 63:129-152.

Morton, R.A., and Miller, T.L., 2005. National assessment of shoreline change: Part 2: Historical shoreline changes and associated coastal land loss along the U.S. Southeast Atlantic Coast: U.S. Geological Survey Open-file Report 2005-1401.

Muhs, D.R., Wehmiller, J.F., Wehmiller, K.R., and York, L.L., 2004. Quaternary sea-level history of the United States.

Developments in Quaternary Science, 1:147-183.
Murphy, C.H., 1995. Carolina Rocks! Sandlapper Publishing Co., Inc., Orangeburg, SC, 261 pp.
Murray, G.E., 1961. Geology of the Atlantic and Gulf Coastal Province of North America. Harper & Bros., New York, 692 pp.
Nairn, R., and Hayes, M.O., 1997. Large-scale coastal evolution in the vicinity of Keta Lagoon, Ghana. (Abstract) International Conference on Coast Research through Large-Scale Experiments, University of Plymouth.
National Wildlife Federation, 2008. Global warming and Georgia. Available at: docs.google.com/viewer?a=v&q=cache:pZI98v8ezZEJ:www.nwf.org/Global-Warming/~/media/PDF.
NERRS, 2012. Sapelo Island, Georgia. Available at: http://www.nerrs.noaa.gov/Reserve.aspx?ResID=SAP.
New Georgia Encyclopedia, 2002. History and archaeology. Available at: http://www.georgiaencyclopedia.org/nge/Article.jsp?id=h-679.
New World Encyclopedia, 2010. Pangaea. Available at: http://www.newworldencyclopedia.org/entry/Pangaea.
Nichols, M.M., and Biggs, R.B., 1985. Estuaries. In: Davis., R.A., Jr. (ed.), Coastal Sedimentary Environments. Second edition. Springer-Verlag, New York, p. 77-186.
NOAA/RPI. 1996a. Environmental Sensitivity Index (ESI) Maps of the Shoreline of the Coast of South Carolina. By Research Planning, Inc. under contract to the National Oceanic and Atmospheric Administration, NOS Data Explorer, 63 maps.
NOAA/RPI. 1996b. Environmental Sensitivity Index (ESI) Maps of the Shoreline of the Coast of North Carolina (3 volumes). By Research Planning, Inc. under contract to the National Oceanic and Atmospheric Administration, NOS Data Explorer, 135 maps.
NOAA/RPI. 1997. Environmental Sensitivity Index (ESI) Maps of the Shoreline of the Coast of Georgia. By Research Planning, Inc. under contract to the National Oceanic and Atmospheric Administration, NOS Data Explorer, 39 maps.
nps.gov, 2010. Cumberland Island National Seashore, Fees and Reservations. Available at: http://www.nps.gov/cuis/planyourvisit/feesandreservations.htm.
Nummedal, D., Oertel, G.F., Hubbard, D.K., and Hine, A.C., 1977. Tidal inlet variability – Cape Hatteras to Cape Canaveral. Proc. 5th Symp. Coastal Sediments '77, Am. Society Civil. Eng., Charleston, SC, pp. 543-562.
O'Brien, M.P., 1931. Estuary tidal prisms related to entrance areas. Jour. Civil Eng, 1(8):738-793.
Oertel, G.F., 1971, Sediment-hydrodynamic inter-relationships at the entrance to the Doboy Sound Estuary, Sapelo Island, Georgia. Ph.D. Dissertation, Univ. of Iowa, Iowa City, Iowa, 172 pp.
Oertel, G.F., 1972, Sediment transport on estuary entrance shoals and the formation of swash platforms. Jour. Sed. Petrology, 42:858-863.
Oertel, G.F., 1973a. A sedimentary framework of the substrate adjacent to Georgia tidal inlets. In: Frey, R.W. (ed.), The Neogene of the Georgia Coast, Guidebook for the 8th Annual Field Trip of the Georgia Geological Society, p. 59-66.
Oertel, G.F., 1973b, Examination of textures and structures of mud in layered sediments at the entrance of a Georgia tidal inlet. Jour. Sed. Petrology, 43:33-41.
Oertel. G.F., 1975a. Ebb-tidal deltas of Georgia estuaries. In: Cronin, L.E. (ed.), Estuarine Research, Vol. 2., Geology and Engineering, Academic Press, New York, pp. 267-276.
Oertel. G.F., 1975b. Post Pleistocene island and inlet adjustment along the Georgia Coast. Jour. of Sediment. Petrol., 45:1:150-159.
Oertel, G.F., 1977. Geomorphic cycles in ebb deltas and related patterns of shore erosion and accretion. Jour. Sediment. Petrol., 47:1121-1131.
Oertel, G.F., 1979. Barrier island development during the Holocene recession, Southeastern United States. In: Leatherman, S. (ed.), Barrier islands, from the Gulf of St. Lawrence to the Gulf of Mexico. Academic Press, New York, pp. 273-280.
Oertel, G.F., 1988. Processes of sediment exchange between tidal inlets, barrier island and ebb deltas. Hydrodynamics and Sediment Dynamics of Tidal Inlets. Springer-Verlag, NY, pp. 297-318.
Oertel, G.F., and Howard, J.D., 1972. Water circulation and sedimentation at estuary entrances on the Georgia Coast:

In: Swift., D.J.P., Duane, D.B., and Pilkey, O.H. (eds.), Shelf Sediment Transport. Dowden, Hutchinson, and Ross, New York, pp. 411-427

Oertel, G.F., and Chamberlain, C.F., 1975. Differential rates of shoreline advance and retreat at coastal barriers of Chatham and Liberty Counties, Georgia. Transactions of the Gulf Coast Assoc. of Petroleum Geologists, 25:383-390.

Oertel, G.F., and Larsen, M., 1976. Development sequences in Georgia coastal dunes and distribution of dune plants. Bull. Georgia Acad. Sci., 34:35-48.

Oertel, G.F., and Dunstan, W.M., 1981. Suspended sediment distribution and certain aspects of phytoplankton production off Georgia, U.S.A. Jour. Marine Geology, 40:171-197.

Oertel, G.F., Kraft, J.C., Kearney, M.S., and Woo, S.J., 1992. A rational theory for barrier-lagoon development. In: Quaternery Coasts of the United States: Marine and Lacustrine Systems, SEPM Special Publication No. 48, pp. 77-87.

Oertel, G.F., Allen, T. R., and Foyle, A.M., 2008. The influence of drainage hierarchy on pathways of barrier retreat: An example from Chincoteague Bight, Virginia, U.S.A. Southeastern Geology, 45:179-201.

Olsen, M., 2006. Georgia's Barrier Islands. Sapelo Island National Research Reserve. Available at: http://www.knowtheconnection.com/pdf/CC%20GA%20barrier%20islands.pdf.

OperBBA, 2012. Operation Big Blue Association, Ghost Crab. Available at: http://www.operationbigblue.org/index.php?option=com_content&view=article&id=117%3Aghost-crab&catid=44%3.

Parker, A.W., 1997. Scottish Highlanders in Colonial Georgia, The Recruitment, Emigration and Settlement at Darien, 1735-1748. University of Georgia Press, Athens, GA, 182 pp.

Paull, C.K., and Dillon, W.P., 1982, Reply to Comment on erosional origin of the Blake Escarpment: an alternative hypothesis. Geology, 4:409-463.

Peltier, W.R., 2002. On eustatic sea level history: Last Glacial Maximum to Holocene. Quat. Science Reviews, 21:377-396.

Peterson, C.H., 2000. Short-term consequences of nourishment and bulldozing on the dominant large invertebrates of a sandy beach. Jour. of Coastal Research, 16(2):368-387.

Petit, J.R., Jouzel, J., Raynaud, D., Barkov, N.I., Barnola, J.M., Basile, I., Bender, M., Chappellaz, J., Davis, J., Delayque, G., Delmotte, M., Kotlyakov, V.M., Legrand, M., Lipenkov, V., Lorius, C., Pépin, L., Ritz, C., Saltzman, E., and Stievenard. M., 1999. Climate and atmospheric history of the past 420,000 years from the Vostok ice core, Antarctica. Nature, 399:429-436.

Pfeffer, W.T., Harper, J.T., and O'Neel, S., 2008. Kinematic constraints on 21st century sea level rise. Science, 321(5894):134-143.

Picton, J.A., 1905. Pantheism: its story and significance. Archibald Constable & Co Ltd., Chicago, 48 pp.

Pidwirny, M., 2006. Stream Morphometry. Fundamentals of Physical Geography, 2nd Edition. Available at: http://www.physicalgeography.net/fundamentals/10ab.html.

Pierce, J.W., and Colquhoun, D.J., 1970. Holocene evolution of a portion of the North Carolina Coast. Geol. Soc. Am. Bull., 81:3697-3714.

Pierce, D.W., Barnett, T.P., and Latif, M., 2000. Connections between the Pacific ocean tropics and midlatitudes on decadal time scales. Jour. of Climate, 13:1,173-1,194.

Pilkey, O.H., 1963. Heavy minerals of the U.S. South Atlantic continental shelf and slope. Geol. Soc. Am. Bull., 74:641-648.

Pilkey, O.H., 2007. Beach nourishment is not the answer. South Carolina Business & Economic Review, Moore School of Business, USC, 53(2).

Pilkey, O.H., and Richter, D.M., 1965, Beach profiles of a Georgia barrier island. Southeastern Geology, 6:11-19.

Pilkey, O.H., and Giles, R.T., 1965. Bottom topography of the Georgia continental shelf. Southeastern Geology, 7:15-18.

Pilkey, O.H., and Terlecky, P.M., 1966. Distribution of surface sediments on the Georgia continental shelf. In: Hoyt, J.H., Henry, V.J., and Howard, J.D. (eds.), Pleistocene and Holocene Sediments, Sapelo Island, Georgia, and Vicinity. Geological Society of America Southeastern Section, Field Trip No. 1, pp. 28-39.

Pilkey, O.H., and Fraser, M.E., 2003. A celebration of the world's barrier islands. Columbia Univ. Press, New York, 309 pp.

Pilkey, O.H., and Pilkey-Jarvis, L., 2006. Useless arithmetic: Why Environmental Scientists Can't Predict the Future. Columbia Univ. Press, 230 pp.

Pilkey, O.H., Blackwelder, B.W., Doyle, L.J., Estes, E., and Terlecky, P.M., 1969. Aspects of carbonate sedimentation on the Atlantic continental shelf off the southern United States. Jour. Sed. Petrol, 39:744-768.

Pinet, P.R., Matz, D.B., and Hayes, M.O., 1967. Petrology of the upper division of the Beacon Sandstone, Antarctic Jour., United States, 2:108-109.

Pinet, P.R., Matz, D.B., and Hayes, M.O., 1971. An Upper Paleozoic tillite in the Dry Valleys, South Victoria Land, Antarctica. Jour. of Sed. Research, 41(3):835-838.

Pirkle, E.C., and Yoho, W.H., 1970. The heavy mineral ore body of Trail Ridge, Florida. Econ. Geol., v. 65, pp. 17-30.

Pirkle, F.L., Pirkle, E.C., and Reyolds, J.G., 1993. Yulee heavy mineral deposits. In: Farrell, K.M., Hoffman, C.W., and Henry, V.J. (eds.), Geomorphology and facies relationships of Quaternary Barrier Island Complexes near St. Marys, Georgia. Georgia Geol. Soc. Guidebooks, 1(1):68-73.

Pirkle, F.L., Pirkle, W.A., and Rich, F.J., 2013. Heavy mineral mining in the Atlantic coastal plain and what deposit locations tell us about ancient shorelines. Jour. Coastal Research Special Issue 69.

Playfair, J., 1802. Illustrations of the Huttonian Theory of the Earth. Edinburgh, 528 pp.

Popenoe, P., 1992. Maps and cross sections depicting the shallow seismic stratigraphy of the continental shelf and slope off Georgia from interpretation of high-resolution seismic-reflection data. Project Rept. No. 15, Dept. of Nat. Resources, Envt. Protection Div., Georgia Geologic Survey, Atlanta, 35 pp.

Postma, H., 1967. Sediment transport and sedimentation in the estuarine environment. Estuaries, 83:158-79.

Pritchard, D.W., 1967. What is an estuary: Physical viewpoint. In: Lauff, G.W. (ed.), Estuaries, Am. Assoc, Adv, Sci. Publ., 83:3-5.

Queram, K.E., 2012. 10 years later, Mason Inlet move considered a success. Star News online. Available at: http://www.starnewsonline.com/article/20120313/ARTICLES/120319907.

Redfield, A.C., 1967. The ontogeny of a salt marsh estuary. In: Lauff, G.H. (ed.), Estuaries. American Assoc. for the Advancement of Science, 83:108-114.

Reel, C.L., 1982. The origin and geologic significance of bubble sand, Bull Island, South Carolina. M.S. thesis, Dept. Geol., Univ. South Carolina, 134 pp.

Rohde, R.A., 2009. Carbon dioxide variations. Available at: commons.wikimedia.org/wiki/File:Carbon_Dioxide_400kyr.png.

Robinson, M.H., Alexander, C.R., Jackson, C.W., McCabe, C.P., and Crass, D., 2010. Threatened archaeological, historic, and cultural resources of the Georgia Coast: Identification, prioritization and management using GIS technology. Geoarchaeology, 25:3:312-326.

Ronline, J.C., Ķeizars, Z., Forrest, B., and Rink, W.J., 2008. Natural residual thermoluminescence as a method of analysis of sand transport along the coast of the St. Joseph Peninsula, Florida. Jour. of Coastal Research, 24:500-507

Ruby, C.H., 1981. Clastic facies and stratigraphy of a rapidly retreating cuspate foreland, Cape Romain, South Carolina. Ph.D. Thesis, Dept. of Geology, Univ. of South Carolina, Columbia, 207 pp.

Rust, B.R., 1978. A classification of alluvial channel systems. In: Miall, A.D. (ed.), Fluvial Sedimentology. Canadian Soc. Petroleum Geology, Mem. 5:187-198.

Sandercock, A.C., and Alhadeff, S.J., 1996. Digital atlas of state of Georgia, physiographic provinces. U.S. Geological Survey, Digital Data Series, Atlanta, Georgia.

Saxena, R.S. (ed.), 1976. Sedimentary Environments and Hydrocarbons. American Association of Petroleum Geologists, and Society of Economic Paleontologists Mineralogists. Annual Meeting Short Course. New Orleans, LA.

Schmitt, M.A., and Haines, A.C., 2003. Beach nourishment: The magic bullet for Georgia's Shore? Available at: http://cms.ce.gatech.edu/gwri/uploads/proceedings/2003/Schmitt%20and%20Haines.PDF.

Sexton, W.J., 1987. Morphology and sediment character of mesotidal shoreline depostional environments. Ph.D. Thesis, Dept. of Geology, Univ. of South Carolina, Columbia, 197 pp.

Sexton, W.J., and Hayes, M.O., 1982. Natural bar-bypassing of sand at a tidal inlet. Proc. 18th Coastal Eng. Conf.,

Capetown , South Africa, Am. Soc. Civil Eng., v. 2, pp. 1,479-1,495.

Sexton, W.J., and Hayes, M.O., 1991. The geologic impact of hurricane *Hugo* and post-storm recovery along the undeveloped coastline of South Carolina, Dewees Island to the Santee Delta. Jour. of Coastal Research, Special Issue 8:275-290.

Sexton, W.J., and Hayes, M.O., 1996. Holocene deposits of reservoir-quality sand along the Central South Carolina coastline. American Assoc. Petroleum Geologists Bull., 80(6):831-855.

Shepard, F.P., 1950a. Longshore Bars and Longshore Troughs. Tech. Memo 15, Beach Erosion Board. U.S. Army Corps of Engineers, Washington, D.C., 38 pp.

Shepard, F.P., 1950b. Beach Cycles in Southern California. Tech. Memo 20, Beach Erosion Board. U.S. Army Corps. of Engineers, Washington, DC., 26 pp.

Shepard, F.P., and Wanless, H.R., 1971. Our Changing Coastlines. McGraw-Hill, New York, 579 pp.

Sherpa Guides, 1999. The Altamaha River Natural Reserve. Available at: http://www.sherpaguides.com/georgia/coast/central_coast/altamaha_river_bioreserve.htm.

Silvester, R., 1977. The role of wave reflection in coastal processes. Coastal Sediments '77, Amer. Soc. Civil Eng., New York, pp. 639-654.

Simons, D.B., and Richardson, E.V., 1962. Resistance to flow in alluvial channels. Am. Soc. Civil Eng. Trans., v. 127, pp. 927-953.

SIWMA, 2010. Sapelo Island Wildlife Management Area. Available at: http://www.ohranger.com/ga/sapelo-island-wildlife-management-area.

Smithsonian Marine Station, 2008. *Mulinia lateralis*. Available at: http://www.sms.si.edu/irlspec/Mulinia_lateralis.htm.

Schmitter, R.D., and Freeman-Lynde, R.P., 1988. Hornblende content of Georgia-South Carolina, U.S.A., nearshore sands: support for shoreward sediment transport in the nearshore zone. Sedimentary Geology, 57:153-169.

Southern Coastal Plain, 2010. Okefenokee Swamp. Available at: http://www1.gadnr.org/cwcs/PDF/13_SouthernCoastalPlain.pdf.

Stalter, R., 1974. Vegetation in the Cooper River Estuary. Cooper River Environmental Study, SC Water Resources Commission, Columbia, SC.

Stephen, M.F., Brown, P.J., FitzGerald, D.M., Hubbard, D.K., and Hayes, M.O., 1975. Beach erosion inventory of Charleston County, South Carolina: A Preliminary Report. South Carolina Sea Grant Technical Report 4, South Carolina Sea Grant Program, Charleston, 79 pp.

Stutz, M.L., and Pilkey, O.H., 2011, Open-ocean barrier islands: Global influence of climatic, oceanographic, and depostional settings. Journal of Coastal Research, 27(2):207-222.

Surfrider Foundation, 2010. Georgia beach fill. Available at: http://surfrider.org/stateofthebeach/05-sr/state.asp?zone=SE&state=ga&cat=bf

Swift, D.J.P., 1975. Barrier-island genesis: Evidence from the central Atlantic shelf, eastern USA. Sedimentary Geology, 14:1-43.

Sydow, J., and Roberts, H.H., 1994. Stratigraphic framework of a late Pleistocene shelf-edge delta, Northeast Gulf of Mexico. Amer. Assoc. Petrol. Geol. Bull., 78(8):1,276-1,312.

Tannehill, I.R., 1956. Hurricanes. Princeton Univ. Press, 308 pp.

Taylor, J.R., Cardamone, M.A., and Mitsch, W.J., 1990. Bottomland hardwood forests: Their functions and values. In: Gosselink, J.G., Lee, L.C., and Mitsch, W.J. (eds.), Ecological processes and cumulative impacts: illustrated by bottomland hardwood wetland ecosystems. Lewis Pub. Co., Chelsea, Mississippi, pp. 13-86.

Teal, J.M., 1958. Distribution of fiddler crabs in Georgia salt marshes. Ecology, 39:185-193.

The Ashantilly Center, 2010. Preserving the legacies of Thomas Spalding and William J. Haynes, Jr. Available at: http://ashantilly.org/blog/?page_id=2.

The Free Dictionary, 2012. Spanish moss. Available at: http://encyclopedia2.thefreedictionary.com/Tillandsia+usneoides.

The Nature Conservancy, 2010. Coastal Georgia. Available at: http://www.nature.org/wherewework/northamerica/states/georgia/features/art28009.html.

Tybee Island Beach Management Plan, 2005. Sponsored by Georgia Coastal Management Program and NOAA. Available at: http://www.cityoftybee.org/Assets/Files/BeachTaskForce/Beachplan.pdf.

Turck, J.A., 2011. Geoarchaeological analysis of two back-barrier islands and their relationship to the changing landscape of coastal Georgia, U.S.A. Ph.D. Dissertation, University of Georgia, Athens, GA., 239 pp.

Turck, J.A., and Alexander, C., 2013. Chapter 7. Coastal landscapes and their relationship to human settlement on the Georgia Coast. In: Thompson, V.D. and Thomas, D.H. (eds.), Life Among The Tides, Recent Archaeology on the Georgia Bight, Anthropology Papers of The American Museum of Natural History, No. 98.

Tye, R.S., 1981. Geomorphic evolution and stratigraphic framework of Price and Capers inlets, South Carolina. M.S. Thesis, Dept. of Geology, Univ. of South Carolina, Columbia. 144 pp.

Tye, R.S., 1984. Geomorphic evolution and stratigraphy of Price and Capers Inlets, South Carolina. Sedimentology, 31:655-674.

UCal Berkeley, 2004. Foraminifera. Museum of Paleontology. Available at: http://www.ucmp.berkeley.edu/index.html.

UGA/Geol., 2009. The geology of Georgia. Available at: http://www.gly.uga.edu/railsback/GAGeology.html.

UGAMI, 2010. University of Georgia Marine Institute on Sapelo Island: Available at: http://www.uga.edu/ugami/visit/about.htm.

U.S. Army Corps of Engineers, 1971. National shoreline study, Washington, D.C., 62 pp.

U.S. Army Corps of Engineers, 2002. Coastal Engineering Manual. Engineer Manual 1110-2-1100, U.S. Army Corps of Engineers, Washington, D.C. Accessible at: http://bigfoot.wes.army.mil/cem026.html.

U.S. Army Corps of Engineers, 2009. Tybee Island, Georgia Shore Protection Project. Available at: http://www.cityoftybee.org/Assets/Files/BeachTaskForce/TybeeIslandOM20090706.pdf.

U.S. Fish and Wildlife Service, 2009a. Blackbeard Island National Wildlife Refuge. Available at: http://www.fws.gov/blackbeardisland/.

U.S. Fish and Wildlife Service, 2009b. Wolf Island National Wildlife Refuge. Available at: http://www.fws.gov/wolfisland/.

U.S. Fish and Wildlife Service, 2010a. Wassaw National Wildlife Refuge. Available at: http://www.fws.gov/refuges/profiles/index.cfm?id=41628.

U.S. Fish and Wildlife Service, 2010b. Okefenokee National Wildlife Refuge. Available at: http://www.fws.gov/okefenokee/.

USGS, 2005. Diagram of the interior of the Earth. Available at: http://pubs.usgs.gov/gip/dynamic/inside.html

USGS, 2012. Earthquake facts and statistics. Available at: http://earthquake.usgs.gov/earthquakes/eqarchives/year/eqstats.php.

Valverde, H.R., Trembanis, C., and Pilkey, O.H., 1999. Summary of beach renourishment episodes on the East Coast Barrier Islands. Jour. of Coastal Research, 15(4):1,100-1,118.

Van Straaten, L.M.J.U., 1950. Environment of formation and facies of the Wadden Sea sediments. Koninkl. Ned. Aardrijkskde Genoot., 67:94-108.

Walton, T.L., and Adams, W.D., 1976. Capacity of inlet outer bars to store sand. Proc. 15th Conf., Coastal Eng., American Society of Civil Engineers, pp. 1919-1937.

Wang, P., Ebersole, B.A., and Smith, E.R., 2002. Longshore sand transport – initial results from large-scale sediment transport facility. ERDC/CHL CHETN-II-46, 10 pp.

Ward, L.G., and Domeracki, D.D., 1978. The stratigraphic significance of backbarrier tidal channel migration (Abstract). Program, Geol. Soc. Am. Meeting (SE Sec.), v.10, p. 20.

Wegener, A., 1912. Die Entstehung der Kontinente. Peterm. Mitt. 185–195, 253–256, 305–309.

Wehmiller, J.F., and Miller, G.H., 2000. Aminostratigraphic dating methods in Quaternary geology. In: Noller, J.S., Sowers, J.M., and Lettis, W.R. (eds.), Quaternary Geochronology, Methods and Applications. Amer. Geophys. Union Reference Shelf, 4:187-222.

Wehmiller, J.F., Simmons, K.R., Cheng, H., Edwards, L., Martin-McNaughton, J., York, L.L., Krantz, D.E., and Shen, C., 2004. Uranium-series coral ages from the US Atlantic Coastal Plain—The "80 ka problem" revisited. Quaternary International 120:3–14.

Weimer, R.J., and Hoyt, J.H., 1964. Burrows of *Callianassa major* Say, geologic indicators of littoral and shallow neritic environments. Jour. Paleontology, 38:761-767.

Wentworth, C.K., 1922. A scale of grade and class terms for clastic sediments. Jour. Geol., 30(5):377-392.

White Oak Plantation, 2010. A History of White Oak Plantation: Available at: http://home.ncifcrf.gov/ccr/lgd/congen2010/WhiteOak/History/HistoryOfWhiteOakPlantation_1768_to_2008.pdf.

Wiedemann, H.U., 1972. Application of red-lead to the detection of dissolved sulfide in waterlogged soils. Zeitschr. Pflanzenernahrung Bodenkunde, 133:73-81.

Willis, R.A., 2006. Genetic stratigraphy and geochronology of last interglacial shorelines on the Central Coast of South Carolina. M.S. Thesis, Louisiana State Univ., Dept. Geology and Geophysics, 134 pp.

Winker, A.D., and Howard, J.D., 1977. Correlation of tectonically deformed shorelines of the southern Atlantic coastal plain. Geology, 5:123-127.

Wood, F.J., 1982. Tides. In: Schwartz, M.L. (ed.), The Encyclopedia of Beaches and Coastal Environments. Hutchinson Ross Pub. Co., Stroudsburg, Pennsylvania, pp. 826-837.

Woodroffe, C.D., 2002. Coasts. Cambridge University Press, Cambridge, 623 pp.

Wunderlich, F., 1972. Georgia coastal region, Sapelo Island, U.S.A.: Sedimentology and Biology. III. Beach dynamics and beach development. Senckenbergiana Maritima, 4:47-79.

Zachos, J., Pagani, M., Sloan, L., Thomas, E., and Billups, K., 2001. Trends, rhythms, and aberrations in global climate 65 Ma to present. Science, 292:5517:686-693.

Zarillo, G.A., Ward, L.G., and Hayes, M.O., 1985. An illustrated history of tidal inlet changes in South Carolina. South Carolina Sea Grant Consortium, Charleston, SC, 76 pp.

INDEX

Q

R

S

ABOUT THE AUTHORS

Miles O. Hayes

Dr. Miles O. Hayes is a coastal geomorphologist and sedimentologist with over 50 years of research experience. He has authored over 250 articles and reports and seven books on numerous topics relating to tidal hydraulics, river morphology and processes, beach erosion, barrier island morphology, oil pollution, and petroleum exploration. Based on extensive field experience throughout the world, he has developed innovative techniques regarding environmental protection, oil-spill response, and shoreline processes. Three of the original concepts proposed and developed by him are: 1) Importance of hurricanes to barrier island and nearshore shelf sedimentation; 2) The effect of tides on shoreline morphology and sedimentation patterns; and 3) The environmental sensitivity index (ESI) for mapping shorelines (with co-author Michel), which has been applied worldwide. Hayes' teaching experience includes a range of both undergraduate and graduate courses while a Professor at the Universities of Massachusetts and South Carolina. Seventy-two graduate students received their degrees under his supervision, most of whom are now leaders in their respective academic, government and industry positions. He is presently Chairman of the Board of Research Planning, Inc. (RPI), a science technology company located in Columbia, SC.

Jacqueline Michel

Dr. Jacqueline Michel is an internationally recognized expert in oil and hazardous materials spill response and assessment with a primary focus in the areas of oil fates and effects, non-floating oils, shoreline cleanup, alternative response technologies, and natural resource damage assessment. As of this time, she has participated in research projects in 33 countries. Much of her expertise is derived from her role, since 1982, as part of the Scientific Support Team to the U.S. Coast Guard provided by the National Oceanic and Atmospheric Administration (NOAA). Under this role, she is on 24-hour call and provides technical support for an average of 50 spill events per year. She leads shoreline assessment teams and assists in selecting cleanup methods to minimize the environmental impacts of the spill. She has written over 150 manuals, reports, and scientific papers on coastal resource impacts, mapping, and protection. As a member of the Ocean Studies Board at the National Academy of Sciences for four years, she served on four National Research Council committees (chairing two), and is a Lifetime Associate of the National Academies. One of the original founders of RPI, which started in 1977, she now serves as the company President.